Oilfield Microbiology

Oilfield Microbiology

Edited by

Torben Lund Skovhus and Corinne Whitby

CRC Press
Taylor & Francis Group
Boca Raton London New York

CRC Press is an imprint of the
Taylor & Francis Group, an **informa** business

CRC Press
Taylor & Francis Group
6000 Broken Sound Parkway NW, Suite 300
Boca Raton, FL 33487-2742

Printed on acid-free paper

International Standard Book Number-13 978-1-138-05775-3 (Hardback)

Library of Congress Cataloging-in-Publication Data

Names: Skovhus, Torben Lund, author. | Whitby, Corinne, author.
Title: Oilfield microbiology / Torben Lund Skovhus, Corinne Whitby.
Description: Boca Raton : CRC Press, Taylor & Francis Group, [2019] | Includes bibliographical references and index.
Identifiers: LCCN 2018047254 | ISBN 9781138057753 (hardback ; alk. paper) | ISBN 9781315164700 (ebook)
Subjects: LCSH: Petroleum—Microbiology.
Classification: LCC QR53.5.P48 S56 2019 | DDC 547/.83—dc23
LC record available at https://lccn.loc.gov/2018047254

Visit the Taylor & Francis Web site at
http://www.taylorandfrancis.com

and the CRC Press Web site at
http://www.crcpress.com

Contents

Foreword

Meeting the Evolving Needs of the Petroleum Industry

Since the time of Louis Pasteur, advances in microbiology have enabled the management of practical problems, and the development of new opportunities in both the public and private sectors. The importance of microorganisms in the petroleum industry was established in the first half of the twentieth century. Microorganisms were found to play a key role in the formation of oil and gas as well as in the degradation of hydrocarbons. New modes of oil exploration and enhanced oil recovery were demonstrated in the field. Biocides were introduced to control microbiologically influenced corrosion (MIC) problems and bioremediation was added to the toolbox for the treatment of contaminated soils and waste streams. The scope of this extensive effort was summarized in a book called *Petroleum Microbiology* by J. B. (Barney) Davis published through Elsevier in 1967. Many, perhaps most, of the topics reviewed remain of vital interest today but much has changed.

The present book, *Oilfield Microbiology*, presents a timely insight into the rapidly developing world of petroleum microbiology. Results, progress, and prospects are presented for selected areas by international experts and practitioners from academia and industry. While the topics and issues addressed find their roots in the pioneering days of the last century, much of the material is new. Advances in the technology available to petroleum microbiologists and a shift in business priorities within the oil and gas sector have driven innovation. Today, the oil industry is faced with the depletion of easily accessible reservoirs, challenging global economics, and a shift in public opinion with respect to the environmental impact of oil and gas production and products.

Advances in the tools available for characterizing microbial populations in the field are changing the practical landscape. Today, molecular microbiological methods (MMM) based on the inexpensive and rapid analysis of nucleic acids in field samples are revolutionizing the approach to both microbial problems and opportunities. Current activity is focused on the development of robust protocols for sampling, sample processing, and data interpretation that can be used across the petroleum industry. Future development of field competent assay devices able to provide "real time" results on-site could shift the monitoring and assessment process in production systems and the environment from occasional sampling followed by off-site analysis to a regime of flexible, ongoing surveillance in the field. Success awaits advances in the assay technology itself as well as the development of a strong knowledge base that can connect a catalogue of microorganisms and other bioindicators to processes of interest. Traditional microbial assays and screening procedures are also targets for innovation. The development of relatively inexpensive high-throughput methods for biocide screening, enumeration, and corrosion rate testing are very much part of the current advance. Speed, reliability, and cost are the major drivers.

Exploration today has shifted to the discovery of unconventional sources of oil in tight source rocks and in geological traps hidden deep under the ocean floor. The discovery of new offshore fields in particular is a complex problem that requires a multidisciplinary approach that must glean as much information as possible from difficult to obtain geological samples to feed the selection and decision-making process. New technology is now adding microbiological data to the mix.

The exploitation of tight formations came of age with the recent widespread implementation of technology that can fracture the host rock to enable the release of oil and gas. The injection and back production of large volumes of water associated with this process brings with it microbiological issues in water handling, disposal, and reuse. MIC of steel infrastructure continues to plague production operations of all sorts. New insights into the links between metabolic processes and corrosion offer opportunities to develop improved predictive models that will allow more certain estimates of the remaining life of specific assets and enable timely interventions to prevent mishaps. Microbial sulfate reduction in the oil reservoir can lead to the production of noxious hydrogen sulfide, which can threaten worker safety and increase the risk of corrosion and cracking in susceptible steels. Souring tends to become apparent overtime and is often associated with the water or seawater injection strategies that are frequently employed to maintain production levels in aging fields. Finding better biocides for specific applications and using novel approaches such as nitrate injection to shift microbial populations away from sulfate reduction *in situ* can extend the life of affected assets by reducing maintenance costs and the risks posed by potential leaks. Because of the long lifetime required of offshore facilities and the difficulty and cost involved in repairing damage, optimizing management programs are of critical interests in ongoing operations.

Public scrutiny of oil patch operations from field to distribution has never been higher and is set against a background of global concerns about climate change and environmental impact. New molecular methods allow characterization of the microbial population in a target location and can be used to assess its response to an accidental release of hydrocarbons. Understanding the fate and impact of oil and new oil products, such as biodiesel in specific environments, is key to managing the risk posed by such events.

The state of the art revealed in this new book indicates that a multidisciplinary effort in petroleum microbiology is underway internationally to meet the evolving needs of the petroleum sector in a demanding time. This effort holds the promise of moving the management and manipulation of microbial activities to a new level of sophistication with benefits accruing from reduced uncertainty in exploration, reduced costs and risk in operations, extension of the life of assets, and improved environmental protection.

Thomas R. Jack
University of Calgary
Canada

Preface

This new book brings together contributions from leading scientists, academics, and experts from the oil and gas industry. The book contains 16 peer-reviewed chapters, split over six parts, with contributors sharing their experiences and research findings from working within the sector.

A key challenge for the oil and gas industry is Microbiologically Influenced Corrosion (MIC), which is responsible for billions of dollars in damage annually. Part I and Part II of the book are devoted to the advancing field of MIC research both in terms of MIC modeling (Part I) and MIC and reservoir souring (Part II). Whilst Part III deals with biocides and biofilms in the oil and gas industry, Part IV presents hydrocarbon biodegradation. Part V covers the rapidly growing field of "-omics" and novel technologies. Finally, Part VI reviews the developments in microbial exploration and enhanced oil recovery.

For many years, our knowledge of microorganisms in petroleum reservoirs was largely based on culture-based methods. Advances however, in molecular microbiological methods (MMM) over the last 25 years, particularly Next Generation Sequencing (NGS) approaches, have provided a deeper understanding of the microorganisms in these natural systems. This has allowed petroleum microbiologists to more readily characterize the *in situ* microbial community in 'real time' including the 'rare' biosphere. These advances have led to opportunities to improve operations and profitability in the industry. From an industry perspective, an increase in our fundamental knowledge of the *in situ* microbial communities alongside the emergence of novel technologies will bring with it more discoveries contributing to how our fossil fuel resources are discovered, produced, transported, and managed in the future.

Chapter 1 by *Voordouw* gives a comprehensive overview of the highlights of 10 years of research in petroleum microbiology. The over-arching aim of his research was to decrease the environmental impact of fossil fuel production and transport through sulfur cycle management, corrosion prevention and improved production.

In order for operating engineers to safely and effectively manage steel infrastructure, it is important that tools are developed to accurately and quantitatively predict the location and severity of corrosion that is likely to occur in the future. **Part I** of the book deals with the topic of modeling microbiologically influenced corrosion (MIC). MIC modeling can provide several benefits to the sector, including guidance on MIC mitigation selection and prioritization of inspection and repair based on a more robust scientific basis for risk-based inspections. Despite these benefits, there currently remain large gaps in our knowledge on MIC modeling research.

Chapter 2 (*Skovhus, Taylor, and Eckert*) provides a unique end-user perspective to describe the trends in MIC modeling; as well as the different types of models, future needs and utility of MIC models for the sector. **Chapter 3** (*Melchers*) considers marine corrosion development in coastal seawaters, under the influence of various environmental factors such as water temperature, velocity, wave action, and steel type, etc. The potential influence of microorganisms in causing increased corrosion in marine environments is also addressed.

Part II continues the analysis of MIC, with a focus on oil and gas reservoir systems. **Chapter 4** (*An and Voordouw*) describes the microbial communities involved in reservoir souring and MIC of shale oil and gas fields. The authors characterize the *in situ* microbial communities obtained from injection, produced and facility waters from a Canadian shale oilfield reservoir to determine souring risks at various salinities. They demonstrate that when injected with nitrate at high salinity, souring can be prevented and so reinjection of high salinity produced water may be a preferable option for the sector over current practices.

A key group of microorganisms in MIC are sulfate-reducing prokaryotes (SRP). One important consideration is that small numbers of metabolically active SRPs could potentially accelerate the corrosive process more than a larger number of less-active microorganisms. To better understand the activities of the *in situ* microbial community, particularly SRPs, RNA (rather than DNA) analyses are applied. However, there are inherent problems associated with RNA-based work including contamination from RNAses, which readily degrade the RNA. It is therefore important to collect, transport, and store environmental samples in a manner that ensures that the integrity of the RNA is maintained. **Chapter 5** (*Coutinho, Pinto, Oliveira, Barboza, Abrantes, and Lutterbach*) describes a standardized method of RNA collection and extraction from environmental samples obtained from the oil sector.

In practice, operators require the ability to detect the microorganisms implicated in MIC in a timely and accurate manner to allow for the implementation of appropriate treatment regimens to mitigate costs and downtime. **Chapter 6** (*Sharma and Huang*) describes an easily transportable 'lab-in-a-box' testing system, which can provide results in a matter of hours.

Reservoir souring is an extensive and costly problem for the industry. Nitrate addition is commonly used as a souring inhibitor but often it requires high concentrations and has variable success. Recently, perchlorate has displayed promise as a more potent inhibitor to sulfide biogenesis. **Part III** of the book is devoted to the importance of biocides and biofilms in the oil and gas industry. **Chapter 7** (*Liu, Engelbrektson, Stoeva, Barnum, Reyes-Umana, and Coates*) describes the use of perchlorate and its effects on the *in situ* microbial community. The authors demonstrate that concurrent nitrate and perchlorate treatment is not more effective than perchlorate treatment alone but can be more effective than nitrate treatment. These findings may influence future treatment decisions for the industry. **Chapter 8** (*Kotu, S. Han, A. Han, Mannan, and Jayaraman*) presents a 'lab-on-a-chip' model for investigating biocides on co-culture biofilms, whilst **Chapter 9** (*Markfoged, Aggerholm, Holsten, Struve, Trénel, Poulsen, and Steinhauer*) considers the efficacy of biocide treatments on biofilm formation of oilfield microorganisms.

Part IV of the book is devoted to hydrocarbon biodegradation and the important role that microorganisms play in the biotransformation of petroleum hydrocarbons. Whilst it has been previously shown that hydrocarbon biodegradation occurs under methanogenic conditions, the metabolic pathways, and key methanogenic hydrocarbon-degrading microorganisms involved are currently unknown. Using a case study, **Chapter 10** (*Toth and Gieg*) identified the metabolic reactions and the key microorganisms driving methanogenic crude oil biodegradation. In particular, *Desulfotomaculum* and *Smithella* are implicated as important hydrocarbon activators in crude oil-containing methanogenic ecosystems. **Chapter 11** (*Matos, Oliveira, Coutinho, Oliva, Junior, Stenuit, Agathos, Lutterbach, and Sérvulo*) is focused on biodiesel, specifically the intrinsic bioremediation of soil contaminated with pure biodiesel (B100) and its low-level blends. The authors showed that whilst biodiesel B5 and B100 were readily biodegraded, Total Petroleum Hydrocarbon (TPH) and Fatty Acid Methyl Esters (FAME) biodegradation were impaired when presented in blends. The authors also showed that biodegradation was largely mediated by autochthonous fungi and bacteria. In **Chapter 12** (*Chen, Min, and Hu*) quantitative PCR (qPCR) and genomic sequencing identified *Acinetobacter* sp. HC8-3 and *Dietzia* sp. CN-3, degraded >90% of the saturated hydrocarbons in crude oil from the sediments in Bohai bay.

It is well known that dispersants can aid in the breakup of crude oil in the marine environment. Dispersants work by increasing the available interfacial surface area for microorganisms to degrade the immiscible hydrocarbons. However, this common view neglects key aspects of the microscale interactions between bacteria and oil droplets that are required for degradation to occur. **Chapter 13** (*Fernandez, Stocker, and Juarez*) discusses a biophysical model for hydrocarbon consumption of suspended oil droplets under conditions of rapid dilution in natural environments. This mechanistic model provides a basis for better understanding of microscale biodegradation in dilute oil environments and can help inform the design of mitigation strategies in marine systems.

Part V is focused on "omics" and novel technologies for the oil and gas industry. Research in the marine environment traditionally involved collecting discrete samples from shipboard surveys and transporting them to the laboratory for analysis, which was time-consuming and costly. Furthermore, it did not allow for rapid decision-making for offshore oil production management. Advances in genomics and metagenomics have unraveled the importance of microbial processes in the oceans, including the microbial responses to contaminants such as oil spills. **Chapter 14** (*Bagi, Knapik, Nilsen, Krolicka, Boccadoro, and Baussant*), describes the detection of microorganisms for use as genosensors (biosensors) for monitoring natural marine environments. The authors describe how the integration of genosensors into autonomous sampling and sample processing platforms can enable the identification and/or quantification of specific microbial genes of interest, providing timely and valuable insights into marine ecosystem processes.

In the final part of this book, **Part VI**, microbial exploration and enhanced oil recovery are described. Nitrate injection into oil reservoirs is often used to prevent reservoir souring by stimulating the growth of heterotrophic nitrate reducing bacteria (hNRB), which couple nitrate reduction to the oxidation of oil organics. **Chapter 15** (*Suri and Voordouw*) showed that by providing high concentrations of alkylbenzenes and nitrate in heavy oils, oil production could be increased by up to 20% in laboratory test systems.

In the past, offshore hydrocarbon exploration campaigns often relied on the geochemical analysis of piston cores extracted from marine sediments. The cores were sent to specialized onshore laboratories. This approach is often logistically difficult, expensive, and time consuming. The authors of the final chapter, **Chapter 16**, (*Miranda, Seoane, Esteban, and Espí*), propose the use of a new generation of

Microbial Exploration Techniques. Using a combination of Next Generation Sequencing (NGS) and machine learning classifiers, they provide a rapid assessment of the probability of hydrocarbon presence in a piston core, based on the microbial DNA fingerprint. The chapter also presents a set of bioindicators of hydrocarbon presence. The authors suggest that these data can be used to build a robust machine learning classification model to categorize new samples in a human independent manner.

The editors wish to thank all of the authors and reviewers who contributed their time, knowledge, and expertise to this book, making it an invaluable resource for many years to come. The editors hope that this book will stimulate further research, discussions, and developments in the field of *oilfield microbiology* and its importance to the oil and gas sector.

Torben Lund Skovhus
VIA University College
Denmark

Corinne Whitby
University of Essex
United Kingdom

Editors

Torben Lund Skovhus, PhD, is Associate Professor and Project Manager at VIA University College in the Centre of Applied Research and Development in Building, Energy, Water and Climate (Horsens, Denmark). He graduated from Aarhus University, Denmark, in 2002 with a master's degree (cand. scient.) in biology. In 2005, he earned a PhD in microbiology at Aarhus University. In 2005, he was employed at Danish Technological Institute (DTI) in the Centre for Chemistry and Water Technology, where he was responsible for the consultancy activities for the oil and gas industry around the North Sea. Dr. Skovhus headed the DTI Microbiology Laboratory while he developed several consultancy and business activities with the oil and gas industry. He founded DTI Oil & Gas in both Denmark and Norway, where he was Team and Business Development Leader for five years. Thereafter he worked as Project Manager at DNV GL (Det Norske Veritas) in the field of corrosion management in both Bergen and Esbjerg.

Dr. Skovhus is currently chair of NACE TEG-286X and ISMOS TSC, an organization he cofounded in 2006. He is an international scientific reviewer and the author of 60+ technical and scientific papers and book chapters related to industrial microbiology, applied biotechnology, corrosion management, oilfield microbiology, water treatment and safety, reservoir souring, and biocorrosion. He is editor of the book, *Microbiologically Influenced Corrosion in the Upstream Oil and Gas Industry* (Boca Raton, FL: CRC Press, 2017).

Corinne Whitby, PhD, is a Senior Lecturer in Environmental Microbiology at the University of Essex, United Kingdom. She earned a PhD in environmental microbiology at the School of Biological Sciences, University of Liverpool, in 1998. After several postdoctoral positions, Dr. Whitby was appointed a Lecturer at the University of Essex in 2006 and promoted to Senior Lecturer in 2012. She has considerable expertise in molecular microbial ecology, and her research is focused in four key areas across terrestrial and aquatic environments: (1) microbial biodegradation of hydrocarbons, specifically naphthenic acids (NAs); (2) microbial-driven biogeochemical cycling; (3) the effect of nanoparticles/nanoplastic pollutants on microbial communities, and (4) characterization of microbial communities in bioaerosols. Her research involves next-generation sequencing (NGS) "omic" approaches to analyze complex microbial communities in the environment. She is an expert in applied microbiology and works closely with end-user beneficiaries, particularly industry, government and environmental agencies. She is also an international scientific reviewer and the author of several scientific papers and book chapters related to molecular microbial ecology, environmental microbiology, applied biotechnology, oilfield microbiology, and bioaerosols.

Contributors

Thais Abrantes
National Institute of Technology
Rio de Janeiro, Brazil

Spiros N. Agathos
Earth and Life Institute (ELI)
Catholic University of Louvain
Louvain-la-Neuve, Belgium
School of Biological Sciences and Engineering
Yachay Tech University
Dan Miguel de Urcuquí, Ecuador

Steffen Larsen Aggerholm
Danish Technological Institute
DTI Oil and Gas
Aarhus, Denmark

Biwen Annie An
Federal Institute for Materials Research and Testing
Berlin, Germany
University of Calgary
Calgary, Alberta, Canada

Andrea Bagi
IRIS Environment
Stavanger, Norway

Hazel Barboza
National Institute of Technology
Rio de Janeiro, Brazil

Tyler Barnum
University of California
Berkeley, California, USA

Thierry Baussant
IRIS Environment
Stavanger, Norway

Catherine Boccadoro
IRIS Environment
Stavanger, Norway

Weiwei Chen
Yantai Institute of Coastal Zone Research, Chinese Academy of Sciences
Yantai, China
University of Chinese Academy of Sciences
Beijing, China
Laboratory for Marine Biology and Biotechnology, Qingdao National Laboratory for Marine Science and Technology
Qingdao, China

John D. Coates
University of California
Berkeley, California, USA

Diogo Coutinho
National Institute of Technology
Rio de Janeiro, Brazil

Richard B. Eckert
DNV GL
Dublin, Ohio, USA

Anna Engelbrektson
University of California
Berkeley, California, USA

Enrique Espí
Repsol Technology Center
Madrid, Spain

Ángela Esteban
Repsol Technology Center
Madrid, Spain

Vicente I. Fernandez
ETH Zurich
Zurich, Switzerland

Lisa M. Gieg
Department of Biological Sciences
University of Calgary
Calgary, Alberta, Canada

Arum Han
Texas A&M University
College Station, Texas, USA

Song-I Han
Texas A&M University
College Station, Texas, USA

Lea Holsten
Schülke & Mayr GmbH
Microbiology Department
Norderstedt, Germany

Xiaoke Hu
Yantai Institute of Coastal Zone Research, Chinese Academy of Sciences
Yantai, China
Laboratory for Marine Biology and Biotechnology, Qingdao National Laboratory for Marine Science and Technology
Qingdao, China

Wei Huang
LuminUltra Technologies, Inc.
Baltimore, Maryland, USA

Arul Jayaraman
Texas A&M University
College Station, Texas, USA

Gabriel Juarez
University of Illinois
Urbana, Illinois, USA

Clenilson Junior
Rio de Janeiro Federal Institute
Rio de Janeiro, Brazil

Kamila Knapik
IRIS Environment
Stavanger, Norway

Susmitha Purnima Kotu
Texas A&M University
College Station, Texas, USA

Adriana Krolicka
IRIS Environment
Stavanger, Norway

Yi Liu
University of California
Berkeley, California, USA

Márcia Lutterbach
National Institute of Technology
Rio de Janeiro, Brazil

Sam Mannan
Texas A&M University
College Station, Texas, USA

Rikke Markfoged
Danish Technological Institute
DTI Oil and Gas
Aarhus, Denmark

Guilherme Matos
Federal Centre for Technological Education CSF
Rio de Janeiro, Brazil

Robert E. Melchers
Centre for Infrastructure Performance and Reliability
The University of Newcastle
Newcastle, New South Wales, Australia

Jun Min
Yantai Institute of Coastal Zone Research, Chinese Academy of Sciences
Yantai, China
Laboratory for Marine Biology and Biotechnology, Qingdao National Laboratory for Marine Science and Technology
Qingdao, China

Juan de Dios Miranda
Repsol Technology Center
Madrid, Spain

Mari Mæland Nilsen
IRIS Environment
Stavanger, Norway

Thais Oliva
Rio de Janeiro Federal Institute
Rio de Janeiro, Brazil

Ana Oliveira
National Institute of Technology
Rio de Janeiro, Brazil

Viviane Oliveira
National Institute of Technology
Rio de Janeiro, Brazil

Luiz A.L.T. Pinto
National Institute of Technology
Rio de Janeiro, Brazil

Morten Poulsen
Danish Technological Institute
DTI Oil and Gas
Aarhus, Denmark

Victor Reyes-Umana
University of California
Berkeley, California, USA

José Miguel Seoane
Repsol Technology Center
Madrid, Spain

Eliana Sérvulo
Federal University of Rio de Janeiro
Rio de Janeiro, Brazil

Neil Sharma
LuminUltra Technologies, Inc.
Baltimore, Maryland, USA

Torben Lund Skovhus
Centre of Applied Research and Development in Building, Energy, Water and Climate
VIA University College
Horsens, Denmark

Katrin Steinhauer
Schülke and Mayr GmbH
Microbiology Department
Norderstedt, Germany

Ben Stenuit
University of Montpellier
Montpellier, France
Catholic University of Louvain
Louvain-la-Neuve, Belgium

Roman Stocker
ETH Zurich
Zurich, Switzerland

Magdalena Stoeva
University of California
Berkeley, California, USA

Lisa Struve
Schülke and Mayr GmbH
Microbiology Department
Norderstedt, Germany

Navreet Suri
Petroleum Microbiology Research Group
Department of Biological Sciences
University of Calgary
Calgary, Alberta, Canada

Christopher Taylor
DNV GL
Dublin, Ohio, USA

Courtney R.A. Toth
Department of Biological Sciences
University of Calgary
Calgary, Alberta, Canada
Currently:
Department of Chemical Engineering and Applied Chemistry
University of Toronto
Toronto, Ontario, Canada

Philipp Trénel
Danish Technological Institute, Field Trials, Technology and Analysis
Aarhus, Denmark

Gerrit Voordouw
Petroleum Microbiology Research Group
Department of Biological Sciences
University of Calgary
Calgary, Alberta, Canada

1 Highlights of 10 Years of Research in Petroleum Microbiology

Gerrit Voordouw

CONTENTS

1.1 INTRODUCTION

The sixth bi-annual meeting of the International Society of Molecular Biology in Oilfield Systems (ISMOS-6) was held June 6–9, 2017, in San Diego, California, 10 years after the launch of ISMOS-1. This also marked the 10th and final year of the NSERC Industrial Research Chair in Petroleum Microbiology, a research program sponsored by the Natural Science and Engineering Research Council of Canada (NSERC), Alberta Innovates, and industry partners, which started in October 2007. Hence two 10-year anniversaries converged and, since I was asked to give the closing address at ISMOS-6, I thought it appropriate to present our research highlights of the past 10 years. This chapter reflects this closing address. Of course, progress in petroleum microbiology has resulted from the efforts by many, as reviewed elsewhere (Head et al. 2003; Ollivier and Magot 2005; Youssef et al. 2009; Gieg et al. 2011).

In order to explain why this research is needed we may consider that despite known environmental problems, society continues to rely heavily on hydrocarbon fuels for its energy needs. Given the decline in conventional hydrocarbon fuel resources (the "low-hanging fruit"), the energy industry turns increasingly to more difficult to produce unconventional resources such as bitumen and shale gas. A common feature of these unconventional resources is that the ratio of energy returned on energy invested (EROEI) is lower than for the depleting conventional resources that we used to enjoy. Because the energy invested gives rise to production of CO_2, it follows that the amount of CO_2 emitted per unit of energy produced is higher for current unconventional production than for past conventional production. More broadly stated, the EROEI of a given energy source is likely to slide with time. This slide can be halted or reversed when a completely new source of energy is discovered, as in the transition from wood and peat to coal and then to oil and gas, which started some 300 years ago.

With the possible exception of nuclear energy, energy alternatives such as wind and solar do not offer the large EROEI step change that was realized in the wood and peat to hydrocarbon fuel transition. This unprecedented step change has allowed today's high-energy-use society to emerge within a very short time. In view of this it is important that our current resources are extracted as efficiently as possible to decelerate the EROEI slide. Petroleum microbiology can contribute to this not only by improving production, but also by solving production-related problems in the areas of sulfur cycle management and corrosion. Therefore, we launched a program in petroleum microbiology in October 2007 that focused on the roles of microorganisms in these three interconnected areas (see Figure 1.1). The highlights of this program are presented below.

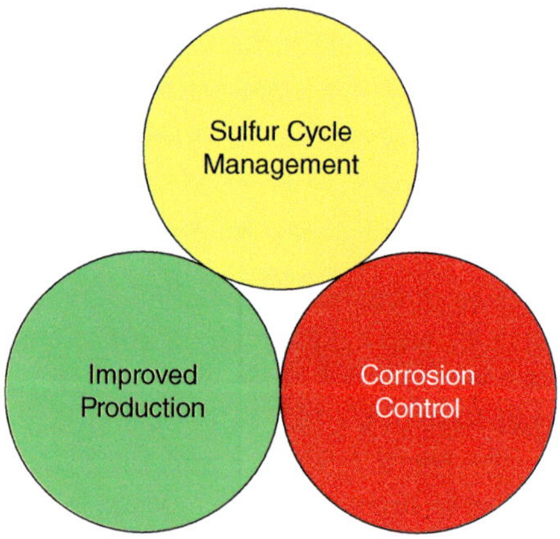

FIGURE 1.1 Overview of research areas, targeted by the NSERC IRC in Petroleum Microbiology.

1.2 RESEARCH HIGHLIGHTS

1.2.1 Sulfur Cycle Management: Emergence of Zones of NRB and SRB

Souring, the reduction of sulfate to sulfide through the activity of sulfate-reducing bacteria (SRB), can be remediated by the injection of nitrate, which stimulates the activity of nitrate-reducing bacteria (NRB). Unfortunately, this technology is not universally successful, and one of the goals of our project was to find out why. Souring is unwanted because the concentration of sulfide allowed in oilfield operations is subject to stringent controls. At neutral pH sulfide is present as HS^- and H_2S. H_2S partitions in aqueous and in oil and gas phases. Gas phase H_2S is often measured on site with Dräger tubes as a proxy for the presence of sulfide. Concentrations of H_2S need to remain below mandated thresholds because increased H_2S results in increased corrosion threat, deteriorated product quality, and compromised worker safety. When remedial measures fail to decrease sulfide and if the existing steel infrastructure is not licensed for sour service, expensive retrofitting is needed. The need to control the activity of SRB, which increase sulfide in fluids and gas produced from water-injected oilfields, is thus clear.

In the past 10 years we have monitored the effect of nitrate injection in the Medicine Hat Glauconitic C (MHGC) field. This is a low-temperature (26° C) field, producing heavy oil with an API gravity of 16°. Souring in this field was caused by the use of 1000 m^3/day of make-up water with a sulfate concentration of 4 mM. This was combined with 3000 m^3/day of produced water (0 mM sulfate) to give injection water of 1 mM of sulfate. Amendment of this injection water with 2–3 mM nitrate decreased the sulfide concentration in produced waters from 0.15 mM to 0.05 mM within a few months, but this was followed by recovery to initial sulfide concentrations within one year. Our explanation for these field observations was that prior to nitrate injection, SRB live close to the injection wellbore where the injected sulfate comes in. When the injection water is amended with nitrate, these SRB are strongly inhibited and this zone is soon dominated by NRB. However, because oil organics are in excess this leads to complete reduction of nitrate to nitrite and then to di-nitrogen (N_2), which is no longer inhibitory and allows the SRB to regrow in a zone located farther away from the injection wellbore. Thus, once sequential zones of NRB and SRB have emerged along the water injection flow path, nitrate injection is no longer effective (Voordouw et al. 2009). Zonation of NRB and SRB has been demonstrated in sand pack bioreactors injected with sulfate, nitrate, and excess volatile fatty acids (VFA, a mixture of acetate, propionate and butyrate) or in sand pack bioreactors with oil (Callbeck et al. 2011 and 2013). Under these conditions nitrate and sulfate are sequentially reduced along the injection water flow path with methanogenesis becoming the dominant microbial process once both of these electron acceptors have become depleted. This holds true both for bioreactors (Chen et al. 2017) and in the field, where methanogens dominate the microbial community composition of produced waters, which lack nitrate and sulfate (Agrawal et al. 2012; Shen et al. 2018).

1.2.2 Sulfur Cycle Management: Nitrate Breakthrough by Toluene-Depletion of Produced Oil

NRB can be subdivided in heterotrophic NRB (hNRB) and sulfide- or sulfur-oxidizing NRB (soNRB). These use organic compounds (including VFA) and sulfur or sulfide as the electron donor for the reduction of nitrate (or oxygen), respectively. Many soNRB are chemolithotrophs, which use only CO_2 as the carbon source for growth. Such microorganisms are ideal to control souring, since they use the injected nitrate solely for the oxidation of sulfide. In contrast hNRB use the injected nitrate to oxidize oil. Although they can control SRB by production of nitrite, this is only transient when nitrite is further reduced to N_2. Heterotrophic NRB and SRB use selected oil components as electron donor for the reduction of nitrate to nitrite and N_2 and of sulfate to sulfide, respectively. VFA can be formed as intermediates in anaerobic metabolism of these oil components (Callbeck et al. 2013). Although SRB appear to use many oil components, including aliphatic and aromatic compounds, the hNRB appear to prefer aromatic hydrocarbons with a further preference for alkylbenzenes, especially toluene (Lambo et al. 2008; Agrawal et al. 2012). Hence, the toluene content of the oil is an important factor to consider when evaluating the potential of nitrate injection to prevent souring of an oilfield.

The toluene concentration in a variety of oils was found to vary by almost 100-fold. This was unrelated to API gravity, e.g., light oil from a field in Papua New Guinea (API gravity of 46°) had 77 mM toluene, whereas a light oil from the Bakken formation (API gravity 38°) had 0.8 mM toluene (Suri et al. 2017). Heavy MHGC oil contains only 1–3 mM of toluene. As a result, continued injection of nitrate resulted in depletion of toluene from this oil. The zone occupied by hNRB expanded gradually from the injection well to the production well when nitrate was continuously injected, leading eventually to the production of water with nitrate, nitrite, and sulfate but no sulfide and the production of toluene-depleted oil (Agrawal et al. 2012). Hence, microbial zonation offers an explanation of why nitrate injections may fail. Analysis of oil composition prior to the start of nitrate injection may indicate how severe this problem will be.

1.2.3 Sulfur Cycle Management: Souring Control by Limiting Sulfate Dispersal

Souring in the MHGC field was first detected six years after the use of make-up water with 4 mM sulfate. This long delay was caused by the sulfide-binding capacity of reservoir rock, e.g., by conversion of iron carbonate to iron sulfide. Make-up water is needed to replace the volume of 1000 m^3/day of produced oil. Souring could have been prevented by the use of sulfate-free make-up water. However, beggars can't be choosers! The South-Eastern part of Alberta is dry, with a limited supply of fresh water. The available water source was one of the best available from a souring control point of view. Although further decrease of the sulfate concentration by nanofiltration is possible, this option was considered too expensive when compared to nitrate injection.

As indicated already, nitrate injection was very successful initially, but sulfide concentrations rebounded within one year due to emerging zonation. Continued nitrate injection gave souring control with nitrate breakthrough in some producing wells after two to three years due to toluene limitation. Interestingly, after five years of nitrate injection the sulfide concentrations in all producing wells of area I in the MHGC field, serviced by water plant 1WP, briefly hit zero. These rebounded somewhat but stayed at 30–50% of pre-nitrate injection levels. What was going on?

This large decrease in sulfide production was caused by a change in water management, associated with the start-up of a polymer pilot plant for enhanced oil recovery (EOR). The start-up of the EOR pilot caused all sulfate-containing make-up water to be used for dissolution of polymer. This polymer-amended water was then injected into a limited region of the field. Produced water from this pilot contained 10% of the injected sulfate concentration (4 mM) as sulfide (0.4 mM), but was otherwise free of sulfate. Its use as the make-up water in 1WP, the main water plant of the field, caused injection water sulfate to drop to zero. This, in turn, strongly decreased produced sulfide concentrations in producing wells from area I and allowed a decreased use of nitrate. Hence, limiting the dispersal of sulfate-containing make-up water decreased souring and should be considered as a souring control strategy, whenever possible (Shen et al. 2018).

1.2.4 Sulfur Cycle Management: Lack of Nitrite Reduction under Extremophilic Conditions

NRB inhibit the activity of SRB through the production of nitrite, which is a strong SRB inhibitor. Nitrite is a magic bullet, which targets dissimilatory sulfite reductase (Dsr), the SRB enzyme that catalyzes the six-electron reduction of sulfite to sulfide. Dsr also catalyzes the slow six-electron reduction of nitrite to ammonium. The kinetic binding constant (K_m) and the rate constant for conversion (k_{cat}) of the enzyme for nitrite are much lower than for sulfite. This means that nitrite is bound more tightly and is converted more slowly, making it a strong competitive inhibitor of Dsr (Haveman et al. 2004). Because nitrate reduction and sulfate reduction can occur in close proximity in nature, many mesophilic SRB have a Nrf-type nitrite reductase, which reduces nitrite to ammonium. This enzyme is located on the outside of the cell to prevent inhibition of the sulfate reduction pathway by the inward diffusion of nitrite (Greene et al. 2003).

Nitrate inhibits SRB only after its conversion to nitrite by NRB. The effectiveness of nitrate injection can be improved by inhibiting the subsequent reduction of nitrite by chemical inhibitors or by existing extremophilic conditions. Economically usable chemical inhibitors of the enzyme nitrite reductase are currently unknown. Although synergy between biocides and nitrite in the control of souring is well-documented (Greene et al. 2006), synergy between biocides and nitrate has not yet been clearly established (Xue and Voordouw, 2015). However, studies with a variety of oilfield–produced waters have indicated that the reduction of nitrite is inhibited by high temperature or high salt. Nitrite is reduced poorly at temperatures in excess of 50° C, whereas the reduction of nitrate occurs at temperatures of up to 70° C. This causes accumulation of nitrite at temperatures of 50–70° C (Fida et al. 2016; Okpala et al. 2017). Oil production from hot offshore oil reservoirs is routinely done by injection of cold seawater, which has a high concentration of sulfate (25 mM). This generates a near-injection wellbore region (NIWR) in which the temperature increases as a function of distance from that of the incoming seawater (30° C), to that of the bulk of the reservoir (95° C in the case of the Terra Nova field; Okpala et al. 2017). This creates mesophilic (30–45° C), thermophilic (45–85° C), and abiotic (85–95° C) zones. Souring control by injection of nitrate in such reservoirs can be improved if the temperature is not allowed to drop below 50° C, for instance by re-injecting hot produced water to eliminate the mesophilic zone (Fida et al. 2016; Okpala et al. 2017). Interestingly, this principle applies also to shale oilfields in which oil is produced through water injection. Produced waters from such fields often have high salt concentrations (in excess of 2 M of NaCl). Nitrate is reduced to nitrite, but nitrite is poorly reduced further under these conditions (An et al. 2017). At low salinity (below 1 M of NaCl) nitrite is rapidly reduced. Hence, nitrate-mediated souring control would benefit from maintaining a high salt concentration by re-injecting produced water. Injecting low-salinity water from another source should be avoided.

1.2.5 Corrosion Control: Microbial Metabolism of Bisulfite

Oilfield produced waters are kept anaerobic to decrease corrosion, i.e., the corrosion rate of iron is higher under aerobic as compared to anaerobic conditions. Anaerobic conditions are created by adding an oxygen scavenger. Bisulfite, which rapidly reacts with O_2 to form sulfate and water, is most commonly used. We determined the effects of bisulfite addition in pipelines through which water was transported from 500 m below the surface to a plant, which generated steam for the production of bitumen by steam-assisted gravity drainage (SAGD). Bisulfite was routinely added to remove O_2 to decrease corrosion. However, pipeline failures were still common. It was noted that the pipeline scale downstream from bisulfite injection sites often contained sulfur. Because this can be formed by reaction of sulfide with oxygen, its presence was taken as evidence that insufficient bisulfite was added and the dose was therefore increased. Microbial community compositions were determined by extraction of DNA, PCR amplification of 16S rRNA genes, and pyrosequencing of the resulting amplicons. This indicated that samples of pipeline scale obtained upstream of bisulfite addition had high fractions of methanogens (*Methanobacteria*) and acetogens (*Acetobacterium*), which could contribute to corrosion, whereas samples of pipeline scale downstream of bisulfite addition had high fractions of *Deltaproteobacteria* of the genera *Desulfocapsa* and *Desulfomicrobium* (Park et al. 2011; Mand et al. 2014). *Desulfomicrobium* is an SRB, which reduces sulfite or sulfate in the presence of hydrogen or an organic electron donor. *Desulfocapsa* derives energy for growth from dismutating bisulfite into sulfate and sulfide. The presence of these microorganisms indicated that excess (not too little) bisulfite was used. The possible conversion of bisulfite to sulfate and sulfur by *Desulfocapsa* could explain the presence of sulfur in pipeline scale downstream from the bisulfite injection point. Formation of sulfur is unwanted because it is very corrosive toward iron. These results indicated that bisulfite was being overdosed and also demonstrated the value of next-generation sequencing methods in monitoring the effects of chemicals added routinely to decrease corrosion.

1.2.6 Corrosion Control: Corrosion of Shale Gas Coiled Tubing by Biocide

Coiled tubing is used in shale oil and shale gas operations for milling plugs separating the multiple stages of hydraulic fractures and removing these from horizontal wells. Coiled tubing is exposed to injection waters internally and produced waters containing plug debris externally. Produced waters are stored on site prior to reuse for milling through successive plugs. Two cycles per day for three days are typically done. Multiple chemicals including biocides are used to limit corrosion of coiled tubing. Biocides are used to decrease Microbiologically Influenced Corrosion (MIC). Because these can be dosed prior to every 12-hour cycle of reuse, there is a potential for overdosing. Decisions on whether more biocide is needed are based on rapid assays for estimating active microorganisms, like the ATP assay. We received field samples from a coiled tubing operation in a shale gas field representing source water and produced waters, which had been reused up to six times. The average general corrosion rates of these samples were evaluated by incubating 50 mL of sample with five carbon steel ball bearings under anaerobic conditions. We have advocated the use of carbon steel ball bearings as probes for measuring general corrosion by weight loss methods (Voordouw et al. 2016). Ball bearings come in different sizes; those of a given size class are all very similar. We used ball bearings of 55.0 ± 0.3 mg and a diameter of 0.238 cm. Samples had low general corrosion rates of on average 0.2 mm/yr, but one had a much higher general corrosion rate of 1.25 ± 0.07 mm/yr. This sample had no detectable culturable APB or SRB. Passage through a 0.2 μm filter to remove microorganisms did not decrease the general corrosion rate. These results suggested that microorganisms were not responsible for this high rate. HPLC indicated a very high concentration of the biocide bronopol in this sample (13.8 mM, 2760 ppm). Incubation of ball bearings with bronopol indicated a linear increase of the general corrosion rate with bronopol concentration. Moreover, similar corrosion rates were obtained as observed for field samples (Sharma et al. 2017). These studies indicated that overdosing of bronopol by addition at multiple cycles of water reuse led to significant corrosion and should be avoided. Erroneous decisions to increase the biocide dose may be made if the assays used to determine residual live bacteria give false positive results. Direct assay of the biocide used (e.g., by HPLC) may be better to determine whether more is needed.

1.2.7 Corrosion Control: Control of Corrosion under Flow Conditions

Corrosion in pipelines occurs under flow conditions. Biocides, used to limit the MIC contribution to this corrosion, are usually dosed in high concentration for a limited time, e.g., 300 ppm for two hours once per week. We packed 60 ball bearings in 1 mL syringe columns to study the effect of such biocide treatments under flow conditions. These columns were continuously injected with the effluent of a continuous culture bioreactor in which oilfield SRB were cultivated in medium with formate and excess sulfate. The system thus mimicked an upstream pipeline, which continuously received SRB that may subsequently contribute to pipeline corrosion. Ball bearings near the column inlet, which were most exposed to the sulfide produced in the continuous culture bioreactor, experienced more severe corrosion than more distant ball bearings. Periodic treatment with biocide decreased the general corrosion rate of ball bearings by up to 50%, relative to untreated controls. Benzalkonium chloride, which also acted as corrosion inhibitor, was most effective in preventing corrosion under flow conditions. However, glutaraldehyde and THPS, which were most effective in inhibiting or killing SRB, decreased the general corrosion rate by only 5–10%. Pre-treatment of ball bearings with oil-dissolved corrosion inhibitors decreased the general corrosion rate by 98–99%, i.e., these were more effective than biocides under the chosen experimental conditions. Hence, the use of ball bearings in small columns allowed the comparative evaluation of multiple treatment scenarios in a relatively short period of time (Pinnock et al. 2018).

1.2.8 Improved Production: Alkylbenzene- and Nitrate-Mediated MEOR

As indicated in section 1.2.2, reduction of nitrate by hNRB using reservoir oil as the electron donor depends on its toluene content. Although a high toluene content of reservoir oil is problematic when nitrate is injected in order to control souring (i.e., to oxidize sulfide), this is desirable if one wants to boost the activity of hNRB for microbially-enhanced oil recovery (MEOR) purposes. In order to access oil-bound toluene, hNRB may attach to oil to achieve emulsification through cell-bound biosurfactants (Kryachko et al. 2012). The toluene content of low toluene-reservoir oil can be increased by amending injection water with toluene. Toluene has a water solubility of 0.5 g/L (5.6 mM). The distribution coefficient of toluene between oil and water phases favors its presence in oil by several thousand-fold. Hence, water-injected toluene would transfer to the oil phase. In sand pack bioreactors, containing 0.5-pore volumes (PV) of residual MHGC oil, the toluene content of the oil was increased from 1 to 10 mM by continuous injection of several pore volumes of 3 mM aqueous toluene. This was then followed by injection of a high concentration of nitrate (80 mM) and an hNRB consortium grown on toluene-amended MHGC oil and nitrate. Following incubation of these sand packs for two weeks, flow was resumed and an additional 0.1 PV (20%) of residual heavy MHGC oil was produced (Gassara et al. 2015). In field applications of this technology, a period of aqueous toluene injection (e.g., of several months) would be followed by injection of nitrate. Flow would be halted to allow hNRB to catalyze toluene-dependent reduction of nitrate and associated oil emulsification. Resumption of flow would then produce additional oil.

Injection of an hNRB consortium would not usually be required, as toluene degrading hNRB of the genera *Thauera* and *Pseudomonas* are commonly found in low temperature reservoirs. The use of water-soluble, oil-insoluble electron donors such as molasses in sand pack bioreactors with 0.5 PV of residual MHGC oil gave similar increases in oil production. However, in this case molasses, nitrate, and hNRB needed to be injected simultaneously (Gassara et al. 2017). Instead, toluene and nitrate can be injected sequentially because water-injected toluene is retained by the oil. Because toluene is a cheap commodity chemical, this technology is in principle cost effective and may work well for late-life, low-temperature heavy oil reservoirs for which more commonly used EOR technologies like polymer injection are becoming too expensive.

1.2.9 Improved Production: Microbial Communities in Deep Alberta Oil Sands

The oil sands of northern Alberta hold vast amounts of bitumen, which is produced by mining or steam-assisted gravity drainage (SAGD), depending on whether or not the resource is located close to the surface. Bitumen is a highly degraded oil with a high viscosity of 10^5–10^6 cP at room temperature and an API gravity of 6–12°. Its low molecular weight components have been removed by anaerobic microbial degradation (Head et al. 2003). In this process oil components react with water to form methane and CO_2 in reactions catalyzed by syntrophic bacteria and methanogenic Archaea. Bitumen is therefore enriched in polyaromatic hydrocarbons (PAHs) and more polar, structurally diverse components such as resins, asphaltenes, and naphthenic acids. We received samples of SAGD cores as a function of depth from 230 to 280 m below the surface and have used these to determine microbial community composition with next-generation sequencing of PCR-amplified 16S rRNA genes. One of the goals of this research was to evaluate the potential for biotechnologies to aid in the production of bitumen. At the *in situ* temperature of approximately 10° C bitumen is extremely viscous and unlikely to be amenable to microbial processes on a reasonable time scale. Hence, microbial transformations of bitumen would have to be catalyzed at higher temperature, e.g., at 60° C where its viscosity has decreased to that of heavy oil. Microbial community composition determination can also indicate whether further conversion of bitumen components by syntrophs and methanogens is still occurring, and this would offer an alternative way of producing energy from the resource by conversion to methane.

The microbial communities in oil sands cores varied on the cm scale. Although anaerobic communities with high fractions of methanogens were found, communities that were dominated by aerobic microorganisms were more common. Interestingly, these could be located in close proximity to sections of core, which harboured mostly anaerobic communities (An et al., 2013). These results have indicated that harvesting energy from subsurface bitumen reservoirs, which are too deep to mine, by microbial conversion to methane is unlikely to succeed. Methanogenesis is very common in tailings ponds, which result from surface mining. The hydrocarbon used for methanogenesis in these ponds is not the residual bitumen, but rather the low molecular weight hydrocarbon diluent used in the bitumen extraction process. A comparative analysis of six tailings ponds indicated all but the Aurora pond to be methanogenic and harbour a large fraction of methanogens (Wilson et al. 2016). Aurora was the only mine in which no diluent was used in the bitumen extraction process. These results support the notion that oil sands do not harbour high fractions of active methanogens, although some active pockets remain. Syntrophy and methanogenesis can apparently not be supported by the more structurally diverse components, which predominate in bitumen.

1.2.10 Improved Production: Role of Fungi and Thermophiles in Oil Sands Outcrops

The oil sands reach the surface in outcrops located in river valleys. Outcrop samples were collected from the Athabasca oil sands from outcrop cliffs and from the river bottom. Cliff samples are exposed to more variable conditions than experienced in river samples, with respect to temperature (–40 to 60° C), water content, and pH with many cliff samples being acidic. Perhaps as a result, the microbial population in cliff samples was found to be significantly smaller than that in river samples (Wong et al. 2015). The microbial community in cliff samples had high fractions of *Eukarya*, mostly fungi, which were largely absent from river samples. Some cliff samples also had high fractions of thermophilic bacteria. Sun-exposed cliffs were found to reach temperatures of up to 57° C during hot summer days when the air temperature was 30° C. Bitumen is more fluid under these conditions with a viscosity of the order of 10^4 cP, as compared to 10^6–10^7 cP at lower temperatures. Hence bitumen degradation would be more feasible at high temperatures, even though such conditions would be transient and restricted to summer months. Interestingly, many of the thermophiles were anaerobic and found in the center of cliff bitumen tablets. Fungi can potentially degrade molecularly diverse hydrocarbons through random attack with partially reduced reactive oxygen species (ROS), using enzymes such as cytochrome P450 monooxygenases. Metagenomic analyses indicated that these were more abundant in cliff than in river communities.

These results indicated that thermophilic bacteria and aerobic fungi in outcrop communities may contribute to bitumen degradation. Whether these can achieve significant upgrading by decreasing viscosity is being actively investigated.

1.3 POSTSCRIPT

In the previous sections, I have sketched some of the research highlights from our 10-year research program in petroleum microbiology. These results are of interest from both fundamental and applied points of view and will contribute to improved extraction and transport of oil. Much more can be done and many more discoveries will be made in this exciting field in the future, which will further decrease the environmental impact of hydrocarbon fuel extraction.

ACKNOWLEDGMENTS

This work was supported by an NSERC Industrial Research Chair Award to GV from 2007 to 2017, as well as by Baker Hughes, BP, Computer Modeling Group Limited, ConocoPhillips, Dow Microbial Control, Enbridge, Enerplus, Intertek, Oil Search Limited, Saudi Aramco, Shell, Suncor Energy, Yara, and YPF as well as by Alberta Innovates. I thank members of the Petroleum Microbiology Research Group for their input, hard work, and dedication in the past 10 years.

REFERENCES

Agrawal, A., H.S. Nathoo, S. Gieg, et al. 2012. Toluene depletion in produced oil contributes to souring control in a field subjected to nitrate injection. *Environ. Sci. and Technol.* 46:1285–1292.

An, B., Y. Shen, and G. Voordouw. 2017. Control of souring in high salinity Bakken shale oil reservoirs by the production of nitrite by halophilic nitrate-reducing bacteria. *Front. Microbiol.* 8: article 1156:1–14.

An, D., S.M. Caffrey, J. Soh, et al. 2013. Metagenomics of Hydrocarbon Resource Environments Indicates Aerobic Taxa and Genes to be Unexpectedly Common. *Environ. Sci and Technol.* 47:10708–10717.

Callbeck, C.M., X. Dong, I. Chatterjee, et al. 2011. Microbial community succession in a bioreactor modeling a souring low temperature oil reservoir subjected to nitrate injection. *Appl. Microbiol. Biotech.* 91:799–810.

Callbeck, C., A. Agrawal, and G. Voordouw. 2013. Acetate production from oil under sulfate-reducing conditions in bioreactors injected with sulfate and nitrate. *Appl. Environ. Microbiol.* 79:5059–5068.

Chen, C., Y. Shen, D. An, et al. 2017. Use of acetate, propionate and butyrate for reduction of nitrate and sulfate and methanogenesis in microcosms and bioreactors simulating an oil reservoir. *Appl. Environ. Microbiol.* 83(7):e02983-16.

Fida, T.T., C. Chen, G. Okpala, et al. 2016. Implications of limited thermophilicity of nitrite reduction for control of sulfide production in oil reservoirs. *Appl. Environ. Microbiol.* 82:4190–4199.

Gassara, F., N. Suri, P. Stanislav, et al. 2015. Microbially enhanced oil recovery by sequential injection of light hydrocarbon and nitrate in low- and high-pressure bioreactors. *Environ. Sci. and Technol.* 49:12594–12601.

Gassara, F., N. Suri, and G. Voordouw. 2017. Nitrate-Mediated Microbially Enhanced Oil Recovery (N-MEOR) from Model Upflow Bioreactors. *J. Hazard Mater.* 324:94–99.

Gieg, L.M., T.R. Jack, and J.M. Foght. 2011. Biological souring and mitigation in oil reservoirs. *Appl. Microbiol. Biotechnol.* 92:263–282.

Greene, E.A., C. Hubert, M. Nemati, et al. 2003. Nitrite reductase activity of sulfate-reducing bacteria prevents their inhibition by nitrate-reducing, sulfide-oxidizing bacteria. *Environ. Microbiol.* 5:607–617.

Greene, E.A., V. Brunelle, G.E. Jenneman, et al. 2006. Synergistic inhibition of biogenic sulfide production by combinations of the metabolic inhibitor nitrite and biocides. *Appl. Environ. Microbiol.* 72:7897–7901.

Haveman, S.A., E.A. Greene, C.P. Stilwell, et al. 2004. Physiological and gene expression analysis of inhibition of Desulfovibrio vulgaris Hildenborough by nitrite. *J. Bacteriol.* 186:7944–7950.

Head, I.M., D.M. Jones, and S.R. Larter. 2003. Biological activity in the deep subsurface and the origin of heavy oil. *Nature.* 426 (6964):344–352.

Kryachko, Y., X. Dong, C.W. Sensen, et al. 2012. Compositions of microbial communities associated with oil and water in a mesothermic oilfield. *Antonie van Leeuwenhoek.* 101:493–506.

Lambo, A.J., K. Noke, and S.R. Larter. 2008. Competitive, microbially mediated reduction of nitrate with sulfide and aromatic oil components in a low temperature, western Canadian oil reservoir. *Environ. Sci. Technol.* 42:8941–8946.

Mand, J., H.S. Park, T.R. Jack, et al. 2014. The role of acetogens in microbially influenced corrosion of steel. *Front. Microbiol.* 5: article 268:1–14.

Okpala, G., C. Chen, T. Fida, et al. 2017. Effect of thermophilic nitrate-reduction on sulfide production in high temperature oil reservoir samples. *Front. Microbiol.* 8: article 1573:1–13.

Ollivier, B., and M. Magot. 2005. *Petroleum Microbiology*. Washington, DC: ASM Press.

Park, H.S., I. Chatterjee, X. Dong, et al. 2011. Effect of sodium bisulfite injection on the microbial community composition in a brackish-water-transporting pipeline. *Appl. Environ. Microbiol.* 77:6908–6917.

Pinnock, T., J. Voordouw, and G. Voordouw. 2018. Use of carbon steel ball bearings to determine the effect of biocides and corrosion inhibitors on microbially-influenced corrosion under flow conditions. *Appl. Microbiol. Biotechnol.* 102:574–5751.

Sharma, M., D. An, T. Liu, et al. 2017. Biocide-mediated corrosion of coiled tubing. *PloS ONE* 12(7): e0181934.

Shen, Y., A. Agrawal, N. Suri, et al. 2018. Souring Control by Limiting Sulfate Dispersal in a Water-Injected Oilfield. *J. Biotechnol.* 266:14–19.

Suri, N., J. Voordouw, and G. Voordouw. 2017. The effectiveness of nitrate-mediated control of the oilfield sulfur cycle depends on the toluene content of the oil. *Front. Microbiol.* 8: article 956:1–13.

Xue, Y., and G. Voordouw. 2015. Control of microbial sulfide production with biocides and nitrate in oil reservoir simulating bioreactors. *Front. Microbiol.* 6: article 1387:1–11.

Voordouw, G., A. A. Grigoryan, A. Lambo, et al. 2009. Sulfide remediation by pulsed injection of nitrate into a low temperature Canadian heavy oil reservoir. *Environ. Sci. and Technol.* 43:9512–9518.

Voordouw, G., P. Menon, T. Pinnock, et al. 2016. Use of homogeneously-sized carbon steel ball bearings to study microbially-influenced corrosion in oilfield samples. *Front. Microbiol.* 7: article 351:1–9.

Voordouw, G., T. Pinnock, and J. Voordouw. 2017. Effect of biocides and corrosion inhibitors on SRB-mediated MIC under flow conditions. NACE International CORROSION 2017 Conference & Expo, New Orleans, LA, March 27–30, 2017. Paper 9650.

Wilson, S.L., C. Li, E. Ramos-Padrón, et al. 2016. Oil sands tailings ponds harbour a small core prokaryotic microbiome and diverse accessory communities. *J. Biotechnol.* 235:187–196.

Wong, M.L., D. An, S. Caffrey, et al. 2015. Roles of thermophiles and fungi in bitumen degradation in mostly cold oil sands outcrops. *Appl. Environ. Microbiol.* 81:6825–6838.

Youssef, N., M.S. Elshahed, and M.J. McInerney. 2009. Microbial processes in oilfields: culprits, problems and opportunities. *Adv. Appl. Microbiol.* 66:141–251.

Part I

Modeling of Microbiologically Influenced Corrosion (MIC)

2 Modeling of Microbiologically Influenced Corrosion—Limitations and Perspectives

Torben Lund Skovhus, Christopher Taylor, and Richard B. Eckert

CONTENTS

2.1 INTRODUCTION

Microbiologically influenced corrosion (MIC) research in the oil and gas industry has seen a revolution over the past decade with the increased application of molecular microbiological methods (MMM) and new industry standards; however, MIC prediction and modeling are areas that have not been fully developed.

Models are widely used in the oil and gas industry, from predicting new oil reserves in the subsurface to predicting the future oil price and choosing the most optimal materials when designing new facilities. Modeling of corrosion rates from specific corrosion mechanisms has also been successfully applied (Nesic 2007). These corrosion mechanisms include stress corrosion cracking, erosion, corrosion fatigue, and CO_2 corrosion. Models can provide numerous benefits, e.g., guidance on MIC mitigation selection and prioritization, identification of data gaps, a scientific basis for risk-based inspections (RBI), and technical justification for asset design and life-extension.

In the case of managing the threat of MIC and establishing the potential rate of degradation, several models have been developed over the past 15–20 years. However, there are few available methods to systematically map the threat of MIC. This is due to several reasons; including the highly unpredictable nature of microbiological growth, disagreement among scientists and operators regarding which groups of microorganisms influence corrosion the most, and limited knowledge about which (physical, chemical, and biological) factors drive the MIC attack and development over time.

2.1.1 Types of MIC Models

Several types of MIC models exist, but they can generally be classified into four main model categories: empirical models, mechanistic models, risk-based models, and molecular models. For a brief overview of these modeling classes, see Table 2.1.

2.1.1.1 Empirical Models

Empirical MIC models predict corrosion rates based solely on a best-fit or multivariate correlation to experimental datasets (either laboratory or field-based). Commonly measured input parameters include measured biological activity, fluid chemistry,

TABLE 2.1
Classification of Models into Four Categories: Empirical, Mechanistic, Risk-Based, and Molecular

Model Class	Definition	Relevant References
Empirical	Utilize best-fit or multivariate correlation to experimental database (lab or field); can be developed intuitively or data-driven (machine-learning)	(Fatah et al. 2013, Grzelak 2006)
Mechanistic	Simulate processes through development and solution of equations expressed from the basic physics, chemistry, or biological phenomena underlying the corrosion mechanism	(Wolodko et al. 2018; Skovhus et al. 2012; Sørensen et al. 2012; Larsen et al. 2013)
Risk-based	Based on probability of incidence of microbiologically influenced corrosion (MIC) and consequence of failure for a given location and/or set of environmental or material parameters; used to plan maintenance and inspection activities	(Skovhus et al. 2017a)
Molecular	A subset of mechanistic models but aimed at understanding fundamental molecular processes involved in interaction of chemical species with one another and material surfaces	(Irrera et al. 2009; Thomsen et al. 2009; Gece and Bilgiç 2010; Masoud et al. 2014)

process conditions (e.g., temperature, pressure, and flow rate), and physical properties (e.g., material and geometry). Examples include the prediction of corrosion rates from sulfate-reducing bacteria (SRB) as a function of sulfate and sulfide concentration (Fatah et al. 2013) and the statistical analyses made to isolate the most important environmental factors by Grzelak (2006).

2.1.1.2 Mechanistic Models

Mechanistic models also have been developed to predict corrosion rates by attempting to simulate actual physical, chemical, and/or biological processes within a MIC environment, such as within a biofilm or under solid deposits. These mechanistic models are semi-empirical in nature (often calibrated with laboratory or field data), and can provide the user with an improved understanding of the effect of selected input variables and an estimated corrosion rate.

Several mechanistic MIC models have been developed, but very few are actually applied by oilfield operators (Wolodko et al. 2018).

An area of growing interest is the potential use of data from MMM as input to mechanistic MIC models. Currently, there is only one model presented in the public domain that uses MMM to predict MIC. Skovhus et al. (2012) and Sørensen et al. (2012) originally developed this model, which is divided into two steps. The first step estimates the processes that occur prior to initiation of MIC and the second calculates the corrosion rate when MIC has or is actively occurring. The model is based on the use of qPCR (quantitative polymerase chain reaction) to measure the abundance of the *in situ* microbiological consortia of sulfate-reducing bacteria, sulfate-reducing archaea, and methanogens, and does not include physical or chemical parameters. Since 2012, the MIC model has been implemented in pipeline operations in the Danish sector of the North Sea by Maersk Oil (Larsen et al. 2013; Thomsen et al. 2016).

2.1.1.3 Risk-Based Models

Risk-based MIC models are used to predict and identify the potential magnitude and location of MIC threats in oil and gas infrastructure, such as production facilities or pipeline systems. Risk-based methods are most often used in industry for planning inspection and maintenance activities of assets to ensure optimal safety and reliability in resource-constrained operations. Risk-based approaches can be classified as qualitative, quantitative, and semiquantitative. Several approaches have been developed and have been incorporated into industry recommended practices to help corrosion engineers make risk-based decisions about aging equipment (e.g., DNV-RP-G101).

Semiquantitative is a term that describes any methodology that has aspects derived from both the quantitative and qualitative approaches, and it aims to combine the major benefits of each approach (e.g., speed of the qualitative and rigor of the quantitative). Typically, most of the data used in a quantitative analysis are often required for semiquantitative approaches, however in less detail. The models also may not be as rigorous as those used for the quantitative approaches. The results are usually given in consequence and probability categories or as risk numbers, but numerical values may be associated with each category to permit the calculation of risk and the application of appropriate risk acceptance criteria.

An example of how quantitative and qualitative input can be combined to assess the threat of MIC is given in the newly proposed model on how MIC assessment is implemented into the RBI analysis (Skovhus et al. 2017a). The first stage of the model is a screening of the probability of failure (PoF) (significant vs. negligible) based on both quantitative and qualitative input e.g. mitigation efforts, historical data, past failures, temperature, pH, details of microbiological monitoring programs, microbiological groups identified, etc. If the PoF is significant, the second stage of the model comes into play. The second stage of the model aims to rank the PoF from quantitative measures alone; e.g. settlement potential, nutrients availability, mitigation effectiveness, and potential oxygen ingress. The PoF is ranked from 1 to 24 in five different PoF ranking groups from very high to very low (see Figure 2.1).

The procedure will not give a direct measure of the loss of containment of the corrosion circuit due to MIC; however, the procedure provides a *MIC susceptibility measure*. In order to establish the total PoF of a corrosion circuit, the outcome of the MIC assessment must be combined with other credible degradation mechanisms within the corrosion circuit.

The PoF ranking is essential for calculating the risk of the system when combined with consequence of failure (CoF). CoF can be calculated based on known methodologies (DNV-RP-G101). CoF is the consequence given a failure, independent of the failure mechanism, meaning that the consequence of leakage caused by MIC will be the same as the consequence of leakage caused by other mechanisms, such as CO_2 corrosion, for example. There are, however, some differences in consequence modeling for different failure mechanisms, but this is related to leak distribution and is covered by the CoF modeling.

One more thing that needs to be included in the CoF evaluation, which is not relevant for degradation mechanisms other than MIC, is the cost of inspecting and repairing, for example, an offshore production facility after the discovery of MIC. The finding of MIC damage in an oil and gas processing facility often results in the

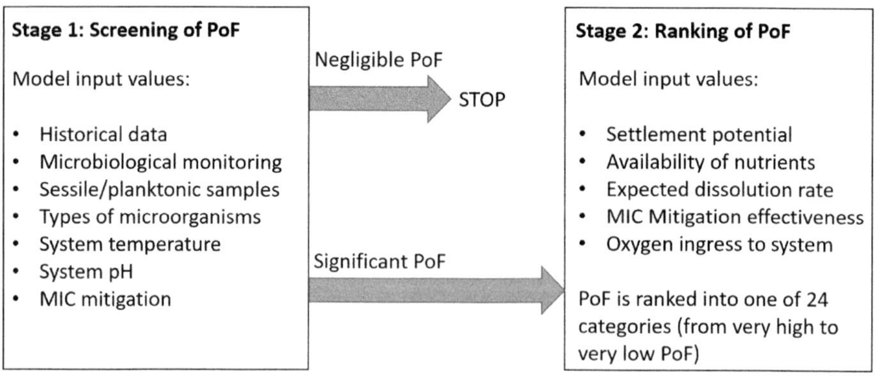

FIGURE 2.1 Overview of a novel procedure for qualitatively assessing the probability of failure (PoF) of topside production facility caused by microbiologically influenced corrosion (MIC). The procedure was developed by Skovhus et al. (2017a).

need for extensive additional inspection because the location or severity of MIC is difficult to predict (as compared to abiotic mechanisms), leading to additional downtime for the facility.

2.1.1.4 Molecular Models

Increasingly, molecular (sometimes also called atomistic when applied to solid materials) modeling has been used to improve the understanding of the mechanisms of those reactions that occur between molecules and also between molecules and surfaces (Broadbelt and Snurr 2000; Richon 2008). This advance in high-fidelity simulation of chemical and catalytic reactions can offer tremendous benefits in the field of corrosion science, including microbiologically influenced corrosion, although these efforts have yet to be fully realized. The most fundamental methods directly solve the quantum mechanical equations that govern the distributions of electrons across a series of atomic centers, thus determining from first principles the patterns of bonding in a molecule or material, which may be expressed as molecular orbitals or electron density probability distributions (Norskov et al. 2010; Levine 2000). Atomistic methods, or molecular mechanics techniques, instead apply a parameterized approach to the strength of bonds between atoms (i.e., a ball and spring model), in order to explore how the energy of a system of atoms changes as the atoms reconfigure during chemical reactions or processes such as diffusion or electromigration (Deshmukh and Sankaranarayanan 2012). Both sets of techniques often require the use of high-performance computer resources (Panchal et al. 2013).

Outputs of molecular modeling methods include electron density distribution (i.e., bonding patterns), the potential energy of an atomistic or molecular system in a given configuration or set of configurations, normal mode analysis (i.e., prediction of vibrational spectroscopy), band structure (i.e., prediction of electronic spectroscopy), equilibrium atomic geometries, location of energy minimal and saddle points (i.e., transition states to chemical reactions, thus allowing prediction of reaction rates from activation energies), solvation energies, stress tensors (for solids), and forces on atoms that can be used for performing molecular dynamics simulations (Taylor and Marcus 2015).

Molecular models have been applied to systems in corrosion, such as the oxidation of metal surfaces (Deshmukh and Sankaranarayanan 2012; Sankaranarayanan and Ramanathan 2008; Jeon et al. 2011; Subbaraman et al. 2013), the hydrogen embrittlement of steels and nickel-based alloys (Lu et al. 2001; Lu et al. 2002; Lu and Kaxiras 2005), the study of the interactions between chemical corrosion inhibitors and metallic or oxidized surfaces (Gece 2008; Gece 2011), and the interactions between amino acids and materials surfaces (Irrera et al. 2009; Thomsen et al. 2009; Gece and Bilgiç 2010; Masoud et al. 2014). Although it has not yet been directly applied to MIC, molecular modeling should have the potential to bring some clarity to the understanding of mechanisms of chemical reactions related to this mode of corrosion.

While overall research in MIC has seen a steady increase since the 1980s, there has been a substantial increase in MIC modeling efforts over the past decade. A number of recent model developments and detailed reviews on the topic highlights this (Skovhus et al. 2012; Skovhus et al. 2017b; Andersen et al. 2017; Singh and Pokhrel 2018; Hashemi et al. 2018; Eckert and Skovhus 2018; Wolodko et al. 2018; Taleb-Berrouane et al. 2018).

2.2 END USER NEEDS FOR MIC MODELS

2.2.1 Process of Managing Corrosion

MIC is a corrosion threat, along with many others, that oil and gas asset owners must address to achieve the design life of their equipment and realize the benefits of their investments. The process of managing corrosion threats falls into a repeating process consisting of three main activities, i.e., threat assessment, identification of mitigation (barriers to the threats), and monitoring the effectiveness of mitigation, as described by Skovhus and Eckert, and others (Skovhus and Eckert 2014; Eckert and Koch 2016; Eckert et al. 2015) (see Figure 2.2).

In the assessment step, asset operators may perform testing and analysis to determine the types of corrosion threat mechanisms that are present. This work could include cathodic protection surveys to assess conditions related to external corrosion or liquid sampling for chemical and microbiological analysis to assess the potential for internal corrosion threats. Assessment often includes some form of inspection to determine whether corrosion has previously caused degradation of the asset. When inspection access is limited, direct assessment protocols, for example, internal corrosion direct assessment for wet gas pipelines, may also be employed to focus the areas of evaluation. The mitigation and prevention step uses information obtained in the previous step about corrosion threat mechanisms to establish measures to reduce or prevent degradation resulting from corrosion. Mitigation of external corrosion is often based on the use of protective coatings and cathodic protection, whereas internal corrosion mitigation could involve the use of corrosion-inhibiting chemicals and biocides, along with mechanical cleaning of the internal surface, installation of liners or enforcing tighter control over product input quality specifications, and monitoring. In the initial design stage, engineers may also perform a life-cycle cost analysis to evaluate the economics of using alloy or nonmetallic materials that are more resistant to corrosion, thereby eliminating the need for regular chemical treatment and inspection. Finally, in the monitoring step, asset owners use various methods to assure that the prevention and mitigation measures being applied to address corrosion are working effectively. Monitoring could include the use of electronic probes or test coupons, routine calendar-based or risk-based inspection, and sampling/analysis of biofilms on monitoring coupons to determine if measures to reduce MIC have the intended effect. Monitoring also provides continuous information to use in ongoing threat assessment, which repeats the cycle of corrosion management (Morshed 2013).

FIGURE 2.2 Corrosion management process. (Skovhus and Eckert 2014.)

End users have needs for MIC models to help perform all three of these corrosion management steps, as discussed further in this section, to optimize the outcomes of each step. In the oil and gas industry, owners/operators of pipelines, offshore platforms and vessels, processing plants, and other assets spend billions of dollars per year in managing the effects of corrosion. In a recent study (Koch et al. 2016) conducted by NACE International that was issued in 2016, the global cost of corrosion was estimated to be $2.5 trillion USD. The failure of infrastructure due to corrosion increases the risk of harming the health and safety of society and causing damage to the environment. As discussed elsewhere in this chapter, the maturity of current MIC models is lacking, and even the molecular and genetic mechanisms that support MIC are not yet fully understood. With this overview in mind, the following sections discuss how end users of improved MIC models could better accomplish effective corrosion management, in contrast to models that already exist.

2.2.2 Assessment

For corrosion threat mechanisms, such as CO_2 corrosion in oil and gas production, a number of mechanistic and deterministic models (Pots et al. 2002; Sooknah et al. 2007) exist for predicting corrosion rates based on the operating environment that will be present. These models are used in the design stage to select the proper alloys and to determine mitigation equipment needs for new assets. The existing CO_2 models (NORSOK M-506), are also used in the operations stage of asset life, to evaluate and predict changes in corrosiveness and for the selection of corrosion inhibitors; however, well-tested, generally accepted rate models do not exist for MIC.

Without the ability to quantify the reduction in MIC susceptibility, or damage rate and incidence frequency, it is often difficult to cost-justify design or operations modifications that could reduce the likelihood of MIC, e.g., making the pipeline capable of accommodating cleaning pigs or providing equipment for biofilm sampling. As a result, assets may be commissioned with few or no MIC mitigative or monitoring capabilities, causing serious shortcomings if MIC is later found to be a threat. Therefore, end users of future MIC models are looking for the ability to predict the likelihood, the locations, and the severity of MIC, based on asset design and operating parameters. Because microorganisms are ubiquitous, asset owners also desire to know the extent to which MIC is a threat as compared to abiotic corrosion mechanisms, since the presence of microorganisms alone is not an indicator of MIC.

2.2.3 Mitigation

In the mitigation step of corrosion management, MIC models could be used to optimize, for example:

- The locations of chemical injection
- The type of chemicals being applied for MIC control
- The frequency, concentration, and contact time of chemical treatment
- The timing of treatment application, relative to operating conditions or events
- The cleaning frequency and types of pigs used

MIC models that considered the effects of mitigation measures on MIC rate or incidence would be extremely useful for optimizing mitigation programs. If such models existed, it is likely that the application of biocides would be reduced, since operators would not need to "over treat" as an "insurance measure" to account for unknowns. In addition, it would be beneficial if MIC models incorporated mitigation effectiveness for both biotic and abiotic corrosion, as in some cases it may be sufficient to use only a corrosion inhibitor, for example, to address both biotic and abiotic corrosion at the same time.

2.2.4 Monitoring

The ability to include MIC as a factor in models for establishing or optimizing a RBI strategy would be a significant benefit. An initial effort in this direction was presented by Skovhus et al. (2017a). MIC is a threat that has not previously been included in RBI program development or the related standards, to any meaningful extent, besides a few recent attempts that have just been published (Skovhus et al. 2017b; Singh and Pokhrel 2018).

Another benefit of having a MIC model from a monitoring viewpoint, e.g., with a Bayesian (probabilistic) approach as proposed by Ayello et al. (2014), would be to identify the monitoring data that provides the most value to the outcome of the model. In other words, instead of attempting to collect all possible data about operating conditions, chemistry, and microbiology, a probabilistic model would illustrate a subset of data types that had the greatest effect on the accuracy of the outcome. From this, asset operators could focus on getting the essential data, rather than increasing cost and risk by gathering all data.

Finally, perhaps the most important use for reliable MIC models would be for integration and analysis of data for decision-making (Eckert and Skovhus 2018). Operators are challenged by corrosion data integration, particularly when molecular microbiological data is produced. The types of data that could be used in an integration model include those listed in Table 2.2.

The ability for a model to integrate information about the physical, chemical, and microbiological environment along with the properties of the metal itself, to predict MIC rates and/or incidence (Skovhus et al. 2017b) would be revolutionary. Operators who are collecting "big data" now, using digital sensors and probes that report data continually, would benefit from the ability to get such data into a MIC model that runs nearly real-time. The ability to observe, in real-time, how changes in different parameters are affecting MIC rates, for example, would be a powerful tool for processing facilities and other types of assets. Another major technical challenge here is that no proven commercial sensors for MIC or MIC precursors currently exist, and the ability to collect data using MMM still relies upon collecting and analyzing physical samples. Atomic hydrogen probes, for example, could serve to provide direct monitoring data for integration into a MIC model, assuming that the role of hydrogen as an intermediary between microbiological metabolism and corrosion was better understood and the nuances affecting hydrogen diffusion through the metal were taken into account. The use of atomic hydrogen as a direct input into a MIC model is of course also challenged by the fact that hydrogen sulfide from abiotic

TABLE 2.2
Data Integration for Microbiologically Influenced Corrosion (MIC) Modeling

Measure Related to Corrosion	Measured Parameters
Fluid composition	Total suspended solids (TSS), salts, metals, volatile fatty acids (VFA), nitrate, sulfate, sulfide, basic sediment and water (BS&W), hydrocarbons, viscosity, etc.
Materials of construction	Grade of carbon steel, stainless steel, corrosion resistant alloys (CRAs), weld metal; nature of seals and gaskets, coating, polymers, liners, etc.
Gas composition	Corrosive gases such as CO_2, H_2S, O_2, etc.
Solids from the system	Solid chemistry, mineralogy, corrosion products, particle-size distribution
Fluid velocity	Liquid and gas velocity, erosion potential, flow regime
Pipeline topography	Low points, dead legs, stagnant zones, slack flow
Corrosion coupons	Corrosion rate (mm/y), pitting rate (mm/y), pit initiation modes, pit propagation, pit density, pit distribution, and pit depth; surface roughness, mechanical damage, erosion
Corrosion measures by online electrochemical sensors	Corrosivity of fluid as monitored by Linear Polarization Resistance (LPR), electrical resistance (ER), hydrogen flux probes, internal fouling probes; dissolved oxygen in treated seawater systems
Inspection data by nondestructive testing	Pitting depth and metal loss by nondestructive scanning techniques; distribution and severity of damage throughout the asset; in-line inspection (ILI) results; permanent ultrasonic sensors
Operating history of the system	Change of fluid type, addition of new wells with new fluid composition, new materials, change of temperature regimes, increase in water production
Chemical treatment of system	History of applied chemicals (e.g. biocide, nitrate, corrosion inhibitor, H_2S scavenger, emulsion breaker, scale inhibitor, etc.); means of application, treatment rates, etc.
Mechanical cleaning of system	Type of pigs, frequency of pigging, combination with biocides or inhibitors, evaluation of effectiveness; analysis of pig returns
Internal/external inspection of system	Visual inspection, camera inspection by remotely operated vehicle (ROV), sensors connected to intelligent pigging, downhole mechanical measures of loss of tubing material (calipers), etc.
Leak/failure history	Failure analysis, root cause analysis (RCA) of previous corrosion failures
Maintenance history	Flow restriction or injectability problems due to solids and fouling; discovery of internal corrosion during routine maintenance

or biotic sources can result in atomic hydrogen generation at the cathode of the half-cell reaction (NACE 2012).

Most likely, a hybrid model that uses some data produced on-stream and other data collected from sampling intermittently would be achievable sooner than a model based on fully automated data sources and sensors.

2.3 INTEGRATED COMPUTATIONAL MATERIALS ENGINEERING

A framework for research in materials science and engineering that is seeing increasing uptake and success is integrated computational materials engineering (Allison et al. 2006). As the name suggests, integrated computational materials engineering (ICME) involves the closely aligned use of computer modeling alongside experimental design and testing to expedite the engineering of new materials with tailored properties. Initial efforts in ICME have claimed to drastically reduce the expense and concept-to-market time period for new materials development and adoption. Recent initiatives, such as the materials genome initiative (Holdren 2011; Warren 2013; de Pablo et al. 2014) and materials informatics (Rajan 2005), are loosely related to ICME, and could be brought into an ICME framework by providing databases and machine learning tools to accelerate the pace of materials engineering projects. Although ICME has been primarily applied in the aerospace industry, there are efforts underway to apply it to corrosion engineering and lifetime prediction (Taylor 2015). In order to do this, the materials modeling processes that have been utilized in ICME need to be augmented by additional modeling of the environmental and interfacial effects that lead to corrosion.

There are four categories of effort or discovery that need to be included in any ICME program: (1) informatics elements, (2) scientific elements, (3) empirical elements, and (4) uncertainty quantification, which includes validation and verification. The interrelations between these categories and some of the elements within each category are expanded upon in Figure 2.3.

As seen in the figure, informatics elements include team-building, data handling, machine learning, and selection of multiphysics elements; scientific elements include the various phenomenological models that are appropriate to the problem. For MIC, this may include damage mechanics, electrochemical modeling, fluid dynamics, metabolics, biochemistry, and thermodynamics. Empirical elements are included as a category on their own to leave room for serendipitous discoveries and/or hunches. The uncertainty quantification elements include methods for dealing with uncertainty and statistical variation, benchmarking models against experiments and/or field data, and identification of knowledge gaps.

A process for guiding the development of ICME projects was developed by Cowles et al. (2012) and applied in the aerospace industry. The process was guided through the use of a checklist that establishes a consensus among the stakeholders on the objectives for the ICME project, and then determines the various elements that need to be modeled, along with an integration pathway that requires (a) establishment of assumptions, inputs, outputs, and boundary conditions; (b) analysis of uncertainty regarding inputs, assumptions, internal parameters, and propagated uncertainties across multiphysics models; (c) model and code verification; and (d) experimental testing for model

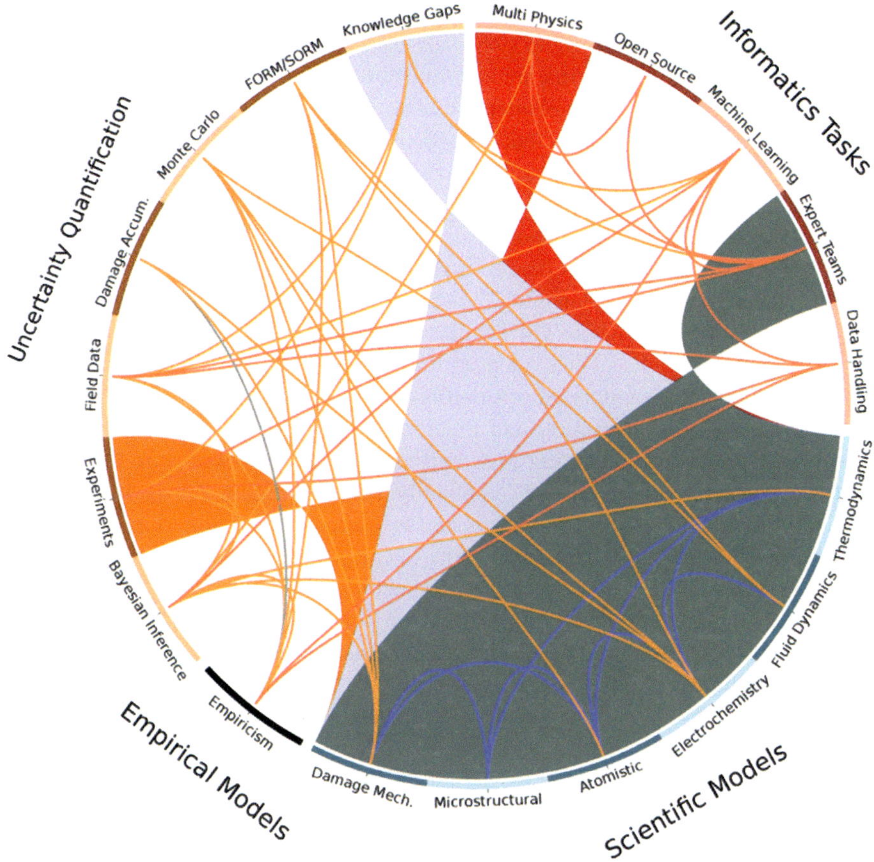

FIGURE 2.3 Activities that might be included in an integrated computational materials engineering project and the linkages between them. Wide ribbons indicate the necessity for coupling scientific models with multi-physics frameworks and experimental testing, as well as their utility in filling knowledge gaps. The lines indicate activities where exchange of information and integration are more direct. (Taylor 2014.)

validation. The checklist approach developed by Cowles et al. (2012) is attractive, as it helps keep a project with many "moving pieces" in scope, and applies a systematic and reproducible approach to quantifying the maturity of any ICME project or tool.

At present, ICME has been utilized as a tool for research projects, although as ICME tools become more mature, it can be expected that the approach will become more mainstream for materials assessments and lifetime prediction projects. The advantage of ICME is the way in which computational modeling is used to decrease the net amount of experiments and lab testing that need to be performed. Modern scientific and engineering computation, while still imperfect, can be used to narrow down the range of experiments that have to be performed. The imperfection of the modeling is gauged by the use of uncertainty quantification. The first widely publicized example of ICME was in the work of Olson et al. (2013) to develop cyber steels that have high strength yet high resistance to hydrogen embrittlement. The Rice-Wang model (Rice and Wang 1989) was used to take first-principles "molecular modeling" databases for the interactions of alloy solutes with grain-boundaries to drive the selection of alloying elements that could promote hydrogen resistance in these steels. Other examples include the computationally assisted design of corrosion inhibitors using molecular modeling (Mondal and Taylor 2014) and the use of Bayesian networks to develop microstructure-performance relationships for lightweight alloys based on an integration of electrochemical modeling, experimental and field studies and microstructural analysis (Taylor et al. 2018).

The current project "Managing Microbiologically Influenced Corrosion in Canada's Offshore & Onshore Oil Production Operations" sponsored by Genome Canada (www.geno-mic.ca) has adopted an ICME-like approach to the study of MIC. An expert team comprised of microbiologists, ontologists, molecular modelers, and risk assessment experts and materials corrosion engineers has joined to examine the processes underlying MIC and to relate those processes to the assessment of the risk of MIC in oil and gas pipeline systems. Figure 2.4 shows some of the flow of information between models that will accelerate the development of the risk assessment tools in the project.

The activities to be performed in the project will include informatics elements (ontologies, aggregation of literature data through text mining, code development), science elements (molecular modeling, biochemical process modeling, electrochemical modeling), empirical studies based on field data and laboratory work, as well as uncertainty quantification in the form of risk assessment and probabilistic analysis. To coordinate such multidisciplinary work, a tool such as an "input-output" matrix can be used that visually charts the transfer of information between the various research groups. Each group engaged in the project is listed as a potential source of data and assigned a "row" in the matrix. Likewise, each group is assigned a "column," which is associated with being a target for data produced by other groups. Each cell in the matrix, then, lists the data elements (if any) that will be transmitted from the "source" group (i.e., the row) to the "target" group (i.e., the column). The cells in the input-output matrix can be attached to an anticipated schedule and include uncertainty quantification and propagation elements.

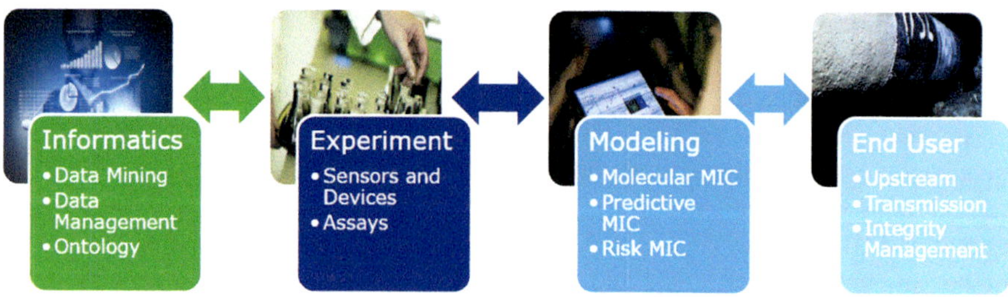

FIGURE 2.4 Synergistic activities and flow of information between the informatics tasks, experimental activities, model construction, and translation to end user within the project "Managing Microbiologically Influenced Corrosion in Canada's Offshore & Onshore Oil Production Operations."

2.4 HOW INDUSTRY COULD BENEFIT FROM MIC MODELS

2.4.1 Application of MIC Models

Application of MIC models to support the three steps of corrosion management was described in section 2.2 and shown in Figure 2.2. MIC models were shown to be relevant and potentially useful for different purposes in asset design and operation. The potential benefits of MIC models include reduction of pipeline, tank, and vessel leaks and failures; optimization of the effectiveness of mitigation measures; support for asset design to extend functional life and increase sustainability; optimization of monitoring measures; and improved, proactive inspection programs to detect MIC before it becomes a serious problem.

Regarding the application of MIC models, there could be different levels of analysis for different end user needs. For example, an end user may only need a high-level understanding of whether MIC is a threat for which to plan overall mitigation measures during the initial stages of asset design. One would anticipate that the data required for this level of modeling would be far less than the data required for a detailed analysis that predicted where MIC would occur and the rate of growth.

A model might also be used on a repeating, periodic basis to assess whether changing conditions are increasing the potential for MIC. In this case, the initial assessment could require more data for detailed analysis and the reevaluation could be performed by simply rerunning the same model using any data that has changed.

Models could also be run in modules that were applicable based on the nature of the environment associated with the asset. For example, a model may have different modules that would be used based on the range of temperatures experienced by the asset or the type of fluid transported in the asset. The different modules would require inputs that are specific to the corrosion threats that are prevalent in each different service type. One example of where this approach is applicable is where a tank carries oil, produced water, and treated seawater at different periods during its lifecycle. Different modules would be used for each service type since the chemical and microbiological conditions are likely to be quite different in each case. Another example is where a pipeline is converted from one service type to another later in its life.

Since MIC results from a combination of chemical, physical, and microbiological conditions, it is certain that information from the process and corrosion monitoring will need to be incorporated into MIC models. Corrosion related data includes coupon and probe results, inspection results, composition monitoring and analysis, product quality monitoring, and so on. Corrosion mitigation measures that are being applied will also influence the outcome of MIC models and will need to be considered and incorporated.

Finally, corrosion and operational data will need to be carefully integrated with validated MMM data from surface biofilms, where possible. Reliable information about the abundance, diversity, and activity of microorganisms in biofilms can be collected using a number of different MMM, such as qPCR, RT-qPCR (reverse transcription qPCR), FISH (fluorescence *in situ* hybridization), ATP (adenosine triphosphate) testing, and next-generation sequencing (NGS), and each method provides a different type of data (Skovhus et al. 2007; Whitby and Skovhus 2011; Juhler et al. 2012; NACE 2018).

In many assets, such as pipelines, surface-related samples are quite difficult to collect. Surface solids and biofilms may be accessible when equipment is removed for service; when a cut out is planned, or when a leak or failure is removed from service. Deposits may also originate as pigging residues from cleaning operations, but these may sometimes be contaminated by cleaning agents or oxygen and are difficult to relate to their point of origin in the system. Asset operators are often unprepared for proper collection and transport of samples intended for MMM analysis. Improved published guidelines are needed by industry for collection of microbiological samples, so reliable data is obtained, integrated, and analyzed properly. In addition, the oil and gas industry needs an increased awareness of the importance of proper sample collection and the value of MMM in MIC diagnostics and modeling in combination with traditional forensic analysis.

2.4.2 Which Models Could Provide the Greatest Benefits?

MIC models could be either molecular, mechanistic, deterministic, or predictive (or other); however, from an end-user standpoint, models could be identified based on their intended outputs. For example, models could calculate MIC susceptibility, rate, incidence, or perhaps all three in a fully integrated model. Such terminology clearly identifies the intended application of the model by end users. An example is provided in Table 2.3.

Identifying MIC models from an end-user perspective could have another benefit, which is helping scientists and researchers who are working on various types of models integrate their work to produce models that solve industry problems. Effective models should also produce results that are verifiable using data collected from assets that are in service. From an industry perspective, end users also need to know the boundaries under which the model provides reliable output data. It is unlikely that any one model would fit all situations and conditions. For example, under certain conditions, sulfate reduction by bacteria or archaea that produce sulfide might drive the primary corrosion mechanism; however, this will not be true for all cases, such as where methanogens are also involved. In other environments with different redox conditions and electron acceptors/donors, the primary MIC mechanism could involve synergies with other microorganisms or a completely different mechanism. Since corrosion occurs through a process of initiation and growth, the microorganisms and the manner in which they are involved in the corrosion process at different stages may also change over time.

Finally, the method of reporting modeling data is an important consideration, so that end users grasp the significance of the data relative to existing assets and corrosion management programs. Linking model output with visual GIS or GPS based displays can help end users apply modeling data more effectively and intuitively.

2.4.3 Bridging the Gap: MIC Management Today vs. Future

The fact that MIC mechanisms and the conditions that promote them are not fully understood today is somewhat surprising, given the advances in other areas of science such as human genetics, computer technology, and space exploration. Biofilms and microorganisms that affect human health are better understood than those found in industry (Muyzer and Marty 2014). Although the effects of biofilms

TABLE 2.3
Potential Types of MIC Models and End User Benefits

Potential Model "Types"	Output Provided	End User Benefits
Susceptibility	Likelihood of microbiologically influenced corrosion (MIC) as a threat under given set of conditions	• Assess basic threat during design and operation • Understand influence of various input parameters on susceptibility • Understand which parameters to collect and monitor • Incorporate in risk management • Prioritize condition assessment, inspection
Rate	Damage rate of MIC under a given set of conditions	• Rate-based threat assessment during design and operation • Better for inclusion into risk-based inspection (RBI) programs, timing of interventions • Life-cycle cost analysis if model includes effects of mitigation and materials • Numerical output • Understand which parameters to collect and monitor • Incorporate in risk management
Incidence	Location and frequency of MIC initiation (distribution in the asset) under a given set of conditions	• Traditional inspection and RBI planning • Targeting mitigation to the most susceptible parts of assets • Understand conditions that drive distribution • Design to address areas where MIC is expected • Guidance for monitoring programs • Incorporate in risk management
Integrated: biotic and abiotic	1. Degree to which the most severe corrosion mechanism is abiotic or biotic 2. Influence of physical, operational, and design factors on rate and distribution 3. Unmitigated and mitigated damage rates of singular or combined mechanisms	• Selection of mitigation that will be the most effective • Understand which parameters to collect and monitor • Optimize mitigation and monitoring programs • Guide design and operations to avoid promoting threats • Incorporate in risk management • Provide traditional inspection and RBI program guidance

on engineered assets may not be as obvious as those that can occur in the human body, the long-term effects of MIC have serious consequences. Extending the safe lifecycle of engineered structures benefits society and reduces the use of materials and energy needed to replace corroded assets, increasing the drive toward materials sustainability (Taylor et al. 2016).

MIC models are an important vehicle through which advancements in knowledge can be delivered and applied to real-world problems. Models take scientific and mathematical principles and present them in a way that someone without a deep understanding of the detailed principles can apply. In one sense, MIC is an ideal phenomenon to model since it involves multiple scientific disciplines that interact in dynamic ways; yet at the same time, the multidisciplinary nature of MIC challenges the developers of such models. A high level of cooperation, communication, and teamwork would be required of any multidisciplinary team that worked on developing a MIC model, since social, political, economic, institutional, and technical barriers must often be overcome to conduct interdisciplinary work effectively. Conceptual frameworks can provide a structure to aid communications between those involved in a large-scale MIC project; promoting a coordinated multidisciplinary approach.

In the pipeline industry, in-line inspection (ILI) is generally considered a "gold standard" technology for integrity assessment. During ILI, a tool (i.e., "smart pig") is inserted into the pipeline and travels with the flow of the fluid from point A to point B (i.e. from the pig launcher to pig receiver). The ILI tool uses different technologies, such as magnetic flux leakage and ultrasonics, to detect anomalies in the pipe wall and measure their size. When the ILI run is completed, the inspection data is downloaded from the tool and analyzed to determine if repairs are needed on the pipeline. The industry and regulators place a high value on ILI data to demonstrate that the pipeline is safe to operate.

MIC, on the other hand, does not yet have a "gold standard" technology through which its presence can be reliably verified and on which industry will depend. Numerous indirect forms of data are presently integrated and analyzed to establish some level of confidence that MIC is an active threat or not. Often the absence of obvious abiotic corrosion mechanisms results in the diagnosis of MIC. The process for data integration and analysis for MIC diagnosis is not standardized and there is not a clear scientific consensus on how to make a valid determination of MIC. A multidisciplinary MIC model could provide the means to solve this problem and a platform for collaboration and consensus. Such a MIC model could suit a variety of end uses, e.g., prediction, diagnosis, asset design and operation, and maintenance. Users would also need guidance on the correct sampling methods and analytical tools, e.g., MMM, to collect high quality data for use in the model.

2.5 CONCLUDING REMARKS

A vision for this suite of tools and models is emerging in the project "Managing Microbiologically Influenced Corrosion in Canada's Offshore & Onshore Oil Production Operations" sponsored by Genome Canada with the support of multiple industry and academic partners (www.geno-mic.ca). The project has participation from six academic institutions from North America and Europe alongside

17 companies and eight federal and provincial agencies. The project is being carried out by a multidisciplinary team of researchers with experience in microbiology, genomics, production chemistry, corrosion, risk assessment, and engineering.

REFERENCES

Allison, J., D. Backman, and L. Christodoulou. 2006. Integrated computational materials engineering: A new paradigm for the global materials profession. *JOM*. 58(11):25–27. https://doi.org/10.1007/s11837-006-0223-5.

Andersen, E.S., T.L. Skovhus, and E. Hillier. 2017. Review of Current Models for MIC Management. In *Microbiologically Influenced Corrosion in the Upstream Oil & Gas Industry*, ed. Skovhus, T.L., D. Enning, and J. Lee, 289–305. Boca Raton, FL: CRC Press. ISBN 978-1-49872-656-6.

Ayello, F., S. Jain, N. Sridhar, and G.H. Koch. 2014. Quantitative Assessment of Corrosion Probability—A Bayesian Network Approach. *Corrosion*. (online):1128–1147. http://doi.org/10.5006/1226.

Broadbelt, L.J., and R.Q. Snurr. 2000. Applications of molecular modeling in heterogeneous catalysis research. *Applied Catalysis*. 200(1–2):23–46. https://doi.org/10.1016/S0926-860X(00)00648-7.

Cowles, B., D. Backman, and R. Dutton. 2012. Verification and validation of ICME methods and models for aerospace applications. *Integrating Mater. and Manuf. Innov.* 1:2. https://doi.org/10.1186/2193-9772-1-2.

de Pablo, J.J., B. Jones, C.L. Kovacs, V. Ozolins, and A.P. Ramirez. 2014. The Materials Genome Initiative, the interplay of experiment, theory and computation. *Current Opinion in Solid State and Materials Science*. 18(2):99–117. https://doi.org/10.1016/j.cossms.2014.02.003.

Deshmukh, S.A., and S.K.R.S. Sankaranarayanan. 2012. Atomic scale characterization of interfacial water near an oxide surface using molecular dynamics simulations. *Phys. Chem. Chem. Phys.* 14:15593–15605. https://doi.org/10.1039/C2CP42308A.

DNV-RP-G101. Recommended Practice DNV-RP-G101: Risk Based Inspection of Offshore Topsides Static Mechanical Equipment. Høvik, Norway: Det Norske Veritas, 2010. https://rules.dnvgl.com/docs/pdf/DNV/codes/docs/2002-01/RP-G101.pdf.

Eckert, R.B., B. Graver, and T.L. Skovhus. 2015. Corrosion Management of MIC Contributes to Pipeline Integrity. *CORROSION 2015*, paper no. C2015-5729. Houston, TX: NACE International.

Eckert, R.B., and G.H. Koch. 2016. Improving Corrosion Control Efficiencies using Corrosion Management Principles. SPE-179945-MS. *Society of Petroleum Engineers, International Oilfield Corrosion Conference and Exhibition*. Scotland, UK. https://doi.org/10.2118/179945-MS.

Eckert, R.B., and T.L. Skovhus. 2018. Advances in the Application of Molecular Microbiological Methods in the Oil and Gas Industry and Links to Microbiologically Influenced Corrosion. *International Biodeterioration and Biodegradation* 126:169–176. http://doi.org/10.1016/j.ibiod.2016.11.019.

Fatah, M., M. Ismail and B. Wahjoedi. 2013. Empirical equation of sulphate reducing bacteria (SRB) corrosion based on abiotic chemistry approach. *Anti-Corrosion Methods and Material*. 60(4):206–212.

Gece, G. 2008. The use of quantum chemical methods in corrosion inhibitor studies. *Corrosion Science*. 50:2981–2992. https://doi.org/10.1016/j.corsci.2008.08.043.

Gece, G. 2011. Drugs: A review of promising novel corrosion inhibitors. *Corrosion Science*. 53(12) 3873–3898. https://doi.org/10.1016/j.corsci.2011.08.006.

Gece, G., and S. Bilgiç. 2010. A theoretical study on the inhibition efficiencies of some amino acids as corrosion inhibitors of nickel. *Corrosion Science.* 52(10):3435–3443. https://doi.org/10.1016/j.corsci.2010.06.015.

Grzelak, A. 2006. A statistical approach to determine Microbiologically Influenced Corrosion (MIC) Rates of underground gas pipelines. M.S. Thesis, Delft, Netherlands: Delft University of Technology.

Hashemi, S.J., N. Bak, F. Khan, K. Hawboldt, L. Lefsrud, and J. Wolodko. 2018. Bibliometric analysis of microbiologically influenced corrosion (MIC) of oil and gas engineering systems. *Corrosion.* 74(4):468–486. https://doi.org/10.5006/2620.

Holdren, J.P. 2011. *Materials Genome Initiative for Global Competitiveness.* National Science and Technology Council, Washington, D.C.

Irrera, S., D. Costa, K. Ogle, and P. Marcus. 2009. Molecular Modeling by DFT of 1, 2-diaminoethane adsorbed on the Zr-terminated and O-terminated, anhydrous and hydroxylated ZnO(0001) surface. *Superlatt. Microstruct.* 49:19–24. https://doi.org/10.1016/j.spmi.2008.11.021.

Jeon, B., S.K.R.S. Sankaranarayanan, A.C.T. van Duin, and S. Ramanathan. 2011. Atomistic insights into aqueous corrosion of copper. *J. Chem. Phys.* 134:234–706. https://doi.org/10.1063/1.3599090.

Juhler, S., T.L. Skovhus, and I. Vance. 2012. *A practical evaluation of 21st century microbiological techniques for the oil and gas industry.* Energy Institute. ISBN 978-0-85293-638-2.

Koch, G., J., Varney J., N. Thompson, O. Moghissi, M. Gould, and J. Payer. 2016. IMPACT (International Measures of Prevention, Application, and Economics of Corrosion Technologies) Study. Report, NACE International, Houston, TX.

Larsen, J., K. Sørensen, S. Juhler, and D. Pedersen. 2013. The application of molecular microbiological methods for early warning of MIC in pipelines. *CORROSION 2013*, paper no. 02029. Houston, TX: NACE International.

Lee, J.S., and B.J. Little. 2017. Management of MIC in the Oil and Gas Industry. In *Microbiologically Influenced Corrosion in the Upstream Oil & Gas Industry*, ed. T.L. Skovhus, D. Enning, and J. Lee, 157–175. Boca Raton, FL: CRC Press. ISBN 978-1-49872-656-6.

Levine, I.N. 2000. *Quantum Chemistry*, 5 ed. Upper Saddle River, NY: Prentice Hall. ISBN 978-8-12033-064-1.

Lu, G., and E. Kaxiras. 2005. Hydrogen embrittlement of aluminum: The crucial role of vacancies. *Phys. Rev. Lett.* 94:155501–155504. https://doi.org/10.1103/PhysRevLett.94.155501.

Lu, G., D. Orlikowski, I. Park, O. Politano, and E. Kaxiras. 2002. Energetics of hydrogen impurities in aluminum and their effect on mechanical properties. *Phys. Rev. B.* 65:64102–64109. https://doi.org/10.1103/PhysRevB.65.064102.

Lu, G., Q. Zhang, N. Kioussis, and E. Kaxiras. 2001. Hydrogen-enhanced local plasticity in aluminum: An ab initio study. *Phys. Rev. Lett.* 87:95501–95504. https://doi.org/10.1103/PhysRevLett.87.095501.

Masoud, M.S., M.K. Awad, A.E. Ali, and M.M.T. El-Tahawy. 2014. Molecular structure of amino alcohols on aluminum surface. *Journal of Molecular Structure.* 1063(0):51–59. https://doi.org/10.1016/j.molstruc.2014.01.019.

Mondal, S.K., and S.R. Taylor. 2014. The identification and characterization of organic corrosion inhibitors: Correlation of a computational model with experimental results. *J. Electrochem. Soc.* 161:C476–C485. https://doi.org/10.1149/2.0191410jes.

Morshed, A. 2013. The Evolution of the Corrosion Management Concept. *Materials Performance.* 52(8):66.

Muyzer, G., and F. Marty. 2014. Molecular Methods in Microbiologically Influenced Corrosion Research, Monitoring and Control. In *Applications of Molecular Microbiological Methods*, ed. T.L. Skovhus, S. Caffrey, and C. Hubert, 7–22. Caister Academic Press, Poole, U.K. ISBN 978-1-908230-31-7.

NACE Report 3T199. 2012. Techniques for Monitoring Corrosion and Related Parameters in Field Applications. NACE International, Houston, TX.
NACE Standard Test Method TM0212-2018. Detection, Testing, and Evaluation of Microbiologically Influenced Corrosion on Internal Surfaces of Pipelines. 2018. Houston, TX: NACE International. TM0212-2018. *Detection, Testing, and Evaluation of Microbiologically Influenced Corrosion on Internal Surfaces of Pipelines.* 2018. Houston, TX: NACE International.
Nesic, S. 2007. Key issues related to modeling of internal corrosion of oil and gas pipelines—A review. *Corrosion Science.* 49:4308–4338. https://doi.org/10.1016/j.corsci.2007.06.006.
Norskov, J.K., F. Abild-Pedersen, F. Studt, and T. Bligaard. 2010. Density Functional Theory in Surface Chemistry and Catalysis. *Proc. Natl. Acad. Sci.* 108(3):937–943. https://doi.org/10.1073/pnas.1006652108.
NORSOK Standard M-506. 2005. CO_2 Corrosion Rate Calculation Model (Rev. 2). Lysaker, Norway: NORSOK. https://www.standard.no/en/sectors/energi-og-klima/petroleum/norsok-standard-categories/m-material/m-5061/.
Olson, G.B. 2013. Genomic materials design: The ferrous frontier. *Acta Mater.* 61:771–781. https://doi.org/10.1016/j.actamat.2012.10.045.
Panchal, J.H., S.R. Kalidindi, and D.L. McDowell. 2013. Key computational modeling issues in integrated computational materials engineering. *Computer-Aided Design.* 45:4–25. https://doi.org/10.1016/j.cad.2012.06.006.
Pots, B.F.M., S.D. Kapusta, R.C. John, M.J.J.S. Thomas, I.J. Rippon, T.S. Whitham, and M. Girgis. 2002 Improvements on De Waards-Milliams corrosion prediction and Applications to corrosion management. *Corrosion 2002*, paper no. 02235. Houston, TX: NACE International.
Rajan, K. 2005. Materials Informatics. *Materials Today.* 8:38–45. https://doi.org/10.1016/S1369-7021(05)71123-8.
Rice, J.R., and J.S. Wang. 1989. Embrittlement of Interfaces by Solute Segregation. *Mater. Sci. Eng.* A107:23–40. https://doi.org/10.1016/0921-5093(89)90372-9.
Richon, A.B. 2008. Current status and future direction of the molecular modeling industry. *Drug Discov. Today.* 13:665–669. https://doi.org/10.1016/j.drudis.2008.04.008.
Sankaranarayanan, S.K.R.S., and S. Ramanathan. 2008. Low temperature oxidation and ultrathin oxide growth on zirconium in the presence of atomic oxygen: a modeling study. *J. Phys. Chem. C.* 112:17877–17882. https://doi.org/10.1021/jp804872u.
Singh, M., and M. Pokhrel. 2018. A Fuzzy logic-possibilistic methodology for risk-based inspection (RBI) planning of oil and gas piping subjected to microbiologically influenced corrosion (MIC). *International Journal of Pressure Vessels and Piping.* 159:45–54. https://doi.org/10.1016/j.ijpvp.2017.11.005.
Skovhus, T.L., B. Højris, A.M. Saunders, T.R. Thomsen, M. Agerbæk, and J. Larsen. 2007. Practical Use of New Microbiology Tools in Oil Production. *SPE 109104. SPE Offshore Europe Conference 2007.* Scotland, UK. https://doi.org/10.2118/109104-PA.
Skovhus, T.L., and R.B. Eckert. 2014. Practical Aspects of MIC Detection, Monitoring and Management in the Oil and Gas Industry. *Corrosion 2014* - paper 3920. Houston, TX: NACE International.
Skovhus, T.L., E.S. Andersen, and E. Hillier. 2017a. Management of Microbiologically Influenced Corrosion in Risk Based Inspection Analysis. SPE-179930-PA. *SPE Production & Operations.* https://doi.org/10.2118/179930-PA.
Skovhus, T.L., L. Holmkvist, K. Andersen, H. Pedersen, and J. Larsen. 2012. MIC risk assessment of the Halfdan Oil Export Spool. *SPE International Conference and Exhibition on Oilfield Corrosion.* Society of Petroleum Engineers. https://doi.org/10.2118/155080-MS.
Skovhus, T.L., R.B. Eckert, and E. Rodrigues. 2017b. Management and control of microbiologically influenced corrosion (MIC) in the oil and gas industry—overview and a North Sea case study. *Journal of Biotechnology.* 256:31–45. https://doi.org/10.1016/j.jbiotec.2017.07.003.

Sooknah, R., S. Papavinasam, and R.W. Revie. 2007. Modeling the Occurrence of Microbiologically Influenced Corrosion. *CORROSION 2007*, paper no. 07515. Houston, TX: NACE International.

Sørensen, K., U. Thomsen, S. Juhler, and J. Larsen. 2012. Cost efficient MIC management system based on molecular microbiological methods. *CORROSION 2012*, paper no. 01111. Houston, TX: NACE International.

Subbaraman, R., S.A. Deshmukh, and S.K.R.S. Sankaranarayanan. 2013. Atomistic Insights into Early Stage Oxidation and Nanoscale Oxide Growth on Fe(100), Fe(111) and Fe(110) Surfaces. *J. Phys. Chem. C.* 117:5195–5207. DOI: 10.1021/jp312514m.

Taleb-Berrouane, M., Khan, F., Hawboldt, K., Eckert, R. and Skovhus, T.L., 2018. Model for microbiologically influenced corrosion potential assessment for the oil and gas industry. Corr. Eng. Sci. Tech, 53:5, 378-392, DOI: 10.1080/1478422X.2018.1483221.

Taylor, C.D., L. Bland, G. Frankel, J. Locke, Y. Zhu, J. Garofano, and K. Smith. 2018. Predicting Corrosion Fatigue Behavior using a Bayesian Network that Integrates Microstructure, Electrochemistry and Fracture Mechanics. *CORROSION 2018*, paper no. 011228. Houston, TX: NACE International.

Taylor, C.D., B. Gully, A.N. Sánchez, E. Rode, and A.S. Agarwal. 2016. Towards Materials Sustainability through Materials Stewardship. *Sustainability*. 8(10):1001. https://doi.org/10.3390/su8101001.

Taylor, C.D. 2014. Integrated Multiscale Modeling of Materials, in *DNV GL Strategic Research and Innovation Position Paper*. DNV GL: Hovik, Norway.

Taylor, C.D. 2015. Corrosion informatics: an integrated approach to modeling corrosion. *Corr. Eng. Sci. Tech.* 50:490–508. https://doi.org/10.1179/1743278215Y.0000000012.

Taylor, C.D., and P. Marcus. 2015. *Molecular Modeling of Corrosion Processes*. ECS-Wiley: Hoboken, NJ.

Thomsen, L., M.T. Wharmby, D.P. Riley, G. Heldd, and M.J. Gladys. 2009. The adsorption and stability of sulfur containing amino acids on Cu. *Surface Science*. 603(9):1253–1261. https://doi.org/10.1016/j.susc.2009.03.014.

Thomsen, U.S., R.L.C. Meng, and J. Larsen. 2016. Monitoring and Risk Assessment of Microbiologically Influenced Corrosion in Offshore Pipelines. *CORROSION 2016*, paper no. 7194. Houston, TX: NACE International.

Warren, J.A. 2013. The Materials Genome Initiative at National Institute of Standards and Technology. Cited 2014. April 21, 2014. http://www.nist.gov/mgi/.

Whitby, C., and T.L. Skovhus. 2011. *Applied Microbiology and Molecular Biology in Oilfield Systems*. New York: Springer Publisher.

Wolodko, J., T. Haile, F. Kahn, C. Taylor, R. Eckert, S.J. Hashemi, A.M. Ramirez, and T.L. Skovhus. 2018. Modeling of Microbiologically Influenced Corrosion (MIC) in the Oil and Gas Industry—Past, Present and Future. *CORROSION 2018*, paper no. 011398. Houston, TX: NACE International.

3 Modeling Microbiologically Influenced Marine Corrosion of Steels

Robert E. Melchers

CONTENTS

3.1 INTRODUCTION

The marine corrosion of structural steels is of particular interest for various infrastructures, particularly those along or near sea and ocean shorelines such as rail and road bridges, wharves, jetties, industrial facilities, and for various maritime structural systems including pipelines, offshore platforms, and shipping of various kinds. In all cases the ultimate drivers are the technical safety of the item(s) of infrastructure, the safety to personnel and users, and, as always, the economics involved. Increasing the penalties for failure to achieve adequate safety can be severe, both at various economic levels for organizations responsible for construction, execution, maintenance, and regulatory control and supervision, but also at the personal level for those involved and who could be held responsible.

Whereas once public utilities and many industries thought and acted in terms of "build and forget" without much consideration of eventual increasing maintenance and ultimately replacement or abandonment of infrastructure, increasing emphasis is being given to "lifetime" management and the associated lifetime risks and costs. In part this has come about because of the ever-larger part of infrastructure budgets being devoted to "maintain and repair" instead of new-build. In part also there is increasing realization that abandonment as such as seldom acceptable in modern society and that there may be very high costs associated with demolition and site reinstatement as well as the costs of new infrastructure. To make sensible decisions about lifetime management and the associated economic and engineering aspects (which usually are highly interdependent), the question of expected lifetime of infrastructure arises. For most commercial infrastructure such as buildings, and industrial infrastructure such as pipelines and tanks and manufacturing facilities, the expected life is somewhere in the region of 20–50 years. For public infrastructure such as smaller road bridges on major but not arterial roads it might be 100 years and for major structures 200 years. These numbers pale into insignificance compared to lifetimes of 10,000–100,000 years being considered as desirable for nuclear waste containers and nuclear-critical facilities.

Except for some highly specialized applications, the most common materials for infrastructure are reinforced concrete and mild and low alloy steels. Steels are still predominant in coastal and offshore facilities, in part because of ease and speed of construction and hence lower costs in what are often hostile construction and operational environments. Typically these are thought of as having protective coatings (paints) or cathodic protection (sacrificial anodes or impressed current). This applies also to reinforced concrete structures. In fact, these protective systems themselves tend to have relatively short lives compared with the infrastructure they are meant to protect and this then raises another economic aspect. In practice the protective systems often are not well maintained. In other cases they are technically infeasible. In yet other cases the infrastructure was built many years ago without cathodic protection (CP) and it cannot be sensibly retrofitted. For these reasons interest remains in the prediction of the expected future rate of corrosion.

The motivation for the research summarized below derives from infrastructure under the action of deterioration caused by corrosion, and in the context considered herein, marine corrosion. The next section provides more background about the very specific context in which infrastructure corrosion must be considered, the need for prediction of likely future scenarios, and the need to have, for that purpose, mathematical models based, as best as practically possible, on science and calibrated, again as best as possible, to real-world data. This includes, particularly for marine corrosion, the need to include microbiological effects. While often mentioned, and sometimes assigned the "explanation of last resort," microbiologically influenced corrosion (MIC) and microbiological effects more generally have been given scant quantitative attention, including in the corrosion literature. The manner in which this has been and is being addressed is reviewed further below, and because longer term corrosion is of most practical interest, drawing heavily on the author's research. Several application examples also are given.

3.2 MOTIVATION—PREDICTION AND AVOIDING *HOCUS POCUS*

The context for quantitative inputs of MIC into infrastructure design and assessment techniques is important. Particularly for infrastructure systems, such techniques rely heavily on quantitative models to, in essence, predict whether a structure will perform adequately and remain sufficiently safe for a defined or expected lifetime. Proof testing for simple items such as light bulbs, furniture, and even automobiles has long been used but is not acceptable for more complex and expensive infrastructure, such as a bridge, a residential or commercial building, or a nuclear facility, etc. For these it is not economically feasible, socially acceptable, and perhaps ethically responsible to rely on repeated failures to estimate the failure rate. Typically only one structure is built and the risk of failure must be expected to be extremely low. Hence the underlying philosophy is prediction of the likely scenario, and utilization for that of mathematical models, based on engineering science and physics and calibrated to a wider range of experience and very specialized testing in laboratories. Anything else runs the risk of being no more than *hocus pocus*.

The necessary predictions for infrastructure safety and serviceability must be made at the time of design. Evidently, this involves prediction of the degree to which the design itself is satisfactory, the degree of success of the construction process(es), the quality and adequacy of the materials actually supplied, and, importantly, the loading systems and their severity applied to the structure over its complete expected life. Since the latter may be 50–100 years, a high degree of accuracy in prediction is necessary. Modern practice and theory dictates that such prediction is based on a substantial use of applied probability theory (Melchers and Beck 2018). Estimates must be made of the probability of failure or success over the expected lifetime. Simply ranking one design ahead of another is not enough; estimates must be available to say the probability of failure (in whatever mode is of interest) is less than, for example, 10^{-6}/year (Melchers and Beck 2018). Such estimates are a long way from what normally is done in corrosion science and corrosion engineering (Melchers 2005a). It is within the above scenario that any model for material deterioration such as for marine corrosion and pitting of steel must fit.

A suitable model for corrosion progression, including pitting, is outlined in the next section. First it is noted that because of the extended time horizons involved, the data necessary for model calibration necessarily must be long-term data. Similarly, the microbiological effects of most interest for infrastructure applications are the longer term effects such as MIC. These longer term effects require understanding of the way the corrosion process and hence the conditions for MIC at the corroding surface change with the progression of corrosion with time. These are unlikely to be obtained in short-term tests under perhaps highly artificial or selective conditions as usual in laboratory conditions, even though these are essential for fundamental microbiological studies.

For MIC the model uses the concept that the development of MIC is likely to be governed by those nutrients that are critical to microbiological metabolism. In its simplest form it can be considered simply as an *environmental input–corrosion output* model. At the present time this is the most useful surrogate available for allowing for the effect of microbiological activity (metabolism) on MIC. It is a broad-brush approach that has had some success, but it is incapable of distinguishing which microorganisms contribute most to corrosion under specific circumstances. To develop such understanding requires considerably more research in the microbiological arena and its link to corrosion mechanisms.

3.3 MODEL FOR MARINE CORROSION

As noted, for infrastructure purposes models are required that represent longer term corrosion and pit depth development. They also must reflect what is likely to occur under actual marine environments, rather than in laboratory conditions. Although there is a considerable amount of corrosion data, as represented in corrosion handbooks (e.g., Schumacher 1979; Kreysa and Eckermann 1992), much of this is for short-term observations (one year or less) and often obtained in laboratory trials. For marine conditions probably the most extensive data are those obtained in the Panama Canal Zone (PCZ) as a result of a very extensive, and presumably very expensive, US experimental program (Southwell et al. 1958; Forgeson et al. 1960). Figure 3.1 shows an example of datasets for average corrosion loss as obtained from mass loss as functions of exposure time, for four different exposure environments, together with an interpreted complex nonlinear trend for each dataset (Melchers 2003a). Trends with similar data nonlinear behavior have been shown to exist in data for a range of other steels as well as aluminum alloys and copper-bearing alloys, all for a wide variety of exposure environments. Clearly they show there is no such thing as a "corrosion rate" and that short-term behavior is very different to longer term trends. This means that test results obtained over, say, one year, and projected in time can be highly misleading. It also can be shown that some other models, such as the pervasive "power law" are poor fits to all or considerable parts of the data trends (Hou and Liang 2004) and that, as a consequence, they can be quite misleading if projected beyond the data base.

The character of the trends shown in Figure 3.1 is similar between the exposure environments, and can be idealized as consisting of two modes (Melchers 2003a) (see Figure 3.2). Further, referencing back to the possible mechanisms responsible for corrosion suggests the phases shown in Figure 3.2. As shown, the earliest (and

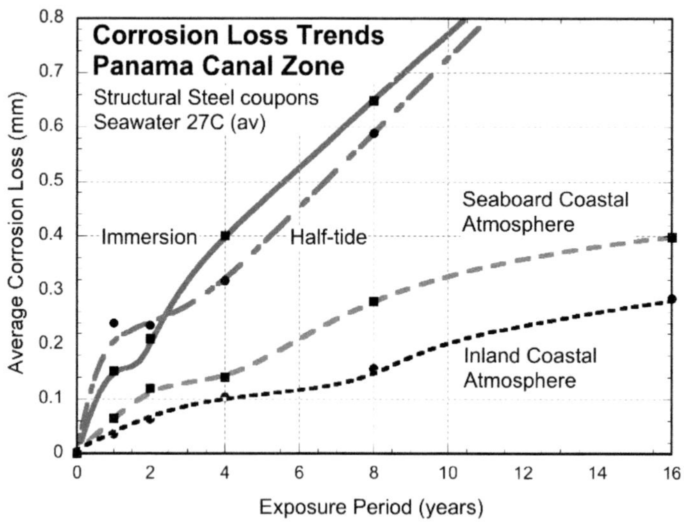

FIGURE 3.1 Longer term corrosion loss data and interpreted trends for the data. (Based on data reported by Southwell et al. 1958; Forgeson et al. 1960.)

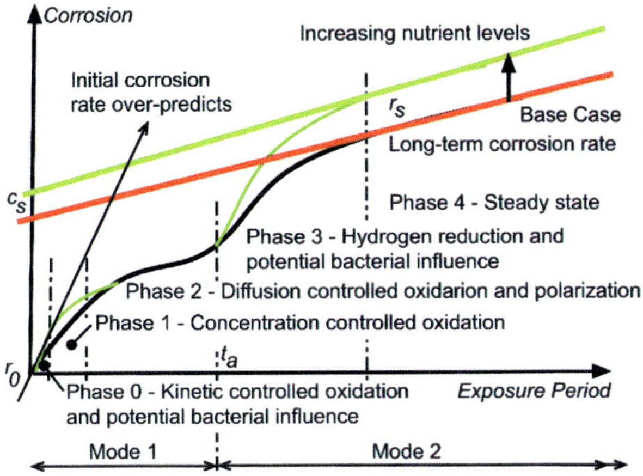

FIGURE 3.2 Bi-modal model for corrosion loss showing the modes and the phases involved, with brief explanation, and also the main model parameters (Melchers 2003a). A generally similar bi-modal trend can be discerned for data for average and for maximum pit depth as a function of time.

quite short) phase involves conventional activation-controlled electrochemical corrosion processes as would be observed for example, in electro-chemical corrosion test in the laboratory. However, it is known from marine microbiology that biofilm attachment also occurs in this very early phase (Little & Lee 2007). This may have an effect on the electro-chemistry. The corrosion kinetics is considered to continue into Phase 1, with the result that under low velocity conditions the weak corrosion products will be formed at cathodic areas on the metal surface, resulting in a slightly nonuniform corrosion pattern on the metal surface. Indeed, this is usually observed under the microscope. However what is also observed is the formation of corrosion pits, already within days of first exposure, as also reported for laboratory experiments (Butler et al. 1972. During this time there may be a MIC effect also. Eventually, however, the corrosion products will commence to interfere with the progression of corrosion. Most easily this can be visualized as the rust layers building up a diffusion barrier that increasingly inhibits the rate of corrosion as the barrier becomes thicker and thus less permeable (Tammann 1923; Evans 1960; Melchers 2003b). This is a useful first-cut model although the precise mechanisms involved are more complex. During this phase too, there may be some MIC effect. Both it and that for phase 2 are shown schematically in Figure 3.2.

A dramatic change in trend is shown at time t_a, at the end of mode 1. At this time the corrosion product build-up has become sufficient to permit less access of oxygen to the corrosion interface (and the pitting there) with the result that the autocatalytic pitting reactions (Wranglen 1974) take over and cause rapid loss of metal and deep pits, as indeed has been observed. In a sense this can be considered as "under-deposit corrosion" long known from practical experience usually to be severe in many different environments. In the present case the deposit is simply composed of the rust layers. The low oxygen environment so created also is suitable for the greater metabolism of microorganisms, provided other conditions such as the availability of essential nutrients for metabolism are available. This is indicated schematically on Figure 3.2. In time, further development of corrosion products and the inevitable balance between building-up of corrosion products and their removal by oxidation (Evans 1960) and perhaps water currents, rainfall, temperature fluctuations, etc., will allow a steady-state scenario to be developed, shown as Phase 4. Indeed data from the PCZ support such a longer term steady state scenario.

The parameters shown as describing the bi-modal model in Figure 3.2 have been the subject of considerable investigation starting with the original concept (Melchers 2003a). The most striking influence is average water temperature as that affects every phase—the rate of chemical reactions and the rates of various diffusion processes such as those in phases 2 and 3 (Melchers and Wells 2006). In addition the effects of alloys in steel composition, long considered to be unsystematic (Schultze and van der Wekken 1976), including carbon content in the steel (Melchers 2003c, 2004a) were given new interpretations although not quantified owing to insufficient data, and the effects of water velocity (Melchers and Jeffrey 2004), water depth (Melchers 2005b), timing of first exposure and effect of oxygen concentration in the water (e.g., Melchers and Chernov 2010) have been considered. This applies to corrosion mass loss as well as to pit depth (Melchers 2004b, 2005c). In addition, the model has permitted various applications, some of which are described later in this chapter.

The effect of microorganisms on longer term corrosion is of particular interest. Figure 3.2 shows the two phases during which MIC can be seen to contribute to corrosion—during phases 1 and 2, and much more so during phase 3, extending in effect to phase 4. Direct evidence for the latter is shown in Figure 3.3 for two nominally similar Pacific Ocean seawater exposure test sites. Although both showed very similar corrosion losses in the first 12 months or so, after that time considerable divergence was found, with site B showing considerably greater corrosion losses compared with site A.

Variables such as water temperature and steel composition that might have caused the difference were discounted. The water temperatures were very similar between the two sites and the exposure coupons were cut from the same sheet. In any case, these and other obvious influences most likely would also have affected mode 1, but this was not the case. Attention then turned to water quality. Analysis showed that the seawaters at site B contained high concentrations of nitrates and phosphates. This was soon attributed to agricultural fertilizer run-off from farming on the adjacent coastline (Melchers 2005d). Detailed analysis of some other cases in the literature, including sewage pollution of the River Thames (Melchers 2007) and the nutrients in the North Sea showed generally similar effects. This led to the hypothesis that elevated concentrations of nutrients could be responsible for the more severe corrosion that was observed, particularly in phases 3 and 4 of Figure 3.2. It also started to explain why earlier field studies in which MIC was suspected but could not be unique isolated, found bacterial profiles that were very similar at sites where corrosion was severe and where it was not severe (Beech and Campbell 2008). Various hypotheses why this might have occurred were subsequently considered, but little or no attention was given to the availability of nutrients.

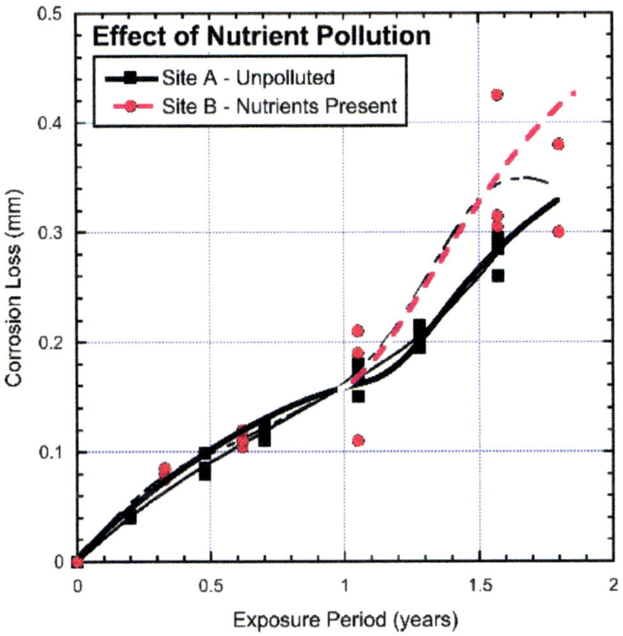

FIGURE 3.3 Corrosion as a function of time for two coastal seawater corrosion testing sites in close proximity and at similar average water temperatures, with one site showing considerably higher longer term corrosion as a result of higher nutrient content.

3.4 SEAWATER DIN AND LONGER TERM CORROSION

It is well-known that for seawaters nearly all nutrients for microorganism metabolism (Little and Lee 2007) are readily available, even in unpolluted seawater. The clear exceptions are ferrous iron as a micronutrient and inorganic nitrogen. The latter, if present, occurs in the form of dissolved inorganic nitrogen (DIN) and may be composed of ammonia, nitrite, and nitrates, the first two usually oxidizing within days to the latter. Ammonia often occurs as part of the effluent from secondary treatment sewage treatment plants as such plants cannot deal with nitrogenous materials. Nitrite in sufficiently high concentrations is known in the corrosion literature as a corrosion inhibitor (Jones 1996), mainly because it competes with the corrosion processes for oxygen. Ironically perhaps, the very process of corrosion supplies the ferrous ions necessary for MIC. (This is the case also for most copper alloys and aluminum alloys (Melchers 2014a), but not for ultra-pure copper (Melchers 2015) which, as a consequence, does not suffer from MIC.)

Based on the notion that DIN is most likely the rate-controlling factor in MIC of steels in seawater, immersion corrosion data for low alloy steels from multiple sources was considered together with data gathered from various reports about the expected or likely annual average DIN concentration in the seawater (Melchers 2014b). The complexity of the bi-modal behavior can be side-stepped for longer term effects by the long-term trend defined by the vertical axis intercept c_s and the long-term rate r_s.

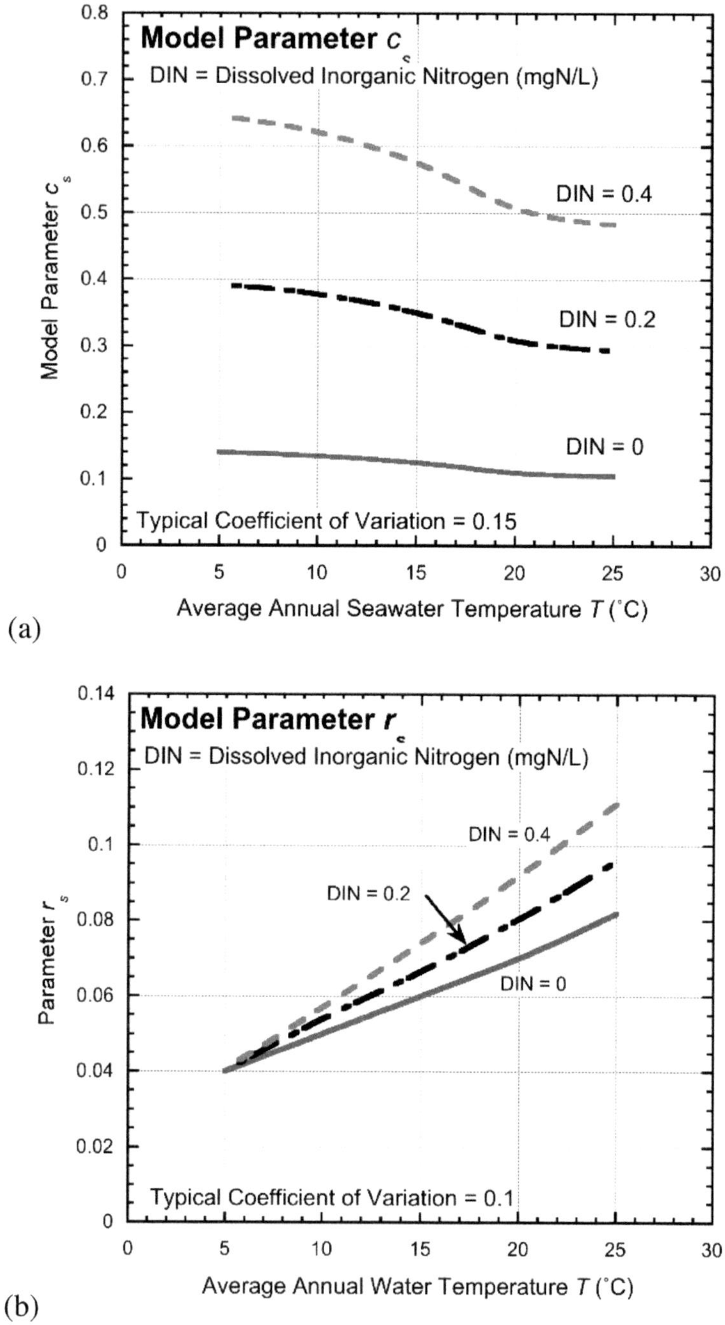

FIGURE 3.4 Influence of dissolved inorganic nitrogen on the seawater corrosion of low alloy steels in terms of the parameters c_s and r_s of the model in Figure 3.2. (Based on Melchers 2014b.)

This opened the way for estimating the effect of DIN on long-term corrosion, allowing also for annual average water temperature. Figure 3.4 shows the results obtained (Melchers 2014b). It shows that DIN has a significant effect on parameter c_s and also a noticeable effect on parameter r_s. It is clear that the effect of water temperature remains important.

3.5 EXAMPLE 1—ACCELERATED LOW WATER CORROSION

Accelerated low water corrosion (ALWC) is the name given to an unusually high rate of corrosion just below the low water mark in tidal seawaters, usually in working harbors. Typically sheet piling or other vertical steel structures in the tidal zone are affected. When first detected in the 1980s in UK harbors, there was considerable concern for the safety and, perhaps more importantly, the remaining economic viability, of some of the ports where it was found. As investigations were conducted in other ports, in the UK and elsewhere, more and more cases were uncovered. However, understanding of the phenomenon remained poor (Beech and Campbell 2008), despite some insights (Gehrke and Sand 2005) indicating the critical role of different species of microorganisms on the process(es) thought to be involved in MIC. Working from the hypothesis that nutrients, and in particular DIN in seawater, held the key to the observations, field exposures of steel strips (and electrically isolated coupons) were conducted at 13 different coastal and harbor sites on the east coast of Australia (Melchers and Jeffrey 2012, 2014a). Rather than sheet piling, mild steel strips 3 mm thick, 50 mm wide, and 3 or 6 m long were exposed for up to 3 years, with periodic recoveries in between. The identification of bacterial species was ignored as these were known, from other studies, to be very similar at all sites. However, care was taken to measure the DIN concentrations at various times at each of the sites. Figure 3.5a shows a typical corrosion profile together with the parameters used to measure the effect of ALWC. When recovered, the strips were cleaned using standard corrosion testing protocols and then cut to exactly 100 mm lengths for which mass losses were determined. The ratio of the maximum corrosion loss for the ALWC region was then compared with that for the immersion region and plotted against the average annual DIN concentration in the local seawater. Figure 3.5b shows one of the results. Also added to Figure 3.5b is the observation that no effect of ALWC was recorded at Kure Beach (NC, USA) in the Atlantic Ocean test site (Humble 1949). The tests were carried out during 1940s in clean Atlantic Ocean seawater on actual steel sheet piling suspended from a jetty (since destroyed by a hurricane) and the resulting profile, showing no ALWC effect, widely reported in the corrosion literature and in corrosion text books. What is less well-known is that these tests were actually a second attempt, and that earlier tests on sheet piling carried out on the inland waterway side of Kure Beach were abandoned after it was found they corroded much faster at the low water level than expected. This was attributed to the polluting effect of the (then) local chemical plant and its sewage and other outfall wastewaters (Humble 1949). This may well be the first observation of ALWC. Similarly, less well-known are observations by Arup and Glantz (1963) that in certain Danish harbors with fish-cleaning facilities corrosion of steel sheet poiling was much greater than elsewhere. This correlates, for example, with the situation at the fishing harbor of Ulladulla (see right side of Figure 3.5b).

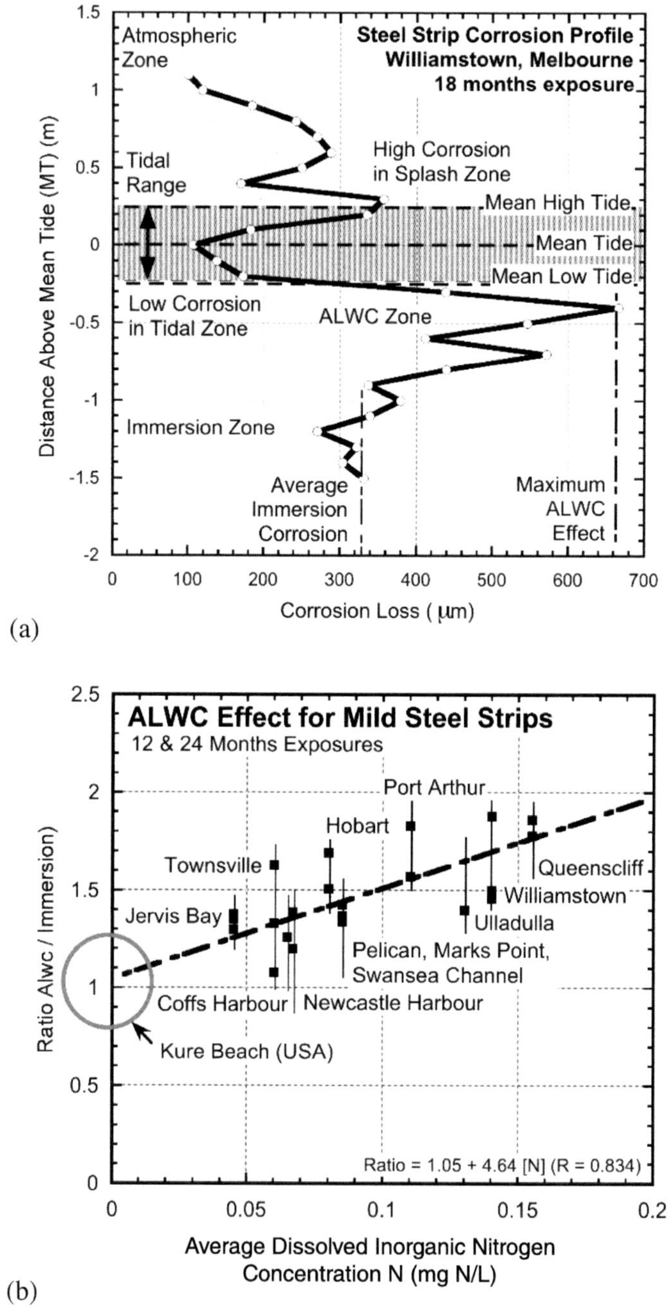

FIGURE 3.5 (a) Typical corrosion profile of steel vertically oriented through the tidal zone showing high level of corrosion just below the low water level and the low corrosion in the inter-tidal zone, and (b) correlation between average DIN concentration at the site and the ratio between peak ALWC and immersion corrosion. (Melchers and Jeffrey 2012, 2014a.)

The trend shown in Figure 3.5b can be extrapolated to much higher DIN values (at least 1.2 mg N/L) as was shown in corrosion profiles derived from data for steel sheet piling located at a number of US Navy bases, and also for a small number of European and Japanese sites, including also much longer exposure periods (Melchers and Jeffrey 2014a). Prompted by these findings closer investigation of the reason why sheet piling typically corrodes through in the central region of the steel profile, irrespective of whether it is a Z or a U profile showed that this has to do with the higher presence of inclusions and the metal grain structure in this location, rather than being caused by any MIC effect (Melchers et al. 2014).

3.6 EXAMPLE 2—MOORING CHAINS FOR OIL AND GAS FLOATING PRODUCTION STORAGE AND OFFLOADING (FPSO) PLATFORMS

FPSOs are floating production storage and offloading vessels increasingly being used for offshore oil and gas resource related activities particularly in very deep waters (currently about 3 km depths). Typically they are specially built vessels or converted oil tankers. For these, remaining "on-station" is a critical aspect, since rupture of production lines can have major environmental as well as production implications. Typically there are 8–12 mooring lines radiating from each vessel, collected at a turret, in many cases located at the bow region. Each mooring line typically consists of a length of chain on the seafloor acting as a drag anchor, a high strength steel wire rope (typically 100 mm diameter), and an upper chain length used for connecting the mooring line to the turret on the vessel, in much the same way as conventional anchor chains are tethered to a vessel. The chains used in the industry are made from high-quality, high-strength steels most commonly made from of 76 mm diameter steel. Chains with steel diameter up to 152 mm also are in use. In each case the length of each individual link is about six times the steel diameter. They are therefore of considerable weight and have a high exposed surface area. Largely for practical (and historical) reasons, conventional practice does not use cathodic protection for the links, although some attempts at this have been made. There also is some evidence of a limited amount of cathodic protection for the upper part of the chains supplied by the CP for the hull of the vessel. Thus, in the majority of cases the chain links are (almost) free to corrode as bare steels exposed to seawater, including in the splash, (limited) tidal zone, the potential ALWC zone, and the immersion zone. Reliance for safety of the chains is placed almost entirely on a sacrificial corrosion allowance. As part of this some operators perform periodic on-site inspections of the chains (usually with personnel climbing on the chains for inspections).

As part of a major Joint Industry Project (Seawater Corrosion of Rope and Chain, SCORCH) sponsored by most of the international oil majors and most of the classification/certification societies and some materials suppliers and operators, the University of Newcastle was asked to investigate the effect of water temperature, on the basis that the industry was moving increasingly into tropical waters whereas most corrosion experience was based on operations in temperate waters (Fontaine et al. 2012; Lee et al. 2015). Of particular interest herein are the observations, by some operators,

FIGURE 3.6 Part of a 76 mm diameter steel mooring chain link showing some of the severe corrosion found as part of the SCORCH project.

of quite severe localized corrosion for vessels operational off the coast of West Africa and also in the Timor Sea. Figure 3.6 shows just one example of such severe corrosion, for a chain exposed for about 8 years. While details of the environments involved and the main findings are still bound by intellectual property constraints, it might be noted that investigation of the water quality immediately around the mooring chains showed very high concentrations of DIN in the seawater, much greater than any recorded in any of the harbors in Figure 3.5 or elsewhere (Melchers and Jeffrey 2014a). Subsequent microbiological analyses showed interesting patterns in the rusts and for the biomass and microorganisms as a function of the (microscopic) distance from the steel (Celikkoi-Aydin et al. 2016). In this case, too, a clear link between species and the higher than expected corrosion could not be established. Nevertheless, the overall results are entirely consistent with what was found for ALWC.

3.7 EXAMPLE 3—CORROSION OF WATER INJECTION PIPELINES

Injection of water (seawater, produced water, waste water, freshwater) into sub-sea (or on-land) oil and gas reservoirs to attempt to drive out the remaining oil or gas is a common practice in industry (see Figure 3.7). The pipes that convey the water to be injected vary in size up to about a 450 mm diameter and they may be many kilometers in length, for example 15 km. Almost without exception the pipelines are

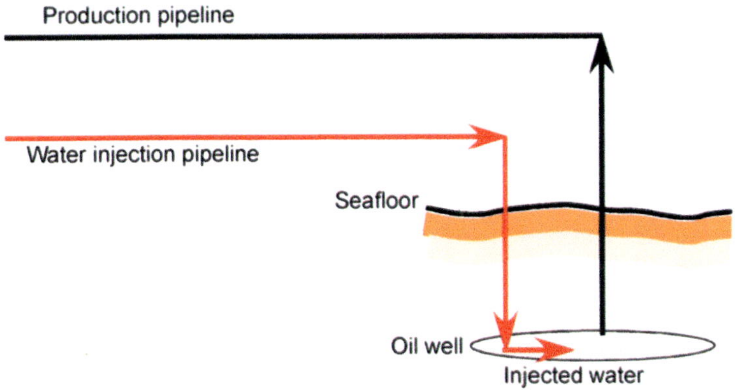

FIGURE 3.7 Typical layout for water injection into oil well, with the water injection pipeline (WIP) usually on the seafloor.

low alloy carbon steels, without interior coatings and without cathodic protection. In practice many of the pipelines show little or no corrosion even after many years of service. Usually this is attributed to de-oxygenation of the injected water (with a typical target remnant dissolved oxygen concentration of around 20 ppm) and for some operators, frequent cleaning of the inside of the pipe by means of pigging—a technique in which something like a robust "bottle-brush" is transported through the pipe to remove marine growth and rusts from the walls of the pipe. This is an expensive operation and alternatives include dosing the injected water with corrosion inhibitors and also with biocides to attempt to control MIC. The latter has the disadvantage that the biocides ultimately leach from the reservoirs and thus into the surrounding waters. Incoming European laws will prevent all but "environmentally friendly" biocides being used (somewhat of a contradiction). This is causing considerable concern for the corrosion management of the water injection pipelines (WIPs). A further factor in the operation of the pipelines is that for some oil or gas reservoirs it is desirable to inject them with nitrates. This is to help the nitrate reducing bacteria present in the reservoir outcompete the sulfate reducing bacteria that also are present, and so suppress the undesirable production of hydrogen sulfide within the reservoir and its release from the reservoir as part of the produced crude oil or gas.

One of the issues the industry has been facing for some years is the severity of the corrosion inside some, but not all, water injection pipelines. The most concerning is so-called channeling corrosion (e.g., Heidersbach and Roodselaar 2012) (see Figure 3.8). There have been a number of investigations based on the concept that this is primarily the result of MIC, and certainly the presence of various microorganisms has been established (Stipanicev et al. 2014). Also, it has been proposed there is a link to nitrate injection (Hubert et al. 2005; Stott 2012; Drønen et al. 2014). However, it might be expected that if the issue is simply MIC, perhaps fostered by nutrients, severe corrosion would have been evident all around the pipe, not just at the bottom. Some have suggested this is the result of the additional effect of erosion along the bottom of the pipe, but this is unlikely as all injected water is filtered to a high degree before being injected. Yet pigging runs along the pipes after some years show that

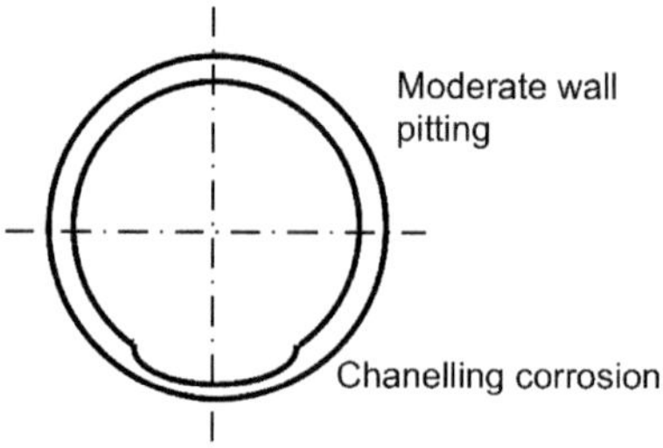

FIGURE 3.8 Schematic view of pipe cross-section showing channelling corrosion.

there is considerable detritus in the pipes, mostly in the form of corrosion products. It is reasonable to assume these are the result of interior corrosion of the WIP.

As part of a new investigation conducted as part of the European Marie Curie BIOCOR project, data became available for five different water injection pipelines operational in the North Sea. Various other data, including pit depth data from a number of intelligent pigging runs became available, as well as various information about the water quality and samples of microbiological activity on coupons exposed for just six months. In short, despite much effort, for a considerable amount of time no coherent picture emerged from analyses of the data. What did emerge, by adopting the model of Figure 3.2, was an interpretation of the pit depth data obtained from the intelligent pigging runs. Two examples are shown in Figure 3.9, the first for one of the pipelines for which nitrate addition was not used, the other for which nitrate injection was used. It is immediately clear that there is a significant difference between them. It is reasonable to conclude from these two, and others not shown here but with similar features, that nitrate had the effect of increasing the pit depths around the walls of the pipes and also added channeling corrosion.

The immediate question to arise form these observations is why specifically the channeling corrosion would occur with the transportation of nitrate in the pipe? Eventually, the following possibility came to be considered—could the material removed periodically by the pigging operation at some stage be from a deposit at the bottom of the pipe, sufficient for under-deposit corrosion to occur in the very low dissolved oxygen (DO) concentration conditions within the pipe? Further enquiries revealed that indeed the velocity in the pipes does vary and from time to time is zero, such as when maintenance activities are carried out. This would allow deposition. The other question was where that deposited material came from. As noted, analysis of the detritus recovered by periodic pigging showed it to be primarily assorted rust products (Comanescu et al. 2016).

From this line of thinking came the need to investigate whether under-deposit corrosion was feasible in the low DO conditions and whether it could be caused by rust products together with some other forms of detritus such as very fine sands. This was investigated as part of an experimental PhD project after the conclusion of the BIOCOR project.

For the experimental project small scale semi-circular pieces of pipe some with added deposits were exposed for up to 6 months in somewhat elevated (30°C) temperature, very low DO concentration, stagnant immersion conditions in four different, physically separated, environments (see Table 3.1). The filtration and UV

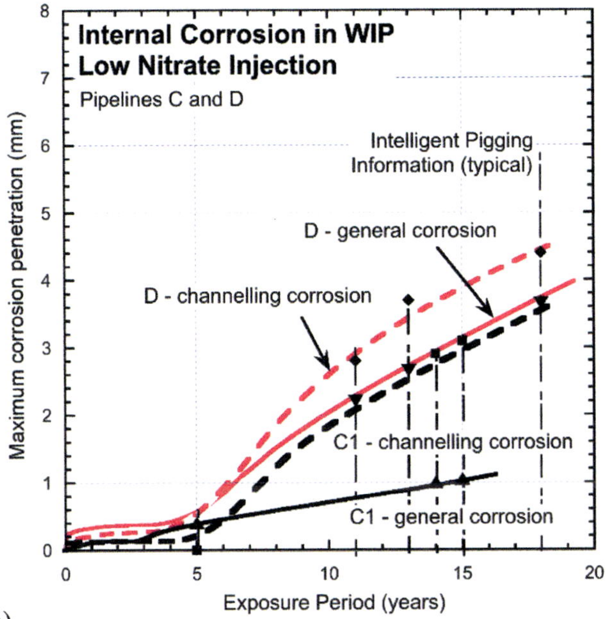

(a)

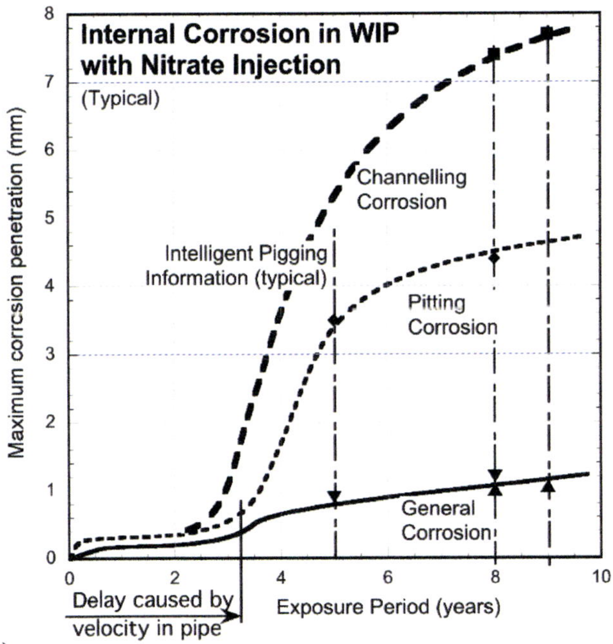

(b)

FIGURE 3.9 Depth of corrosion penetration obtained from intelligent pigging of some WIPs in the North Sea and interpreted corrosion trends consistent with Figure 3.2, showing effect of nitrate addition to injected water on internal corrosion penetration.

TABLE 3.1
Experimental Conditions for WIP Corrosion Simulation and Findings

Case	Deposits	Seawater	Nitrate	Experimental Result
1	no	natural	no	least corrosion pitting
2	yes	filtered and UV treated	no	somewhat greater pitting
3	yes	natural	no	deep pitting
4	yes	natural	yes	deepest pitting

treatment are known to remove virtually all microorganisms. The deposits consisted of magnetite, calcium carbonate, and sea-sand, in proportions to match reasonably closely the typical pigging detritus. The nitrate source was commercial calcium nitrate ($Ca(NO_3)_2$) and the deoxidizer used was sodium hydrogen sulfite ($NaHSO_3$), an industry standard de-oxidizer at the time of the study (Wang and Melchers 2017).

As seen in Table 3.1, the severity of pitting corrosion increased with the addition of deposits, addition of microorganisms, and addition of nitrate. A schematic view of the experimental trends is shown in Figure 3.10 (Wang and Melchers 2017). The fact

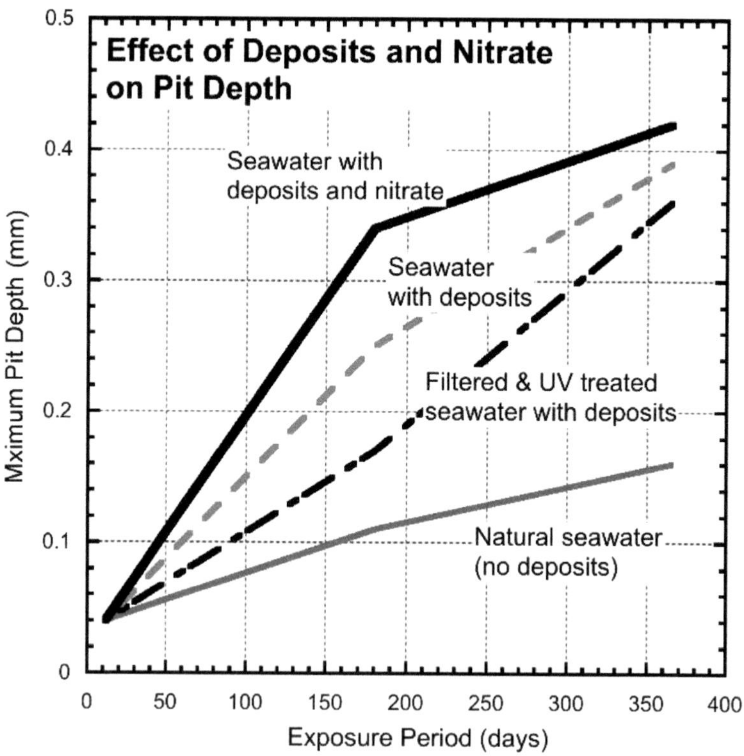

FIGURE 3.10 Summary of experimental results obtained for the corrosion of steel coupons in four different but stagnant, low DO concentration waters, with deposits and without, and the effect of nitrate addition. (Based on Wang and Melchers 2017.)

that nominally "clean" seawater can cause more severe corrosion and pitting than filtered and UV treated seawater was shown earlier (Melchers and Jeffrey 2014b) and was attributed to the latent presence of appropriate nutrients even in nominally "clean" seawater (i.e., seawater with DIN <0.03 mg N/L, much less than is normally permitted in drinking water).

3.8 OUTLOOK AND CLOSURE

An important outcome of the research and the examples outlined herein is that short-term observations of corrosion, including the effect of microorganisms, do not necessarily relate well to the corrosion to be expected in the longer term. As outlined in the model of Figure 3.2, the development of corrosion products changes the local environment in which the corrosion and microbiological processes must occur, and these cannot easily be replicated in short-term, laboratory experiments. For many practical applications, particularly related to infrastructure durability, the longer term behavior is by far of most interest.

The three examples given above, drawn from the author's experience with industrial projects, show clearly the *relative* importance of microbiological influences on corrosion of structural steels in marine environments. In all cases independent evidence was available showing the presence of microorganisms normally associated with MIC, including those that may be grouped as sulfate-reducing bacteria (SRB), iron oxidizing bacteria (IOB), iron reducing bacteria (IRB), etc. Similar evidence has been obtained in other environments such as various atmospheres (Manchuca et al. 2016). However, it remains the case that making a direct association between bacterial species and corrosion, let alone the severity of corrosion, is far from straightforward at the present time. Certainly laboratory based experiments for short-term association between individual species and enhanced corrosion are available, and have been so for a long time, but whether this extends to longer term effects, and how such correlations are influenced by environmental conditions remains unresolved. Only slowly is progress being made from "who is there" (now "easy" to investigate), to "at what rate, and how," and to "what do they do?" The progression of understanding in this direction remains an important research challenge.

The approach described herein is based on the notion that all three questions are related to the metabolism of whatever microorganisms may be involved, without unique identification of the species. In the seawater environment where most nutrients necessary for metabolism are readily available, the focus on DIN as critical when Fe is supplied by the corrosion process (as in steels, commercial Al alloys and commercial Cu alloys) is a reasonable, practical, and cost-effective approach for prediction of likely MIC effects. However, since fundamental microbiological understanding and microbiological sensitivity aspects are bypassed, it clearly does not necessarily assist in proposing remedial or preventative measures. Reducing or eliminating one or more of the critical nutrients is, of course, one possible course of action, paralleling approaches used in some branches of medicine. However, nutrient removal or limitation may not always be practical. For example, it is technically difficult to eliminate Fe even from aluminum and most copper alloys. Similarly, elimination of DIN is unlikely to be practical in harbors and areas of industrial activity, although tertiary treatment can remove them from sewage effluents.

An important observation from all three examples, and in the model of Figure 3.2, is that corrosion is not exclusively a microbiological event. On the contrary, it is essentially an electrochemical process to which microbiological activity may contribute, sometimes significantly. The balance between purely abiotic electrochemical activity and that contributed by biological activity depends entirely on the details of the environment, as should be amply evident from the examples given herein. Because of the necessity of Fe as a micronutrient for microbiological activity, it follows also that MIC without any abiotic corrosion to supply Fe ions is unlikely to occur. This is one reason why cathodic protection, aimed at reducing or eliminating the dissolution of Fe ions (and thus corrosion) often is considered effective also against MIC, typically through providing a slightly greater applied potential (around 100 mV) than would be the case for protection against abiotic corrosion alone (Jones 1996; Little and Lee 2007).

Finally, while MIC often results in quite severe corrosion, as shown for example in Figure 3.10, such corrosion is not necessarily MIC. Under-deposit corrosion also can be quite severe and quite localized, without any involvement of MIC. Again, it depends on the environment in which the corrosion process occurs.

ACKNOWLEDGMENTS

The author acknowledges the support of the Australian Research Council for continued support of basic research, the European Marie Curie BIOCOR project, the SCORCH-JIP and Deepstar projects (Dr. Andrew Potts and Tim Lee, AMOG Consulting), as well as a range of coworkers and colleagues, particularly Drs. Robert Jeffrey, Iulian Comanescu, and Daniel Wang at Newcastle, and Dr. Laura Machuca and Professor Brian Kinsella at Curtin University.

REFERENCES

Arup, H.H., and Glantz, G. 1963. An evaluation of corrosion of sheet steel piling in Danish harbors, Proc. Second Int. Cong. Metallic Corrosion. 11–15 March. New York: National Association of Corrosion Engineers, 424–431.

Beech, I.B., and Campbell, S.A. 2008. Accelerated low water corrosion of carbon steel in the presence of a biofilm harbouring sulphate-reducing and sulphur-oxidising bacteria recovered from a marine sediment. *Elect. Acta.* 54:14–21.

Butler, G., Stretton, P., and Beynon, J.G. 1972. Initiation and growth of pits on high-purity iron and its alloys with chromium and copper in neutral chloride solutions. *British Corros. J.* 7(7):168–173.

Celikkol-Aydin, S., Gaylarde, C.C., Lee, T.M., Melchers, R.E., Witt, D.L., and Beech, I.B. 2016. 16S rRNA gene profiling of planktonic and biofilm microbial populations in the Gulf of Guinea using Illumina NGS. *Marine Environmental Research.* 122:105–112.

Comanescu, I., Melchers, R.E., and Taxén, C. 2016. Corrosion and durability of offshore steel water injection pipelines. *Ships and Offshore Structures.* 11(4):424–437.

Drønen, K., Roalkvam, I., Beeder, J., Torsvik, T., Steen, I.H., Skauge, A., and Liengen, T. 2014. Modeling of Heavy Nitrate Corrosion in Anaerobe Aquifer Injection Water Biofilm: A Case Study in a Flow Rig. *Environmental Science and Technology.* 48:8627–8635.

Evans, U.R. 1960. *The Corrosion and Oxidation of Metals: Scientific Principles and Practical Applications.* London: Edward Arnold (Publishers).

Fontaine, E., Potts, A.E., Kai-tung Ma, Arredondo, A.L., and Melchers, R.E. 2012. SCORCH JIP: Examination and testing of severely-corroded mooring chains from West Africa. *Offshore Technology Conference*, Houston, TX, 30 April–3 May 2012, Paper OTC 23012.

Forgeson, B.W., Southwell, C.R., and Alexander A.L. 1960. Corrosion of metals in tropical environments—Part 3—Underwater corrosion of ten structural steels. *Corrosion.* 16(3):105t–114t.

Gehrke, T., and Sand, W. 2005. Interactions between microorganisms and physiochemical factors cause MIC of steel pilings in harbours (ALWC). Proceedings, *Corrosion* 2005, paper no. 03557. Houston: NACE International.

Heidersbach, K.L., and van Roodselaar, A.C. 2012. Understanding, preventing, and identification of microbial induced erosion-corrosion (Channelling) in water injection pipelines. NACE *Corrosion*, paper no. C2012-0001221. Mar 11–15. Salt Lake City, Utah.

Hou, W., and Liang, C. 2004. Atmospheric corrosion prediction of steels. *Corrosion.* 60(3):313–322.

Hubert, C., Nemati, M., Jenneman, G., and Voordouw, G. 2005. Corrosion risk associated with microbial souring control using nitrate or nitrite. *Applied Microbiology and Biotechnology.* 68:272–282.

Humble, A.A. 1949. The cathodic protection of steel piling in seawater. *Corrosion.* 5(9):292–302.

Jeffrey, R.J., and Melchers, R.E. 2007. The changing topography of corroding mild steel surfaces in seawater. *Corros. Sci.* 49:2270–2288.

Jones, D.A. 1996. *Principles and Prevention of Corrosion*, 2nd ed. Upper Saddle River, NJ: Prentice-Hall.

Kreysa, G., and Eckermann, R. 1992. *Dechema Corrosion Handbook*, vol. 11. New York: VCH Publishers.

Lee, T.M., Melchers, R.E., Beech, I.B., Potts, A.E., and Kilner, A.A. 2015. Microbiologically influenced corrosion (MIC) of mooring systems: diagnostic techniques to improve mooring integrity. *Proc. of the 20th Offshore Symposium*, Feb. 2015, Houston: Society of Naval Architects and Marine Engineers.

Little, B.J., and Lee, J.S. 2007. *Microbiologically Influenced Corrosion.* Hoboken, NJ: Wiley.

Machuca, L.L., Jeffrey, R.J., Bailey, S.I., Gubner, R., Watkin, E.L.J., Ginige, M.P., Kaksonen, A.H., and Heidersbach, K. 2014. Filtration-UV irradiation as an option for mitigating the risk of microbiologically influenced corrosion of subsea construction alloys in seawater. *Corros. Sci.* 79:89–99.

Machuca, L.L., Jeffrey, R.J., and Melchers, R.E. 2016. Microorganisms associated with corrosion of structural steel in diverse atmospheres. *International Biodeterioration & Biodegradation*, 114:234–243.

Melchers, R.E., and Beck, A. 2018. *Structural reliability analysis and prediction*, 3rd ed. Chichester, UK: John Wiley and Sons.

Melchers, R.E. 2003a. Modeling of marine immersion corrosion for mild and low alloy steels—Part 1: Phenomenological model. *Corrosion.* 59(4):319–334.

Melchers, R.E. 2003b. Mathematical modeling of the diffusion controlled phase in marine immersion corrosion of mild steel. *Corros. Sci.* 45(5):923–940.

Melchers, R.E. 2003c. Effect on marine immersion corrosion of carbon content of low alloy steels. *Corros. Sci.* 45(11):2609–2625.

Melchers R.E. 2004a. Effect of small compositional changes on marine immersion corrosion of low alloy steel. *Corros. Sci.* 46(7):1669–1691.

Melchers, R.E. 2004b. Pitting corrosion of mild steel in marine immersion environment - 1: maximum pit depth. *Corrosion* 60(9):824–836.

Melchers, R.E. 2005a. The effect of corrosion on the structural reliability of steel offshore structures. *Corros. Sci.* 47(10):2391–2410.
Melchers, R.E. 2005b. Effect of immersion depth on marine corrosion of mild steel. *Corrosion.* 61(9):895–906.
Melchers, R.E. 2005c. Effect of alloying on maximum depth of pits in mild steel in marine immersion environments. *Corrosion.* 61(4):355–363.
Melchers, R.E. 2005d. Effect of nutrient-based water pollution on the corrosion of mild steel in marine immersion conditions. *Corrosion.* 61(3):237–245.
Melchers, R.E. 2007. The effects of water pollution on the immersion corrosion of mild and low alloy steels. *Corros. Sci.* 49:3149–3167.
Melchers, R.E. 2014a. Bi-modal trend in the long-term corrosion of aluminum alloys. *Corros. Sci.* 82 (2014):239–247.
Melchers, R.E. 2014b. Long-term immersion corrosion of steels in seawaters with elevated nutrient concentration. *Corros. Sci.* 81:110–116.
Melchers, R.E. 2015. Bi-modal trends in the long-term corrosion of copper and copper alloys. *Corros. Sci.* 95:51–61.
Melchers, R.E., and Chernov, B.B. 2010. Corrosion loss of mild steel in high temperature hard freshwater. *Corros. Sci.* 52:449–454.
Melchers, R.E., and Jeffrey, R. 2004. Influence of water velocity on marine corrosion of mild steel. *Corrosion.* 60(1):84–94.
Melchers, R.E., and Jeffrey R.J. 2012. Corrosion of long vertical steel strips in the marine tidal zone and implications for ALWC. *Corros. Sci.* 65:26–36.
Melchers, R.E., and Jeffrey, R.J. 2014a. Corrosion of steel piling in seawater harbours. *Proc. Instn. of Civil Engineers, Maritime Engineering,* 167(MA4):159–172.
Melchers, R.E., and Jeffrey, R.J. 2014b. Long-term corrosion of mild steel in natural and UV-treated coastal seawater. *Corrosion.* 70(8):804–818.
Melchers, R.E., and Wells, P.A. 2006. Models for the anaerobic phases of marine immersion corrosion. *Corros. Sci.* 48(7):1791–1811.
Melchers, R.E., Jeffrey, R.J., and Usher, K.M. 2014. Localized corrosion of steel sheet piling. *Corros. Sci.* 79:139–147.
Schultze, W.A., and van der Wekken, C.J. 1976. Influence of alloying elements on the marine corrosion of low alloy steels determined by statistical analysis of published literature data. *British Corros. J.* 11(1):18–24.
Schumacher, M. (ed.) 1979. *Seawater Corrosion Handbook.* New Jersey: Noyes Data Corporation.
Schwermer, C.U., Lavik, G., Abed, R.M.M., Dunsmore, B., Ferdelman, T.G., Stoodley, P., Gieseke, A., and de Beer, D. 2008. Impact of nitrate on the structure and function of bacterial biofilm communities in pipelines used for injection of seawater into oilfields. *Applied and Environmental Microbiology.* 74(9):2841–2851.
Southwell, C.R., Forgeson, B.W., and Alexander, A.L. 1958. Corrosion of metals in tropical environments—Part 2—Atmospheric corrosion of ten structural steels. *Corrosion.* 14(9):53–59.
Stipanicev, M., Turcu, F., Esnault, L., Rosas, O., Basseguy, R., Sztyler, M., and Beech, I.B. 2014. Corrosion of carbon steel by bacteria from North Sea offshore seawater injection systems: Laboratory investigation. *Bioelectrochemistry.* 97:76–88.
Stott, J.F.D. 2012. Implementation of Nitrate Treatment for Reservoir Souring Control: Complexities and Pitfalls. *SPE International Conference and Workshop on Oilfield Corrosion.* May 28–29. Aberdeen. Paper No. SPE-155155-MS.
Tammann, G. 1923. *Lehrbuch der Metallographie*, 2nd ed. Leipzig: Leopold Voss.
Wang, X., and Melchers, R.E. 2017. Long-term under deposit pitting corrosion of carbon steel pipelines. *Ocean Engineering.* 133:231–243.
Wranglen, G. 1974. Pitting and sulphide inclusions in steel. *Corros. Sci.*, 14:331–349.

Part II

Microbiologically Influenced Corrosion (MIC) and Reservoir Souring

4 Microbial Communities Involved in High Salinity Souring in Shale Oilfields

Biwen Annie An and Gerrit Voordouw

CONTENTS

4.1 INTRODUCTION

Saline and hypersaline (>3 M NaCl) environments can be stressful for non halophilic (salt-loving) microorganisms. Halophilic eukarya, bacteria and archaea were isolated from several saline environments such as the Great Salt Lake, the Dead Sea and petroleum reservoirs. There are two distinct mechanisms used by halophiles for osmotic adaptation. The first mechanism is referred as "high salt-in," which is the accumulation of potassium chloride (KCl) inside the cytoplasm (Oren, 2001). The second mechanism is "high salt-out," through the production and accumulation of compatible solutes for osmotic balance (Oren, 2001). The "high salt-in" strategy requires intracellular accumulation of KCl, until the concentration inside the cell is equivalent to the salt concentration (mostly NaCl) in the extracellular environment. However, this strategy requires an adaptation of the intracellular machineries (e.g., protein synthesis) to high concentrations of KCl (Oren, 2001). For example, halophiles that utilize this strategy for osmoregulation have highly acidic intracellular proteins (Oren, 2001). Microorganisms that use the "high salt-in" strategy do not require osmotic solutes for osmoregulation. In fact some halophiles such as *Halanaerobium*, which accumulate intracellular KCl, can metabolize osmotic solutes synthesized by other microorganisms, e.g., glycine betaine into trimethylamine and acetate (Daly et al., 2016).

The second type of osmotic regulation strategy utilized by halophiles is referred as "high salt-out" and requires the production or accumulation of osmotic/compatible solutes. This is more energetically expensive to maintain, but no modification of intracellular machineries is required (Oren, 2001). Most commonly known compatible solutes include: glycerol (only in halophilic *Eukarya*), glutamate, glutamine, proline, glycine betaine, ectoine, glucosylglycerol and trehalose (Averhoff and Müller, 2010; Oren, 2001, 2013a; Pastor et al., 2010). To achieve osmoregulation, cells must actively synthesize osmotic solutes at the cost of energy production and growth (Oren, 2001). One of the best-studied osmotic solutes is glycine betaine (GB) (Oren, 2001). Some halophiles, such as *Marinobacter* and *Methanohalophilus*, produce GB for osmoregulation that can be broken down to methylated amines by *Halanaerobium* (Oren, 2001; Daly et al., 2016).

Daly et al. (2016) characterized the halophilic syntrophic relationship between the thiosulfate reducer and fermenter *Halanaerobium* and the halophilic methanogen *Methanohalophilus*. Genes associated with GB synthesis were found in the genome of *Methanohalophilus* and *Marinobacter*. Genes associated with the breakdown of GB to trimethylamine (TMA) were found in the genome of *Halanaerobium*. Pathways associated with methanogenesis using TMA were found in the genomes of the methanogens *Methanolobus* and *Methanohalophilus*. However, to maintain continuous methanogenesis using GB breakdown requires high biomass. But if all biomass that can synthesize GB were to be eliminated due to viral activities then long-term methanogenesis cannot be supported. Physical evidences of halophilic interspecies interactions was demonstrated by Cayol et al. (2002), who showed that fermentative *Halanaerobium* can metabolize glycerol and produce hydrogen (H_2) for sulfate reduction by *Desulfohalobium retbaense* (Cayol et al., 2002). Due to the energy limitations at high salinity, it is not surprising that microorganisms might cooperate for energy conservation. In the case of *Halanaerobium* and *Desulfohalobium retbaense*, such cooperation might increase the risks of corrosion and souring. On the other hand, a synergistic relationship between *Halanaerobium* and *Methanohalophilus* might be further explored for gas production and enhanced oil recovery.

Halophiles are often explored for their biotechnological potential, e.g., halophiles produce osmotic solutes, which have a wide range of applications. Solutes such as ectoine, betaine and trehalose enhance refolding of partially denatured proteins and are used in pharmaceuticals to stabilize damaged biomolecules (Roberts et al., 2005). Halophiles also have high resistance to heavy metals by producing exopolysaccharides that can absorb toxic metals (Stan-Lotter and Fendrihan, 2012).

In the oil and gas industry, halophiles may play a significant role in bioremediation and oil production processes. Many oil reservoirs are rich in salts, whether from natural deposits or the injection of saline waters (e.g., seawater) for oil recovery. Hence, the produced waters from these sites often contain high levels of salt and hydrocarbon. Hydrocarbon-contaminated saline waters are best treated by bioremediation using halophiles (Sorokin et al., 2011). In addition to bioremediation, halophiles can also produce biosurfactants that can be used for MEOR in saline oil reservoirs (Zargari et al., 2014).

4.2 SHALE OIL AND SHALE GAS OPERATIONS

Recently, the advancement of horizontal drilling and hydraulic fracturing technologies have facilitated the massive expansions of shale oil and shale gas production. Shale gas accounts for 6% of the global natural gas resources (Mouser et al., 2016). There are also nearly 345 billion barrels of shale oil recoverable worldwide, making shale oil a potentially large component of future crude oil production (Brittingham et al. 2014). The extremely finely grained sedimentary shale rocks require the process of hydraulic fracturing by generating breakages, thus increasing hydrocarbon migration.

Horizontal drilling provides the pathway through which the fracturing fluids can enter to facilitate the hydraulic fracturing process (Shrestha et al., 2017). The basic principle of hydraulic fracturing is to create new fractures or enhance existing rock fractures for hydrocarbon migration (Chen et al., 2014). Fluid flows in conventional fields are typically determined through Darcy's Law (Equation **1**), where q is the flow rate of fluids, A is the cross-sectional area and the total pressure drop (p_a–p_b) divided by viscosity (μ) and the length of interest (L) (equation 1).

$$q = A(k/\mu)(p_a - p_b)/L \tag{4.1}$$

However, fluid flows in shale reservoirs are difficult to predict due to the presence of nanopores (1–100 nm; Lee and Kim, 2016). Thus, the Forchheimer or non-darcy model was developed to predict the effect of high velocity gas flow and pressure drops in hydraulic fracturing on production rate (Lee and Kim, 2016) instead of the traditional Darcy's law.

Fracturing fluids used in hydraulic fracturing are composed of 90% water, 9% sand (proppant) and 1% additives (Chen et al., 2014). Water is used to create thin fractures in the reservoir, while sand holds open the newly created fractures (Chen et al., 2014). The types of additives used in the operation may vary among geographical regions. Additives can include acids that are used to clean wellbore regions by dissolving mineral deposits. Corrosion inhibitors are also added into the fracturing fluids, as some of the acids used are highly corrosive (Chen et al., 2014). Polymers are added to increase the viscosity of the fracturing fluids, and of these guar gum is one of the most commonly used (Chen et al., 2014). Guar gum is used to maintain the suspension of sand or proppant in the hydraulic fracturing process. Breakdown of these polymers by microorganisms is undesirable, as it can compromise the efficacy of the production process.

The salinity of the flowback waters increases over time as are produced oil-water emulsions from the shale oilfields. In shale gas operations, following horizontal drilling and hydraulic fracturing methane gas is recovered without further injection (Cluff et al., 2014), but shale oilfields are continuously injected with clean and often low salinity water to produce oil after hydraulic fracturing, like the waterflooding stage of conventional oil production. The waters introduced during the hydraulic fracturing process and the injection waters used for subsequent oil recovery often contain sulfate leading to potential souring problems. Shale oil production from the Bakken formation, which spans parts of Saskatchewan and Manitoba in Canada and of North Dakota in the US, produces 1.2 million barrels of oil per day in 2015 (Laurenzi et al., 2016). In the Bakken formation 0.5 to 3 million gallons

(1.9 to 11.4 million liters) of water are required per well (Wang et al., 2016). The large quantities of water required by the shale operations may result in biogenic sulfide production, iron sulfide formation, plugging issues and microbiologically influenced corrosion (MIC) of downhole equipment, surface separation facilities, storage tanks and flowlines. Treating the wells with biocides could control potential reservoir souring by microorganisms. In addition to reservoir souring, biodegradation of polymers by microorganisms is also an undesirable outcome. Breakdown of polymers such as guar gum may decrease the efficacy of proppants (sand) in the operation. Thus, understanding the potential impact of microorganisms in shale gas and shale oil operations based on reservoir and site-specific operating conditions will ensure that biocides and other treatments to mitigate microbial risks are applied effectively.

Wastewaters produced from hydraulic fracturing and production are also threats for the environment (Chen et al., 2014). These wastewaters are high in total dissolved solids (TDS, mainly salts) and contain hydrocarbons, additives, heavy metals and toxic compounds. Wastewaters are typically transported away from the operational sites for disposal. Some flowback fluids are disposed by subsurface injection, and there is evidence suggesting that the hydrocarbons can migrate upwards reaching groundwater (Chen et al., 2014; Shrestha et al., 2017). The presences of hydrocarbons and gases in groundwater can cause serious health and safety concerns.

Recycling of wastewaters is a desirable alternative as it can decrease the costs of using fresh water and waste management (Chen et al., 2014; Shrestha et al., 2017). One limiting factor is the high TDS, which can generate scaling, and this is an undesirable outcome of recycling. At the Montney shale formation located in northeast British Columbia and northwest Alberta, Canada, it costs $8/barrel to dispose of flowback waters but only $2.75/barrel to recycle (Boschee, 2014). The Bakken formation located in Saskatchewan, Canada, is currently disposing 12 million barrels of produced water per day while using 9 million barrels of fresh water or source water per day (Boschee, 2014). Hence, industries and operators are faced with decisions of starting an effective recycling system. Overall, developing an effective way to recycle flowback and produced waters can reduce environmental impact and costs.

4.3 MICROORGANISMS IN OIL RESERVOIRS

A diverse microbial community can be found in oil reservoirs. Sulfate-reducing bacteria (SRB) are among the most studied as they have an important ecological role related to reservoir souring and corrosion. SRB are known to be strictly anaerobic and can utilize a range of substrates such as lactate, volatile fatty acids (VFA), alcohols and hydrogen as electron donors (Barton and Fauque, 2009). Some SRB can also utilize hydrocarbons, such as toluene and alkanes, as well as benzoate and phenol for sulfate reduction (Barton and Fauque, 2009).

Phylogenetically, most SRB belong to the class *Deltaproteobacteria* or the phylum *Firmicutes* (Gieg et al., 2011). Most SRB have been isolated from non-saline environments, but halophilic SRB have been found and can thrive at extremely high salinities (Oren, 2001). *Desulfohalobium utahense*, isolated from the Great Salt Lake (Jakobsen et al., 2006) can reduce sulfate up to 270 g/L of NaCl or 4.6 M NaCl. For osmotic regulation, SRB such as *Desulfovibrio halophilus* use organic solutes for osmotic balance (Oren, 2001) while others accumulate inorganic salts intracellularly. At high salinity, the choice of electron donor is dominated by the energy yield. Acetate oxidation can generate −47 kJ of energy per mol of sulfate (Oren, 2001), while sulfate reduction using H_2 can generate −152 kJ per mol of sulfate (Equation 2).

$$4H_2 + SO_4^{2-} + H^+ \rightarrow HS^- + 4H_2O \qquad (4.2)$$

Thus it is not surprising that halophilic SRB such as *Desulfovermiculus halophilus* and *Desulfohalobium utahense* can utilize H_2 but not acetate for sulfate reduction (Belyakova et al. 2006).

In addition to SRB, there are also thiosulfate reducers, sulfur reducers and sulfide oxidizers that contribute to the sulfur cycle in oil reservoirs (Youssef et al., 2009; Gieg et al., 2011). These are stimulated by the water injected into the oil reservoir during oil recovery processes, which often contains sulfate. Concentrations of hydrogen sulfide are strictly regulated in produced oil and gas to protect the safety of humans, product quality and steel infrastructures.

There are several ways to prevent the formation of sulfide by SRB or remove the sulfide already present in oil reservoirs. Biocides such as glutaraldehyde are often used to eliminate the growth of SRB and other bacteria. However, they are costly to use and not environmentally friendly. Nitrate injection is often preferred over biocides for souring control in reservoirs. Injection of nitrate stimulates the growth of heterotrophic nitrate-reducing bacteria (hNRB) and sulfide-oxidizing nitrate-reducing bacteria (soNRB) (Myhr et al. 2002; Bødtker et al. 2009). Nitrate injection can remediate souring through three mechanisms: hNRB can outcompete SRB for the same carbon and energy sources (Callbeck et al. 2011), soNRB can utilize nitrate to oxidize sulfide back to sulfate (Bødtker et al. 2009), and both hNRB and soNRB can produce nitrite which inhibits the dissimilatory sulfite reductase (Dsr) enzyme, which the SRB rely on to reduce sulfite to sulfide (Greene et al. 2003). The Dsr enzyme has a higher affinity to bind nitrite than to bind sulfite, preventing sulfite reduction in the presence of nitrite. However, a problem associated with nitrate injection is that nitrite will not normally accumulate in the reservoir since it is reduced further to ammonia or N_2 (Hubert, 2009). To achieve effective souring control nitrate reduction must terminate at the nitrite stage.

4.4 MICROORGANISMS IN SHALE RESERVOIRS

Recently, several studies have been published on the microbial community compositions of shale gas fields. These studies revealed a change in the microbial composition over time. Shale gas fields are rich in hydrocarbons with high pressure and temperature, which are favorable conditions for extremophiles that are indigenous to the formation. Hydraulic fracturing can introduce freshwater microorganisms into the formation as many of the source waters used are low in salinity. Previous studies on waters produced from hydraulic fracturing at different shale operations showed similar microbial community composition (Liang et al., 2016; Mouser et al., 2016). During the initial flowback stages, the microbial communities were predominately aerobic fresh water microorganisms, which transitioned to anaerobic halophilic communities including *Halanaerobium,* halophilic SRB and methanogenic taxa. Some taxa such as *Marinobacter, Halomonas* and *Halanaerobium* are capable of hydrocarbon degradation and fermentation at high salinity (Daly et al., 2016; Cluff et al., 2014; Liang et al., 2016). *Halanaerobium* in particular have a high potential for MIC as it can reduce thiosulfate to sulfide. They can also breakdown polysaccharides to H_2 and organic acids, which may enhance the growth of other microorganisms. Later flowback stages are also dominated by methanogens, particularly methylotrophic methanogens such as *Methanohalophilus* and *Methanolobus* (Wuchter et al., 2013; Daly et al., 2016).

Shale oilfields are more dynamic and complicated as compared to shale gas fields. The down-hole environment of the field is often high in temperature (>60°C) and salinity (>1 M NaCl), while the waters injected for hydraulic fracturing are low in temperature and salinity. Thus, temperature and salinity of the reservoir may decrease over time with continued oil recovery causing changes in microbial communities. Microbial communities will flourish in low temperature and/or low salinity regions. Bowers and Wiegel (2011) analyzed the optimal salinity and temperature ranges for 110 published *Archaea*. Results showed that most of the halophiles analyzed grow optimally between 30°–40°C and 2.5–3.5 M NaCl.

Extreme environments such as high temperature or salinity are very stressful for the microorganisms to live in. High salinity, in particular high NaCl, limits the survival of the microorganisms by forcing them to spend an excessive amount of energy on osmoregulation (Oren, 2001). Hence the majority obligate halophiles have an optimal temperature around 30°C (Bowers and Wiegel, 2011). The formation and produced waters from shale oilfields are often high in salinity due to the mineral and salt contents released from the shale formation during the hydraulic fracturing process (Waldron et al. 2007). But several bacterial and archaeal taxa can survive under very high NaCl concentrations (Waldron et al. 2007). To cope with these hypersaline environments, the organisms have evolved different strategies as described previously.

In shale reservoirs, it is not known whether biogenic souring is due to indigenous species or due to microorganisms introduced with the injection fluids. In addition, the effect of continuous injection of low salinity water for oil recovery in shale oilfields is currently unknown. High salinity and temperature can inhibit the presence and activity of microorganisms. However decreasing the salinity and temperature of the reservoir may increase biogenic souring potential (Youssef et al., 2009; Gieg et al., 2011). Previously, it was shown that in extreme environments, such as at high temperature, nitrite can accumulate with nitrate injection, thus effectively inhibiting souring (Gieg et al., 2011; Fida et al., 2016). High salinity is an equally extreme condition as high temperature. However, the efficacy of nitrate-mediated souring control at high salinity has not been reported previously. Compared to nitrate-mediated souring control at low salinity conditions, there is a lack of knowledge on its effectiveness at high salinity.

4.5 CASE STUDY—BAKKEN SHALE OIL RESERVOIR

Currently, data on microbial communities residing in shale oilfields, such as the Bakken shale oil reservoir, is extremely limited with little to no research. A Bakken field near Estevan in Saskatchewan, Canada, is a light, shale oil (API 44°) reservoir with a down-hole temperature of approximately 60°C. The reservoir was fractured and

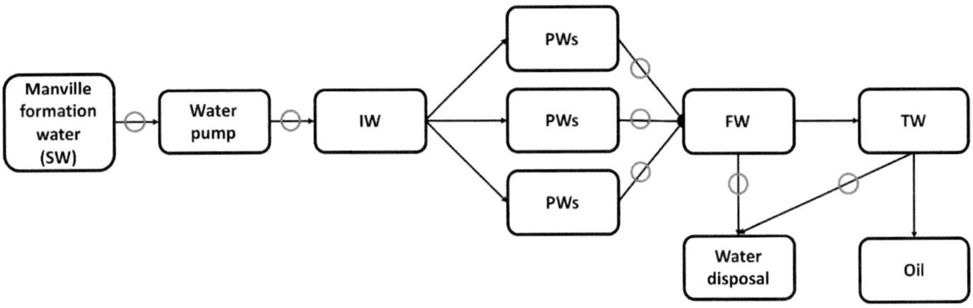

FIGURE 4.1 Schematic diagram of shale oil production from a field in the Bakken formation. Mannville formation source water (SW) is injected at the injection well (IW) to pressurize multiple producing wells (PW). Oil-water emulsions from producing wells enter a free water knockout (FW) facility first and then a treater water facility (TW) to separate oil from water. The produced water is not reinjected but is disposed. Circles indicate the sampling points. (Adapted from An et al., 2017.)

continuously produced through water flooding with source water from the Mannville formation. Five types of sample were received from the field (see Figure 4.1): source water (SW), injection water (IW), produced water (PW), free water knockout water (FW) and treater water (TW) (An et al., 2017).

The SW and IW samples had low salinity (0.7 ± 0.1 Meq of NaCl), high sulfate (29.9 ± 2.6 mM) and low ammonium (2.0 ± 0.9 mM) concentrations. The water chemistry of the SW and IW remained stable between the three sampling periods (between 2013 to 2015), indicating that the water chemistry was relatively constant with time. In contrast, the sampled PWs (~15 wells) showed highly variable salinities (0.6 to 3.7 Meq of NaCl), sulfate (5.1 to 34.8 mM) and ammonium (2.4 to 33.2 mM) concentrations. All produced fluids were converged into the FW and TW facilities (see Figure 4.1). The FW and TW (FW-TW) samples had a salinity of 2.4 ± 0.3 Meq of NaCl, and ammonium and sulfate concentrations of 22.5 ± 3.5 and 17.2 ± 1.6 mM, respectively. These averages were consistent over time and maintained high salinity. The low salinity, high sulfate concentrations in some of the PWs were likely due to a breakthrough of the SW. Increased sulfate concentrations in the reservoir may increase souring downhole by stimulating SRB activities. In addition, of the wells sampled, it indicated a positive relationship between salinity and ammonium concentration. Statistical analyses (JMP®, Version 13.1. SAS Institute Inc., Cary, NC, 1989–2007), with the exception of one PW sample, showed a positive linear relationship ($r^2 = 0.90$), suggesting the increased salinity and ammonium concentration of the PWs were likely contributed by the shale oil formation. Further analyses of the operation indicated another source of ammonium as a product of methylotrophic methanogenesis.

The microbial communities in 2013 samples were dominated by halophilic/halotolerant taxa. High fractions of *Methanohalophilus, Halanaerobium* and *Desulfovermiculus*, as well as of *Flexistipes*, *Halomonas* and *Desulfohalobium* were found in high salinity PW and FW-TW samples, but not in the SW sample. SW was dominated by halotolerant and non-halophilic microorganisms such as *Halomonas, Clostridium-sensu-stricto, Methanolobus, Pelobacter, Desulfovibrio* and *Methanobacterium.* In 2015, the FW-TW samples were mainly dominated by *Halanaerobium, Modicisalibacter, Desulfovermiculus, Methanohalophilus* and *Chromohalobacter,* similar to the 2013 data. The community compositions of low salinity PW (0.5 to 1 M NaCl) samples had very high fractions (45–95%) of *Marinobacter* with smaller fractions of *Halomonas, Thiomicrospira, Methanothermococcus* and *Desulfotignum.*

The dominant microbial communities in SW samples were *Methanolobus Thiomicrospira, Desulfovibrio, Methanobacterium* and *Pelobacter.* The microbial communities of SW were halotolerant not halophilic, with optimal growth ranges between 0.2 to 1.0 M NaCl. Of these members, *Desulfovibrio* are SRB known to be involved in reservoir souring and MIC (Hubert et al., 2009; Kakooei et al., 2012; Guan et al., 2014) at low salinity and low temperature conditions. The presences of sulfur-reducers such as *Pelobacter* can further increase souring risks. Overall, as the low salinity source waters are continuously injected into the reservoir, there is an increasing souring potential associated with it.

Microbial community compositions of PWs were dependent on salinity. Samples from high salinity Bakken PWs were dominated by obligate anaerobes and halophiles, such as *Halanaerobium, Desulfovermiculus* and *Desulfohalobium.* This suggests fermentative and souring potential at high salinity. Archaeal taxa such as *Methanohalophilus Methanothermococcus* and *Methanothermobacter* were also detected in these high salinity PW samples, suggesting potential methanogenic activities. However, PW samples with low salinity (0.7 Meq of NaCl) had decreased fractions of *Halanaerobium* and increased fractions of *Marinobacter,* which grows optimally at lower salinities (Guo et al., 2007) and is known for its fermentative and nitrate-reducing properties in oil reservoirs (Guo et al., 2007; Daly et al., 2016; Rani et al., 2017). Overall, the decrease in PW salinities was reflected in the microbial community compositions. Samples from the FW-TW facility, which processes oil-water emulsions from all (>100) producing wells, had high salinity throughout. These were dominated by the same obligately anaerobic, halophilic taxa as found in high salinity PWs (*Halanaerobium, Methanohalophilus, Desulfovermiculus* and *Desulfohalobium*).

Overall, based on the microbial community analyses, model interactions between key microbial players have been proposed (see Figure 4.2). The dominant SRB *Desulfovermiculus* and *Desulfohalobium* reduce sulfate using oil organics (butyrate and propionate) or H_2 (Belyakova et al., 2006; Jakobsen et al., 2006). The fermentative hydrogen producer *Halanaerobium* could enhance this sulfate reduction by providing H_2 and organic acids (see Figure 4.2) (Cayol et al., 2002). In the presence of thiosulfate, *Halanaerobium* can catalyze the reduction of thiosulfate and produce sulfide leading to further souring risks (Booker et al., 2017). Though it has not been reported extensively, the formation of thiosulfate can occur in oilfields through both abiotic and biotic processes (Bullock and Forbes, 1933; Siu and Jia, 1999; Suh and Akagi, 1969). It has been reported that thiosulfate can be formed from sulfide and sulfite (equation 3) (Siu and Jia, 1999) or from aromatic nitro compounds and sulfide (equation 4) (Bullock and Forbes, 1933). Some SRB also have the capability of thiosulfate production (Suh and Akagi, 1969). Furthermore, thiosulfate can also

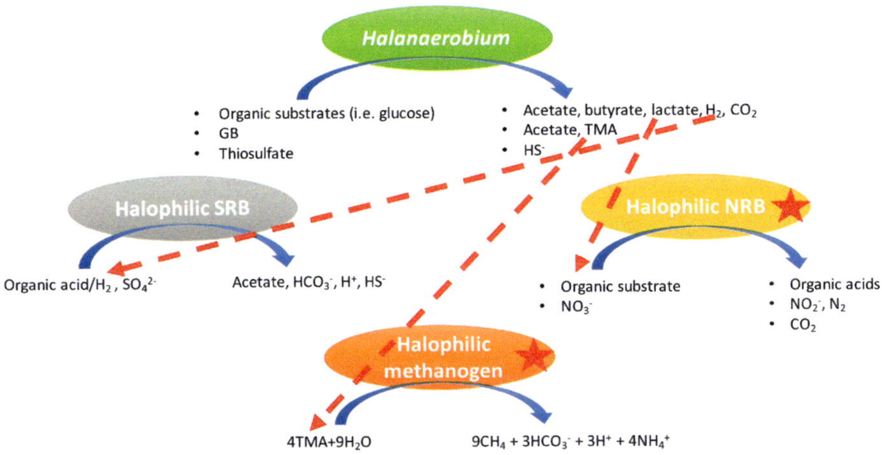

FIGURE 4.2 Proposed interactions between halophilic microbial taxa based on microbial community analyses of field sample and literature. Dashed arrows indicate potential syntrophic relationship shared between microbial community taxa. Star indicates potential producer of glycine betaine (GB), which can be utilized by *Halanaerobium*. *Halanaerobium* can produce trimethylamine (TMA) that can be utilized by halophilic methanogens for methanogenesis. Utilization of organic substrates such as glucose and guar gum by *Halanaerobium* can produce organic acids and H_2 utilized by halophilic NRB and SRB.

be formed as an intermediate in oilfields through the abiotic oxidation of pyrite (equation 5; FeS_2) (Harrold et al., 2016). The presence of thiosulfate could be a main driver of the series of microbial interactions at elevated salinities in shale reservoirs.

$$2HS^- + 4HSO_{3^-} \rightarrow 3S_2O_3^{2-} + 3H_2O \tag{4.3}$$

$$R\text{-}NO_2 + Na_2S_2 + H_2O \rightarrow R\text{-}NH_2 + Na_2S_2O_3 \tag{4.4}$$

$$2FeS_2 + 7.5O_2 + 7H_2O \rightarrow S_2O_3^{2-} \rightarrow 4SO_4^{2-} + 2Fe(OH)_3 + 8H^+ \tag{4.5}$$

Other halophilic heterotrophic taxa such as *Modicisalibacter, Halomonas, Chromohalobacter* and *Marinobacter* were found in FW-TW samples in more variable fractions and several of these are well known NRB (Vreeland et al., 1980; Rani et al., 2017). Notably, *Halomonas* reduces nitrate to nitrite at elevated salinities (Vreeland et al., 1980). The accumulation of nitrite inhibits SRB (Gieg, Jack, and Foght. 2011). In addition, *Halanaerobium* can metabolize the osmotic protectant betaine produced by *Halomonas, Chromohalobacter* and *Methanohalophilus* (Pastor et al., 2010; Lai and Lai, 2011) into methylated amines used by *Methanohalophilus* for methane production (Daly et al., 2016). *Halanaerobium* does not depend on compatible solutes for osmoprotection but instead accumulates K^+ and Cl^- intracellularly. Furthermore, methanogenesis using methylated amines by methylotrophic methanogens such as *Methanohalophilus* would increase the ammonium concentration.

Bakken samples with high proportions of *Methanohalophilus*, were extremely saline (>2 Meq of NaCl) with high concentrations of ammonium (~2 0 mM). Methylotrophic methanogenesis has not been proposed previously as a reason for the high ammonium concentrations in shale oilfields. Further testing is needed to prove that high ammonium in produced waters originates from methylamine addition to fracturing fluids or injection waters. Overall the microbial communities of high salinity PW and FW-TW samples indicated a strong halophilic and halotolerant signature that contrasted with the non-halophilic microbial communities in low salinity SW-IW samples.

4.6 SOURING CONTROL STUDIES USING BAKKEN SAMPLES

High concentrations of sulfate in source water samples can potentially stimulate the growth of SRB such as *Desulfovermiculus, Desulfohalobium, Desulfocella, Desulfotignum* and *Desulfovibrio*, which were all found in the microbial compositions of Bakken field samples. SRB enrichments were set up at both 0.75 M NaCl (43.8 g/L NaCl) and 2.5 M NaCl (146 g/L NaCl) with the field samples. Positive growth was observed with volatile fatty acids (mixture of propionate, acetate and butyrate) or lactate as the electron donor and sulfate as the electron acceptor. Enrichments of PW, FW and TW were dominated by *Halanaerobium* and the SRBs *Desulfovibrio, Desulfohalobium* and *Desulfovermiculus* at both salinities. Microbial community analyses indicated *Halanaerobium*, which produces H_2 through fermentation, can be utilized by SRB for sulfate-reduction (Cayol et al., 2002; Belyakova et al., 2006; Thabet et al., 2011). However, the identified microbial community taxa from field samples and enrichments were not thermophilic, whereas most oilfields are high in temperature. Hence, the identified taxa would be expected to grow in regions of the reservoir and above ground operations where the temperature has cooled from the *in situ* temperature of 60°C. Overall, the identification of halophilic SRB from the Bakken field enrichments further implied the possibility of souring in the field at different salinities. In addition, souring will also occur when the temperature of the field is lowered to below 50°C, which is highly likely due to the continuous injection of low-temperature, low salinity and high sulfate source water for oil recovery.

The presence of high fractions of nitrate-reducers (*Halomonas* and *Marinobacter*) in the microbial community composition of Bakken field samples indicates the possibility for nitrate-mediated souring control. Through our research, it was shown that nitrate reduction at high salinity gave nitrite accumulation, which will prevent SRB from producing sulfide (An et al., 2017). At lower salinities, in primary or secondary enrichments (0, 0.5 and 1.0 M NaCl), nitrite did not accumulate and was likely reduced further to N_2, indicating that souring control through nitrate injection will be less effective under these conditions. Dominant microbial taxa in high salinity hNRB enrichments were *Halomonas, Deferribacteraceae, Halanaerobium, Desulfovermiculus, Geotoga* and *Desulfosalsimonas. Halomonas* and *Deferribacteraceae* are capable of denitrification at high salinity, and *Halomonas* is known for incomplete reduction of nitrate to nitrite (Vreeland et al., 1980). Overall, growth of hNRB at high salinity showed nitrite accumulation, which can be further explored for nitrate-mediated souring control.

Because the outcome of batch cultures can be more variable with some cultures reducing nitrate completely to N_2 and others accumulating nitrite, continuous cultures (chemostats) were established. Chemostats were used to study the dominant microbial taxa involved with sulfate-reduction and nitrate-reduction at low (0.5 M NaCl) and high (2.5 M NaCl) salinities. Continuous cultures showed distinct separation of microbial community compositions between high and low salinity. A complete microbial turnover was observed more rapidly in chemostats than in batch cultures as dead or nongrowing cells are constantly removed, leaving only actively growing cells for DNA analyses. DNA sequencing of batch cultures analyze all biological components in the system, including the inactive or dead cells and therefore is a less accurate representation of active taxa in the system.

Sulfate-reduction and nitrate-reduction were faster at 0.5 M NaCl than at 2.5 M NaCl and nitrite persisted at high salinity. At 2.5 M NaCl, under sulfate-reducing conditions the microbial community was dominated by *Halanaerobium* and *Desulfovermiculus*. Similar results were observed in field samples and batch enrichments suggesting a high possibility of interdependence between the two species. As described earlier, halophilic SRB use H_2 as a primary electron donor due to higher energy yields, and *Desulfovermiculus* is a known H_2-utilizing SRB. It is possible that *Desulfovermiculus* uses the H_2 produced by *Halanaerobium* through fermentation for sulfate-reduction (Belyakova et al., 2006). In contrast, the dominant SRB was *Desulfovibrio* at 0.5 M NaCl, whereas *Halanaerobium* and *Desulfovermiculus* were mostly absent. The complete shift of microbial communities due to continuous injection of low salinity SW-IW can be represented by chemostats and may reflect what might happen in the reservoir at a faster rate.

The microbial community composition in nitrate-reducing chemostats at high salinity were nearly identical with *Halanaerobium, Halomonas* and *Marinobacter* with VFA or lactate as the electron donor. At low salinity, the continuous culture with nitrate was dominated by *Geoalkalibacter*, an anaerobic nitrate-, sulfur- and iron-reducing microorganism. *Geoalkalibacter* grows optimally at 0.3 M NaCl, a similar salinity range as the 0.5 M NaCl medium and SW-IW. Overall, microbial communities were separated based on salinity ranges correlated to observations with field samples. The data also showed that repeated transferring of high salinity SRB enrichments caused a loss in nitrate-reducing taxa. Similarly, a continuous transferring of nitrate-reducing taxa also led to a loss in SRB, while fermentative taxa could persist continuously under both conditions.

The bioreactor studies using medium-injected sand-packed columns aimed to determine the feasibility of nitrate-mediated souring control at high salinity under flow conditions, as in the field. Nitrate reduction started within two weeks of the start of incubation using an active consortium at 2.5 M NaCl and immediately at 0.5 M NaCl. Continuously injecting low concentrations of nitrate at high salinity led to nitrite accumulation and souring control (see Figure 4.3) (An et al., 2017).

The microbial communities in bioreactors continuously injected with lactate and sulfate (LS) only at 2.5 M NaCl (LS_2.5) produced 3 mM sulfide and were dominated by *Halanaerobium* and *Desulfovermiculus*. At 0.5 M NaCl, a similar H_2S production was achieved as in bioreactors at 2.5 M NaCl (see Figure 4.3 A and B), though the dominant microbial community was *Desulfovibrio*. In the presence of

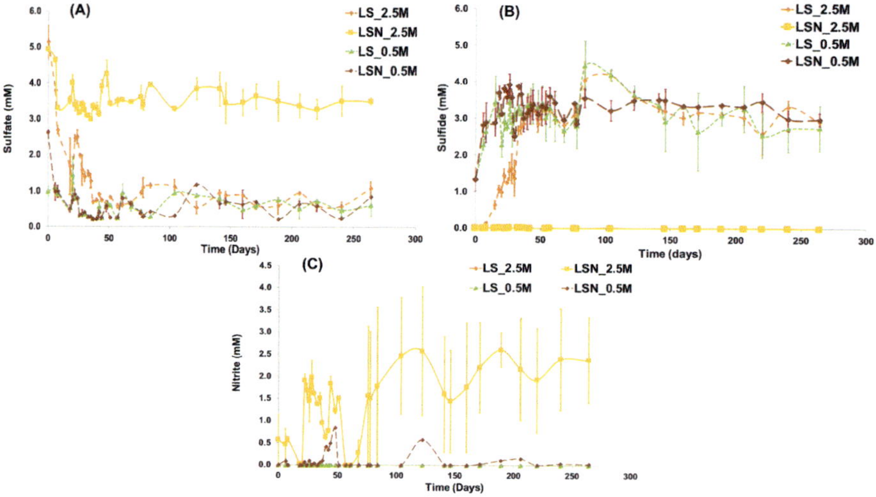

FIGURE 4.3 Effect of nitrate on sulfide concentration in the effluent of bioreactors at low and high salinity. High salinity bioreactors were inoculated with high salinity chemostat cultures and were injected with lactate, sulfate and nitrate at 2.5 M NaCl (LSN_2.5) or with lactate and sulfate at 2.5 M NaCl (LS_2.5). Low salinity bioreactors were inoculated with low salinity chemostat cultures and were injected with lactate, sulfate and nitrate at 0.5 M NaCl (LSN_0.5) or with lactate and sulfate at 0.5 M NaCl (LS_0.5). The concentrations ±SD of sulfate (A), sulfide (B) and nitrite (C) are shown as a function of time. Bioreactors were run at a flow rate of 0.6 PV/day. (Adapted from An et al., 2017.)

nitrate, the bioreactors at 2.5 M NaCl showed no sulfide production throughout the experiment (see Figure 4.3 A and B), whereas sulfide was observed in the bioreactors at 0.5 M NaCl. This was due to the accumulation of nitrite in the bioreactors continuously injected with nitrate at 2.5 M NaCl (see Figure 4.3 C), which was not observed in the bioreactors injected with nitrate at 0.5 M NaCl (see Figure 4.3 B). *Halomonas*, which dominated the community in the high salinity bioreactors, reduce nitrate only to nitrite (Vreeland et al., 1980). Excess concentrations of electron donors (VFA or lactate) were present throughout in these and other bioreactor studies. Thus, complete sulfate and nitrate reduction were possible in the bioreactors continuously injected with nitrate at 2.5 M NaCl. The lack of SRB activities and incomplete nitrate reduction were not caused by a shortage of electron donors but due to salinity restraints.

At low salinity (0.5 M NaCl), sulfide production was comparable between the bioreactors with and without nitrate (see Figure 4.3 A and B). Similar microbial community composition was also observed between the two types of bioreactors. They were primarily dominated by *Desulfovibrio*. Our 16S rRNA sequencing results also indicated the presences of nitrate-reducing communities in the LSN_0.5 bioreactors, suggesting *Desulfovibrio* must reduce sulfate after complete nitrate reduction has occurred. The excess available electron donors allowed a complete nitrate reduction followed by sulfate reduction, similar to the microbial zonation model in low temperature oilfields (Voordouw et al., 2009). In addition, an increase in the microbial diversity was also observed at 0.5 M NaCl compared to 2.5 M NaCl. A diverse halotolerant microbial community dominated at 0.5 M NaCl, which transitioned to a strictly halophilic community harboring mainly *Halanaerobium, Desulfovermiculus* and *Halomonas* at 2.5 M NaCl.

Overall, it has been shown that at high salinity, nitrate-mediated souring control can successfully inhibit the activities of SRB due to nitrite accumulation. The different microbial community compositions at high and low salinities further indicate the possibility of souring in the Bakken field as a low salinity, sulfate-rich source water is continuously injected into the reservoir. Nitrate-mediated souring control technology can only be effective if the reservoir remains at high salinity. Hence, re-injection of high salinity produced water present in treatment and free-water knockout facilities should be considered if nitrate were to be used for souring control in the Bakken field.

ACKNOWLEDGMENTS

This work was supported by an NSERC Industrial Research Chair Award to Dr. Gerrit Voordouw, which was also supported by BP America Production Co., Baker Hughes Canada, Computer Modeling Group Limited, ConocoPhillips Company, Dow Microbial Control, Enbridge Inc., Enerplus Corporation, Intertek Commercial Microbiology, Oil Search (PNG) Limited, Shell Global Solutions International, Suncor Energy Inc. and Yara Norge AS, as well as by Alberta Innovates-Energy and Environment Solutions. Special thanks to Priyesh Menon, who provided us with all the help and support to start this project. Thanks to Chris Farnden, Dominic Cote and Kirk Miner from Baker Hughes Canada for providing field information and sample collections.

REFERENCES

An, Biwen A., Yin Shen, and Gerrit Voordouw. 2017. "Control of Sulfide Production in High Salinity Bakken Shale Oil Reservoirs by Halophilic Bacteria Reducing Nitrate to Nitrite." *Frontiers in Microbiology* 8 (JUN): 1164. doi: 10.3389/fmicb.2017.01164.

Averhoff, Beate, and Volker Müller. 2010. "Exploring Research Frontiers in Microbiology: Recent Advances in Halophilic and Thermophilic Extremophiles." *Research in Microbiology* 161 (6): 506–14. doi: 10.1016/j.resmic.2010.05.006.

Barton, Larry L., and Guy D. Fauque. 2009. "Chapter 2 Biochemistry, Physiology and Biotechnology of Sulfate-Reducing Bacteria." In *Advances in Applied Microbiology*, 1st ed., 68:41–98. doi: 10.1016/S0065-2164(09)01202-7.

Belyakova, E V, E P Rozanova, I A Borzenkov, T P Tourova, M A Pusheva, A M Lysenko, and T V Kolganova. 2006. "The New Facultatively Chemolithoautotrophic, Moderately Halophilic, Sulfate-Reducing Bacterium *Desulfovermiculus Halophilus* Gen. Nov., Sp. Nov., Isolated from an Oilfield." *Microbiology (Mikrobiologiya)* 75 (2): 161–71. doi: 10.1134/S0026261706020093.

Bødtker, Gunhild, Kristine Lysnes, Terje Torsvik, Eva Ø Bjørnestad, and Egil Sunde. 2009. "Microbial Analysis of Backflowed Injection Water from a Nitrate-Treated North Sea Oil Reservoir." *Journal of Industrial Microbiology & Biotechnology* 36 (3): 439–50. doi: 10.1007/s10295-008-0515-6.

Booker, Anne E., Mikayla A. Borton, Rebecca A. Daly, Susan A. Welch, Carrie D. Nicora, David W. Hoyt, Travis Wilson, et al. 2017. "Sulfide Generation by Dominant *Halanaerobium* Microorganisms in Hydraulically Fractured Shales." *Applied and Environmental Microbiology* 2 (4): 1–13. doi: 10.1128/mSphereDirect.00257-17.

Boschee, Pam. 2014. "Produced and Flowback Water Recycling and Reuse Economics, Limitations, and Technology." *Oil and Gas Facilities* February (1): 17–21. doi: 10.2118/0214-0016-OGF.

Bowers, Karen J., and Juergen Wiegel. 2011. "Temperature and PH Optima of Extremely Halophilic Archaea: A Mini-Review." *Extremophiles* 15 (2): 119–28. doi: 10.1007/s00792-010-0347-y.

Brittingham, Margaret C., Kelly O. Maloney, Aïda M. Farag, David D. Harper, and Zachary H. Bowen. 2014. "Ecological Risks of Shale Oil and Gas Development to Wildlife, Aquatic Resources and Their Habitats." *Environmental Science & Technology* 48 (19): 11034–47. doi: 10.1021/es5020482.

Bullock, Jesse Lunt, and George Shannon Forbes. 1933. "The Formation of Sodium Thiosulfate from the Oxidation of Sodium Sulfides by Means of an Aromatic Nitro Compound." *Journal of the American Chemical Society* 55 (1): 232–38. doi: 10.1021/ja01328a025.

Callbeck, Cameron M., Xiaoli Dong, Indranil Chatterjee, Akhil Agrawal, Sean M. Caffrey, Christoph W. Sensen, and Gerrit Voordouw. 2011. "Microbial Community Succession in a Bioreactor Modeling a Souring Low-Temperature Oil Reservoir Subjected to Nitrate Injection." *Applied Microbiology and Biotechnology* 91 (3): 799–810. doi: 10.1007/s00253-011-3287-2.

Cayol, Jean Luc, Marie Laure Fardeau, Jean Louis Garcia, and Bernard Ollivier. 2002. "Evidence of Interspecies Hydrogen Transfer from Glycerol in Saline Environments." *Extremophiles* 6 (2): 131–34. doi: 10.1007/s007920100229.

Chen, J, MH Al-Wadei, and RCM Kennedy. 2014. "Hydraulic Fracturing: Paving the Way for a Sustainable Future?" *Journal of Environmental and Public Health* 2014:1-10. doi: 10.1155/2014/656824.

Cluff, Maryam A, Angela Hartsock, Jean D MacRae, Kimberly Carter, and Paula J Mouser. 2014. "Temporal Changes in Microbial Ecology and Geochemistry in Produced Water from Hydraulically Fractured Marcellus Shale Gas Wells." *Environmental Science & Technology* 48 (11): 6508–17. doi: 10.1021/es501173p.

Daly, Rebecca A., Mikayla A. Borton, Michael J. Wilkins, David W. Hoyt, Duncan J. Kountz, Richard A. Wolfe, Susan A. Welch, et al. 2016. "Microbial Metabolisms in a 2.5-Km-Deep Ecosystem Created by Hydraulic Fracturing in Shales." *Nature Microbiology* 1 (10): 16146. doi: 10.1038/nmicrobiol.2016.146.

Fida, Tekle Tafese, Chuan Chen, Gloria Okpala, and Gerrit Voordouw. 2016. "Implications of Limited Thermophilicity of Nitrite Reduction for Control of Sulfide Production in Oil Reservoirs." *Applied and Environmental Microbiology* 82 (14): 4190–99. doi: 10.1128/AEM.00599-16.

Gieg, Lisa M., Tom R. Jack, and Julia M. Foght. 2011. "Biological Souring and Mitigation in Oil Reservoirs." *Applied Microbiology and Biotechnology* 92 (2): 263–82. doi: 10.1007/s00253-011-3542-6.

Greene, E. A., C. Hubert, M. Nemati, G. E. Jenneman, and G. Voordouw. 2003. "Nitrite Reductase Activity of Sulphate-Reducing Bacteria Prevents Their Inhibition by Nitrate-Reducing, Sulphide-Oxidizing Bacteria." *Environmental Microbiology* 5 (7): 607–17. doi: 10.1046/j.1462-2920.2003.00446.x.

Guan, Jing, Bing Liang Zhang, Serge Maurice Mbadinga, Jin Feng Liu, Ji Dong Gu, and Bo Zhong Mu. 2014. "Functional Genes (Dsr) Approach Reveals Similar Sulphidogenic Prokaryotes Diversity but Different Structure in Saline Waters from Corroding High Temperature Petroleum Reservoirs." *Applied Microbiology and Biotechnology* 98 (4): 1871–82. doi: 10.1007/s00253-013-5152-y.

Guo, B, J Gu, Y G Ye, Y Q Tan, K Kida, and X L Wu. 2007. "*Marinobacter Segnicrescens* Sp Nov., a Moderate Halophile Isolated from Benthic Sediment of the South China Sea." *International Journal of Systematic and Evolutionary Microbiology* 57: 1970–74. doi: 10.1099/ijs.0.65030-0.

Harrold, Zoë R., Mark L. Skidmore, Trinity L. Hamilton, Libby Desch, Kirina Amada, Will Van Gelder, Kevin Glover, Eric E. Roden, and Eric S. Boyd. 2016. "Aerobic and Anaerobic Thiosulfate Oxidation by a Cold-Adapted, Subglacial Chemoautotroph." *Applied and Environmental Microbiology* 82 (5): 1486–95. doi: 10.1128/AEM.03398-15.

Hubert, Casey, Gerrit Voordouw, and Bernhard Mayer. 2009. "Elucidating Microbial Processes in Nitrate- and Sulfate-Reducing Systems Using Sulfur and Oxygen Isotope Ratios: The Example of Oil Reservoir Souring Control." *Geochimica et Cosmochimica Acta* 73 (13): 3864–79. doi: 10.1016/j.gca.2009.03.025.

Jakobsen, Trine Fredlund, Kasper Urup Kjeldsen, and Kjeld Ingvorsen. 2006. "*Desulfohalobium Utahense* Sp. Nov., a Moderately Halophilic, Sulfate-Reducing Bacterium Isolated from Great Salt Lake." *International Journal of Systematic and Evolutionary Microbiology* 56 (Pt 9): 2063–69. doi: 10.1099/ijs.0.64323-0.

Kakooei, Saeid, Mokhtar Che Ismail, Bambang Ariwahjoedi, and Bandar Seri Iskandar. 2012. "Mechanisms of Microbiologically Influenced Corrosion : A Review." *World Applied Sciences Journal* 17 (4): 524–31.

Lai, Shu Jung, and Mei Chin Lai. 2011. "Characterization and Regulation of the Osmolyte Betaine Synthesizing Enzymes GSMT and SDMT from Halophilic Methanogen *Methanohalophilus Portucalensis*." *PLoS ONE* 6 (9): e25090. doi: 10.1371/journal.pone.0025090.

Laurenzi, Ian J, Joule A Bergerson, and Kavan Motazedi. 2016. "Life Cycle Greenhouse Gas Emissions and Freshwater Consumption Associated with Bakken Tight Oil." *Proceedings of the National Academy of Sciences* 113 (48): E7672–80. doi: 10.1073/pnas.1607475113.

Lee, Kun Sang, and Tae Hong Kim. 2016. *Integrative Understanding of Shale Gas Reservoirs. Integrative Understanding of Shale Gas Reservoirs*. Springer Briefs in Applied Sciences and Technology. Cham: Springer International Publishing. doi: 10.1007/978-3-319-29296-0.

Liang, Renxing, Irene A. Davidova, Christopher R. Marks, Blake W. Stamps, Brian H. Harriman, Bradley S. Stevenson, Kathleen E. Duncan, and Joseph M. Suflita. 2016. "Metabolic Capability of a Predominant *Halanaerobium* Sp. in Hydraulically Fractured Gas Wells and Its Implication in Pipeline Corrosion." *Frontiers in Microbiology* 7 (JUN): 1–10. doi: 10.3389/fmicb.2016.00988.

Mouser, Paula J, Mikayla Borton, Thomas H Darrah, Angela Hartsock, and Kelly C Wrighton. 2016. "Hydraulic Fracturing Offers View of Microbial Life in the Deep Terrestrial Subsurface." *FEMS Microbiology Ecology* 92 (August): 1–18. doi: 10.1093/femsec/fiw166.

Myhr, S, B.L.P. Lillebo, E Sunde, J Beeder, and T Torsvik. 2002. "Inhibition of Microbial H_2S Production in an Oil Reservoir Model Column by Nitrate Injection." *Applied Microbiology and Biotechnology* 58: 400–408. doi: 10.1007/s00253-001-0881-8.

Oren, Aharon. 2001. "The Bioenergetic Basis for the Decrease in Metabolic Diversity at Increasing Salt Concentrations: Implications for the Functioning of Salt Lake Ecosystems." *Hydrobiologia* 466: 61–72. doi: 10.1023/A:1014557116838.

Oren, Aharon. 2013. "Life at High Salt Concentrations, Intracellular KCl Concentrations, and Acidic Proteomes." *Frontiers in Microbiology* 4 (NOV): 1–6. doi: 10.3389/fmicb.2013.00315.

Pastor, José M., Manuel Salvador, Montserrat Argandoña, Vicente Bernal, Mercedes Reina-Bueno, Laszlo N. Csonka, José L. Iborra, Carmen Vargas, Joaquín J. Nieto, and Manuel Cánovas. 2010. "Ectoines in Cell Stress Protection: Uses and Biotechnological Production." *Biotechnology Advances* 28 (6): 782–801. doi: 10.1016/j.biotechadv.2010.06.005.

Rani, Sundas, Hyeon-woo Koh, Hongik Kim, Sung-keun Rhee, and Soo-je Park. 2017. "*Marinobacter Salinus* Sp. Nov., a Moderately Halophilic Bacterium Isolated from a Tidal Flat Environment." *International Journal of Systematic and Evolutionary Microbiology* 67 (2): 205–11. doi: 10.1099/ijsem.0.001587.

Roberts, Mary F, EA Galinski, DD Martin, RA Ciulla, MF Roberts, MF Roberts, AD Brown, et al. 2005. "Organic Compatible Solutes of Halotolerant and Halophilic Microorganisms." *Saline Systems* 1 (1): 5. doi: 10.1186/1746-1448-1-5.

Shrestha, Namita, Govinda Chilkoor, Joseph Wilder, Venkataramana Gadhamshetty, and James J. Stone. 2017. "Potential Water Resource Impacts of Hydraulic Fracturing from Unconventional Oil Production in the Bakken Shale." *Water Research* 108: 1–24. doi: 10.1016/j.watres.2016.11.006.

Siu, T, and Charles Q Jia. 1999. "Kinetic and Mechanistic Study of Reaction between Sulfide and Sulfite in Aqueous Solution." *Ind. Eng. Chem. Res.* 38: 3812–16. doi: 10.1021/ie990254x.

Sorokin, Dimitry Y., Ben Abbas, Tatjana P. Tourova, Boris K. Bumazhkin, Tatjana V. Kolganova, and Gerard Muyzer. 2014. "Sulfate-Dependent Acetate Oxidation under Extremely Natron-Alkaline Conditions by Syntrophic Associations from Hypersaline Soda Lakes." *Microbiology (Reading, England)* 160 (2014): 723–32. doi: 10.1099/mic.0.075093-0.

Sorokin, Dimitry Y, Albert J H Janssen, and Gerard Muyzer. 2011. "Biodegradation Potential of Halo(Alkali)Philic Prokaryotes." *Critical Reviews in Environmental Science and Technology* 42 (8): 811–56. doi: 10.1080/10643389.2010.534037.

Stan-Lotter, Helga, and Sergiu Fendrihan. 2012. *Adaption of Microbial Life to Environmental Extremes-Novel Research Results and Application.* Springer Wien New York. doi: 10.1007/s13398-014-0173-7.2.

Suh, B., and J. M. Akagi. 1969. "Formation of Thiosulfate from Sulfite by *Desulfovibrio Vulgaris*." *Journal of Bacteriology* 99 (1): 210–15.

Thabet, Olfa Ben Dhia, Terres Wafa, Khelifi Eltaief, Jean Luc Cayol, Moktar Hamdi, Guy Fauque, and Marie Laure Fardeau. 2011. "*Desulfovibrio Legallis* Sp. Nov.: A Moderately Halophilic, Sulfate-Reducing Bacterium Isolated from a Wastewater Digestor in Tunisia." *Current Microbiology* 62 (2): 486–91. doi: 10.1007/s00284-010-9733-z.

Voordouw, G, Aleksandra Grigoryan, Adewale Lambo, Shiping Lin, Hyung Soo Park, Thomas R Jack, Dennis Coombe, et al. 2009. "Sulfide Remediation by Pulsed Injection of Nitrate into a Low Temperature Canadian Heavy Oil Reservoir." *Environmental Science & Technology* 43 (24): 9512–18. doi: 10.1021/es902211j.

Vreeland, R. H., C. D. Litchfield, E. L. Martin, and E. Elliot. 1980. "*Halomonas Elongata*, a New Genus and Species of Extremely Salt-Tolerant Bacteria." *International Journal of Systematic Bacteriology* 30 (2): 485–95. doi: 10.1099/00207713-30-2-485.

Waldron, Patricia J., Steven T. Petsch, Anna M. Martini, and Klaus Nüslein. 2007. "Salinity Constraints on Subsurface Archaeal Diversity and Methanogenesis in Sedimentary Rock Rich in Organic Matter." *Applied and Environmental Microbiology* 73 (13): 4171–79. doi: 10.1128/AEM.02810-06.

Wang, Lin, Scott Burns, Daniel E. Giammar, and John D. Fortner. 2016. "Element Mobilization from Bakken Shales as a Function of Water Chemistry." *Chemosphere* 149 (April). Elsevier Ltd: 286–93. doi: 10.1016/j.chemosphere.2016.01.107.

Wuchter, Cornelia, Erin Banning, Tracy J. Mincer, Nicholas J. Drenzek, and Marco J L Coolen. 2013. "Microbial Diversity and Methanogenic Activity of Antrim Shale Formation Waters from Recently Fractured Wells." *Frontiers in Microbiology* 4 (DEC): 1–14. doi: 10.3389/fmicb.2013.00367.

Youssef, Noha, Mostafa S Elshahed, and Michael J Mcinerney. 2009. *Chapter 6 - Microbial Processes in Oilfields: Culprits, Problems, and Opportunities. Advances in Applied Microbiology.* 1st ed. Vol. 66. Elesvier Inc. doi: 10.1016/S0065-2164(08)00806-X.

Zargari, Saeed, Amin Ramezani, Sassan Ostvar, Rasool Rezaei, Ali Niazi, and Shahab Ayatollahi. 2014. "Isolation and Characterization of Gram-Positive Biosurfactant-Producing Halothermophilic Bacilli from Iranian Petroleum Reservoirs." *Jundishapur Journal of Microbiology* 7 (8): e10981. doi: 10.5812/jjm.10981.

5 RNA and Biocorrosion—Collection, Transport, and Extraction Standardization of Samples from the Oil Sector

Diogo Coutinho, Luiz A.L.T. Pinto, Viviane Oliveira, Hazel Barboza, Thais Abrantes, and Márcia Lutterbach

CONTENTS

5.1 INTRODUCTION

Corrosion is the process of deterioration of a material (mainly metallic) due to the chemical and/or electrochemical action of the environment (Gentil, 1982). Microbiologically influenced corrosion (MIC), or biocorrosion, is an electrochemical process of metal dissolution influenced and/or accelerated by the presence and activity of microorganisms, which has caused serious damage in different industrial sectors (Videla, 2005). Biocorrosion is directly related to the adhesion of microorganisms and their subsequent colonization on the surface of materials. Structures formed on these surfaces from microbial agglomerates are termed biofilms or biofouling. Biofilms are heterogeneous structures composed of cellular aggregates, extracellular polymer material (resulting from microbial metabolism), organic and inorganic matter, and most significantly, water (Davey and O'Toole, 2000). These agglomerations also present interstitial channels, where the passage of circulating, nutrient-containing fluid occurs (Lewandowski et al., 1995). This conceptual model of the biofilm explains the limitation of access by some oxidizing biocides and their lower efficiency on microorganisms adhered to the metal (Videla, 2003).

Microorganisms are widely distributed in nature, and different microbial groups may be involved in the corrosion process, often including bacteria, fungi and algae that participate in the deterioration process of metal surfaces. Among the microorganisms related to biocorrosion, SRB are indicated as the group responsible for the most serious cases—being the main culprits for the biocorrosion of metallic materials commonly used in several industrial sectors (Cortez, 2009). These bacteria have an anaerobic metabolism and are found in most natural waters as planktonic but are generally active inside deposits and at the interface between the biofilm and metal surface where oxygen is depleted.

Microbiological monitoring of affected sites can be done using various culture media to evaluate microbial growth by detecting the metabolites that are generated or by determining the presence of bacterial genes in the samples studied. Therefore, it is important to study the biocorrosion processes, as well as the various parameters that influence the formation of biofilms in industrial systems, relating them to the harmful effects on materials and the biodeterioration of products.

In science, each methodology or technology used for a given purpose has gaps that must be filled by analysis using another approach. The quantification of microorganisms generated by DNA analysis provides data that culture techniques cannot provide, since only about 1–10% of the microorganisms present in the environment adapt to laboratory conditions (Torsvik et al., 1990; Amann et al., 1995; Osburne et al., 2000). In spite of the quantification of bacteria by qPCR, the results are representative of the number of bacteria present in a sample, since the cultivable and non-cultivable microorganisms are quantified. When qPCR data are compared to that from cultivation techniques, such as the most probable number (MPN) method, qPCR quantification may seem overestimated since the bacterial cells that are alive, in process of death and dead are quantified.

With this in mind, our laboratory used a series of tests on another molecule that may provide insight to a recurring question/hypothesis, which is: Could SRB with high metabolic activity accelerate the corrosion process more than a large number of bacteria that are dead or with low metabolic activity? SRB activity is directly linked to the dsrAB enzyme, which catalyses the last step of anaerobic respiration with sulfate, the reduction of sulphate to sulphide. All SRB analyzed to date have dsrAB (Wagner et al., 1998; Klein et al., 2001; Dhillon et al., 2003) and it has been target in molecular techniques for detection of sulphate reducers in different environments (Castro et al., 2002; Fukaba et al., 2003; Nakagawa et al., 2004; Perez-Jimenez and Kerkhof, 2005; Dar et al., 2007). Studies of the precursor molecule of this enzyme can provide significant information for a better understanding of the sulphate reducing mediated biocorrosion phenomenon.

Quantifying levels of transcripts for key genes in anaerobic sedimentary environments can provide insights into the metabolic state of microorganisms carrying out important processes in those environment (Chin et al., 2004; Chin et al., 2008). Ultimately, our goal at the end of the testing and standardization is to use SRB mRNA for monitoring areas susceptible to biocorrosion. Experiments have already been performed in our laboratory with environmental samples and pure culture, where it was possible to extract mRNA and synthesize complementary DNA (cDNA), a more stable form, in addition to its detection by qPCR (results not shown). mRNA extraction can be combined with reverse transcription of RNA (RT-PCR) to determine the amount of mRNA, allowing quantification of the relative gene expression at a given time or in certain types of tissues or cells (Qin, 2006; Mackay, 2007). mRNA is the intermediate molecule between DNA and proteins, being produced only by metabolically active cells. Some studies have already been published where mRNA was used as a direct measure of the transcription activity of the dsrAB gene (Neretin et al., 2003; Price et al., 2010; Larsen et al., 2013).

The qPCR technique using mRNA may be employed to evaluate microbial activity through the expression of a specific gene over time in response to the use of a biocide, to environmental variations, or to compare environments with different characteristics. This SRB gene expression analysis approach may promote better understanding of the participation of this group of microorganisms in MIC.

The purpose of this work was to carry out a laboratory test using tank water to establish a standard with recommendations for collection, preservation, packaging and transport of samples of produced water for later RNA extraction. An assay was developed that allowed the evaluation of the quality and concentration of RNA obtained from a sample stored under different conditions and storage time.

5.2 METHODS

Because it is extremely labile, it is necessary to take some care when working with mRNA, since the molecule degrades easily by manipulation, temperature, ribonuclease activity and time between collection and processing (Bustin et al., 2005; Fleige and Pfaffl, 2006). Currently there are several commercial kits that extract RNA from different types of samples, in addition to in-house protocols that are efficient, but not as clean as kits. In-house protocols use solutions that are commonly found in research labs, such as detergents and alcohol. In the commercial kits, which are designed specifically for these methodologies, there are affinity mRNA columns in addition to successive washing steps. These procedures make possible, at the end of the process, a mRNA with less degradation and higher purity. As oil and gas industry samples have their peculiarities, it was first necessary to seek the best methodology for extracting RNA that would be able to maintain the RNA's structural integrity at the end of the method. For this, the following strategies were used:

1. Sample centrifugation + RNA extraction by TRIzol™
2. Sample filtration + RNA extraction by TRIzol™
3. Centrifugation + RNA extraction by Maxwell® automatic extractor
4. Filtration + extraction of the RNA by Maxwell® automatic extractor

After the definition of four strategies to extract the mRNA, sample collection was scheduled to begin the test. The sample was collected from a tank of production water and transferred to an opaque blue bottle. It was shipped immediately to the laboratory at room temperature. Shipping time was twenty four hours. Initially, an aliquot of the original bottle was separated. From this aliquot the first extraction of RNA was made, being considered for purposes of identification as "0 Days." All extractions were performed following strategy #1 of RNA extraction in triplicates

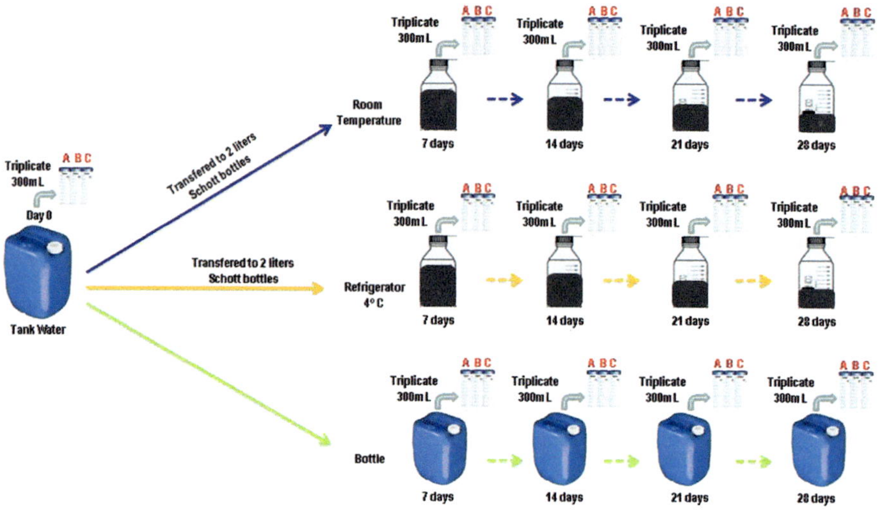

FIGURE 5.1 Experimental design for evaluation of RNA quality under different conditions and storage time.

for each condition. After the initial extraction, aliquots were divided into two other Schott type bottles. One of the samples was stored under refrigeration at 4°C (identified as "refrigerator") and the other sample was kept at room temperature on the lab bench (identified as "room temperature"). In the fourth storage condition, the sample remained in the original opaque blue bottle, at room temperature (identified as "bottle"). Samples were subsequently extracted to compare the yield and quality parameters of the extracted RNA over a total period of 28 days. Extractions occurred at seven-day intervals.

Two assays were performed under the same conditions to verify the replicability of the results. The experimental design can be visualized in Figure 5.1.

After each extraction, the mRNA was quantified and evaluated by absorbance ratios in 260/280 and 260/230 by a Nanodrop™ spectrometer (ThermoFischer). Finally the mRNA was submitted to eletrophoretic migration in agarose gel to verify if the molecules were degraded.

In parallel with the molecular biology assays, the samples were also analyzed using the MPN method to quantify culturable SRB population growth throughout the experiment. Bacterial cultivation in artificial growth media using serial dilution is accepted as a standard technique for the estimation of SRB in oil and gas systems, according to NACE International Standard TM0194-2014.

5.3 RESULTS AND DISCUSSION

Through an agarose gel it is possible to estimate the integrity of the RNA by visualizing the ribosomal mRNA subunits (see Figure 5.2). The presence of trails indicates degradation that may jeopardize subsequent analyses. Therefore, a good quality RNA must

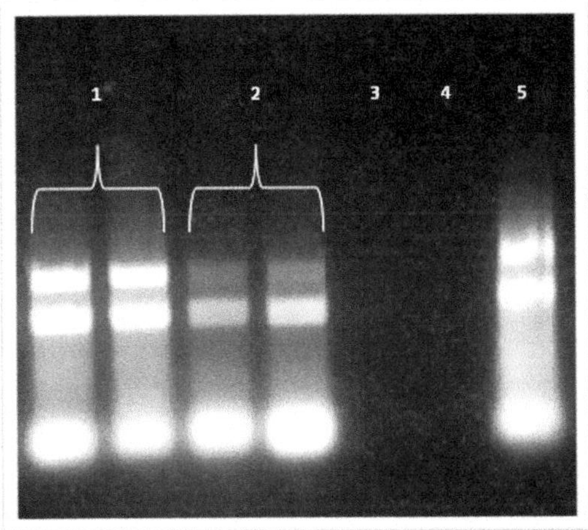

FIGURE 5.2 Electrophoretic migration in agarose gel of different strategies for RNA extraction. 1: Sample centrifugation + RNA extraction by TRIzol™; 2: Sample filtration + RNA extraction by TRIzol™; 3: Centrifugation + RNA extraction by Maxwell® automatic extractor; 4: Filtration + extraction of the RNA by Maxwell® automatic extractor; 5: Degraded mRNA.

have well defined bands, with good intensity and without traces. It was verified that the strategy that presented the best efficiency was strategy #1 (centrifugation + RNA extraction by TRIzol™). In this type of extraction the bands of the RNA were more intense and without trails, which are signs of absence of degradation of the molecule. To visually represent the characteristics of a degraded RNA, Lane 5 included an extracted RNA sample that remained at room temperature for two days.

The 260/280 nm absorbance ratio is often used to evaluate protein contamination, and the reference value (protein-free sample) should be close to 2.0. The absorbance ratio 260/230 nm is used to evaluate the contamination by other compounds, and the value indicated as reference is also around 2.0 (Desjardins & Conklin, 2010). The quality of data obtained for the samples extracted during the experiment can be verified in Figure 5.3.

Analyzing only the data obtained from the nanodrop spectrophotometer, it was not possible to verify a significant difference between the 260/280 ratio during the extractions and that all remained within the ideal range (close to 2.0), suggesting that the extractions were efficient in removal of proteins, regardless of the form of storage and time. For the contamination by chemical compounds, it was possible to observe that there was reduction of the efficiency during the twenty eight days, independent of the form of storage.

The data concerning RNA quantification are shown in Figure 5.4. In both experiments it was possible to extract the molecules at all conditions and times analyzed. Attention was drawn to the higher concentration after twenty one days in the first experiment, which may have been caused by the need for bacteria adaptation to the microenvironment. This may have caused increased mRNA expression under these conditions.

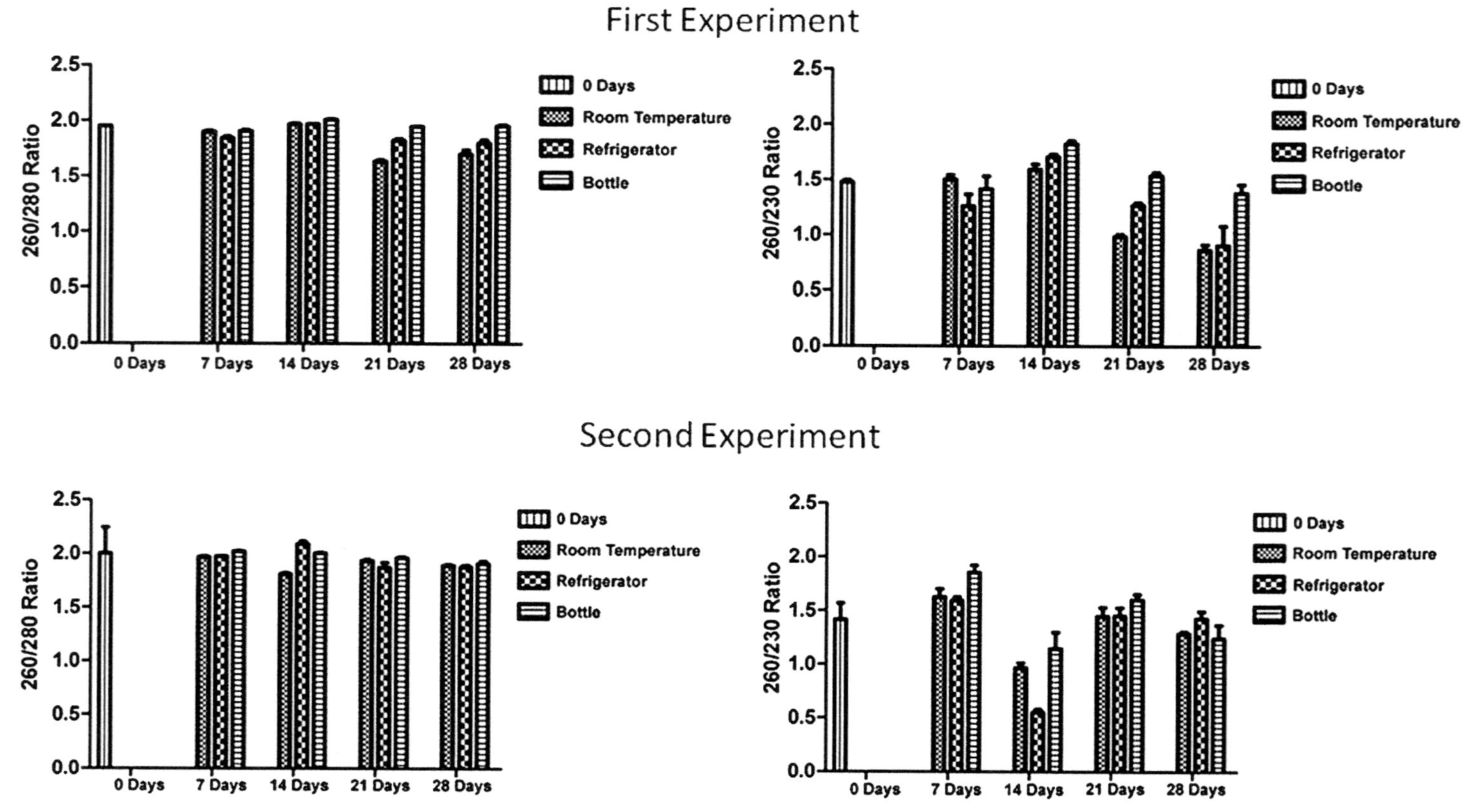

FIGURE 5.3 Graphs with the purity ratios for samples extracted in the 28-day period with extractions at 7-day intervals for the different storage conditions.

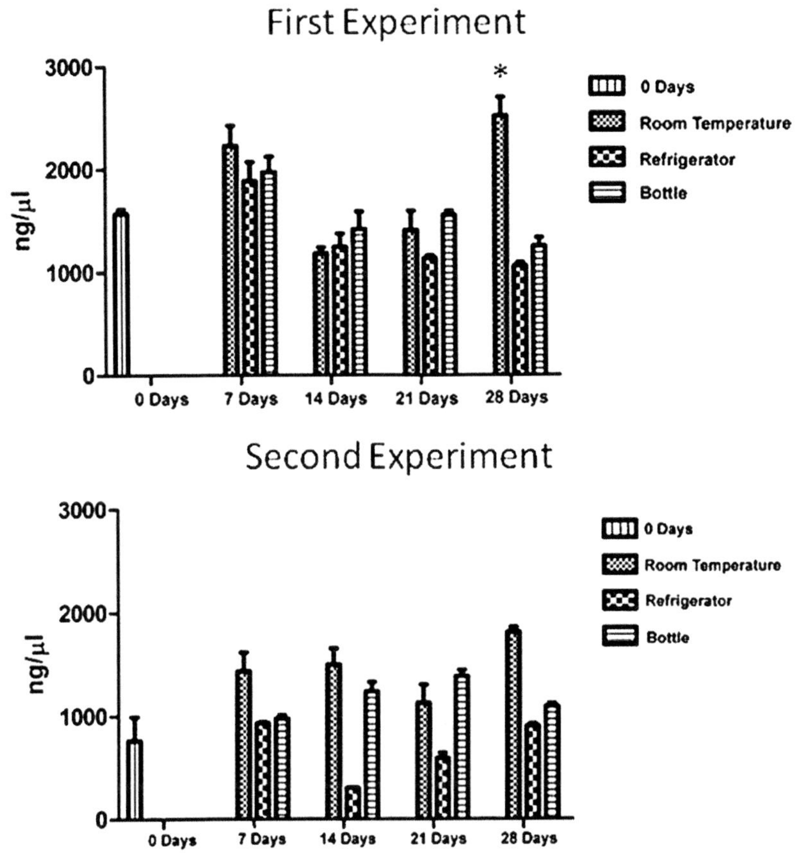

FIGURE 5.4 Quantification of extracted RNA samples during the 28 days of assay under the different storage conditions.

When observing the intensity of the bands, it was verified that for up to seven days there was no significant variation of the quality of the RNA extracted, independent of the storage. For longer periods, a better preservation is observed in the samples kept under refrigeration and in the original opaque bottle. Therefore, it is suggested that twenty one days be set as the maximum time limit for the extraction, since in twenty eight days the degradation of the RNA molecule was evident (see Figure 5.5).

In Figure 5.6 it is possible to visualize the results of agarose gel electrophoresis for each day of extraction in the three conditions evaluated for the second assay. The data corroborate with the previous experiment, where it is possible to once again verify degradation at twenty eight days and the best strategies to prevent the degradation of RNA molecules were to keep the sample in the original opaque bottle or in a refrigerator.

The results of the MPN experiment, as shown in Figure 5.7, indicated that the samples kept under refrigeration maintained SRB numbers closer to those found on

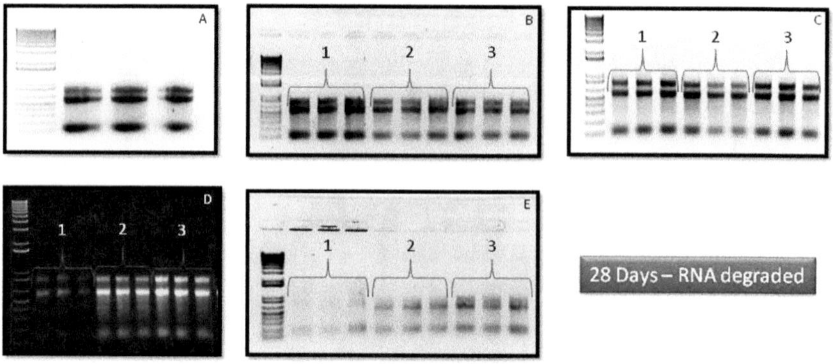

FIGURE 5.5 Photos of the gels corresponding to each day of extraction (A: 0 days; B: 7 days, C: 14 days, D: 21 days, E: 28 days) and for the three storage conditions analyzed (1: Room temperature, 2: Refrigerator, 3: Bottle) in the first assay.

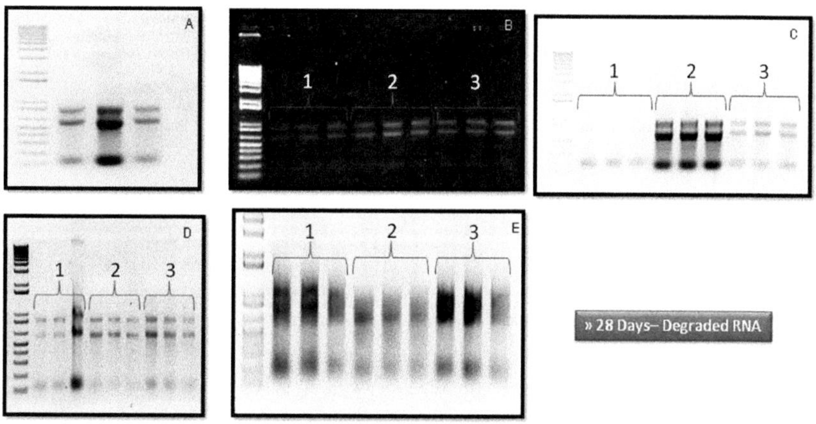

FIGURE 5.6 Photos of the gels corresponding to each day of extraction (A: 0 days; B: 7 days; C: 14 days; D: 21 days; E: 28 days) and for the three storage conditions analyzed (1: Room temperature; 2: Refrigerator; 3: Bottle) in the second assay.

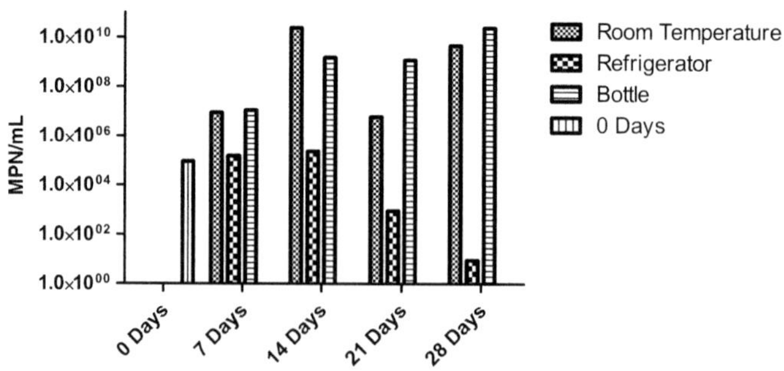

FIGURE 5.7 Quantification of SRB by MPN and for the three storage conditions analyzed (1: Room temperature; 2: Refrigerator; 3: Bottle) in the second assay.

the day of receipt until the fourteenth day. From twenty one days onward there was a reduction of three orders of magnitude in the MPN results until reaching only nine cells per milliliter at the end of the test. For the other two types of storage, a significant increase in SRB MPN was observed, which lasted up to twenty eight days in both samples from the room temperature bottle and the Schott bottle.

5.4 CONCLUSIONS AND PERSPECTIVES

The data obtained from these two assays suggest that within the conditions analyzed for this type of sample, the mRNA remained stable without degradation for up to twenty one days of storage, thus allowing subsequent assays such as gene expression studies.

While the main objectives of this work were the analyses related to RNA extraction and integrity, it would be important to verify the MPN results as, there is a possibility that the gene expression could be influenced by the growth of the bacteria over time, impacting the results of RNA expression.

Based on this work, samples should be transported in Schott flasks or in an opaque bottle under refrigeration. The procedures for extraction and quantification by MPN should be performed within fourteen days from the collection date, as the growth rate reduces due to metabolic deceleration, which probably also reduces the amount of RNA expressed between the collection period and the start of the laboratory analyses.

The next steps in this line of study of RNA are directed to the standardization of the experiments related to the activity of the dsrAB gene associated with its influence on the biocorrosion process itself.

REFERENCES

Amann, R.I., Ludwig, W., and Shleifer, K.H. 1995. Phylogenetic identification and *in situ* detection of individual microbial cells without cultivation. *Microbiology Reviews*. 59:143–169.

Ben-Dov, E., Brenner, E., and Kashmaro, A. 2007. Quantification of sulphate-reducing bacteria in industrial wastewater, by real-time polymerase chain reaction (PCR) using dsrA and apsA genes. *Microbial Ecology*. 54:439–451.

Bustin S.A., Benes V., Nolan, T., and Pfaffl, M.W. 2005. Quantitative real-time RT-PCR a perspective. *Journal of Molecular Endocrinology*. 34:597–601.

Castro, H., Reddy, K.R., and Ogram, A. 2002. Composition and function of sulphate-reducing prokaryotes in eutrophic and pristine areas of the Florida Everglades. *Applied and Environmental Micrbiology*. 68:6129–6137.

Chin, K. J., A. Esteve-Núñez, C. Leang, and D. R. Lovley. 2004. Direct correlation between rates of anaerobic respiration and levels of mRNA for key respiratory genes in Geobacter sulfurreducens. *Applied Environmental Microbiology*. 70: 5183–5189.

Chin, K. J., M. L. Sharma, L. A. Russell, K. R. O'Neill, and D. R. Lovley. 2008. Quantifying expression of a dissimilatory (bi)sulfite reductase gene in petroleum-contaminated marine harbor sediments. *Microbial Ecology*. 55:489–499.

Cortez, M.T.S.C. 2009. Abordagem molecular na resolução de problemas com biocorrosão. Master's Thesis, Department of Biology. University of Aveiro, Portugal.

Dar, S.A., Yao, L., van Dongen, U., Kuenen, J.G., and Muyzer, G. 2007. Analyses of diversity and activity of sulphate-reducing bacterial communities in sulphidogenic bioreactors using 16S rRNA and dsrAB genes as molecular markers. *Applied Environmental Microbiology*. 73:594–604.

Davey, M.E., and O'Toole, G.A. 2000. Microbial biofilms: From ecology to molecular genetics. *Microbiology and Molecular Biology Review*. 64(4):847–867.

Desjardins, P., and Conklin, D. 2010. NanoDrop microvolume quantitation of nucleic acids. *Journal of Visualized Experiments*. 45:2565.

Dhillon, A., Teske, A., Dillon, J., Stahl, D.A., and Sogin, M.L. 2003. Molecular characterization of sulphate-reduction bacteria in the Guaymas Basin. *Applied and Environmental Microbiology*. 69:2765–2772.

Fey, A., Eichler, S., Flavier, S., Christen, R., Höfle, M.G., and Guzmán, C.A. 2004. Establishment of a real-time PCR-based approach for accurate quantification of bacterial RNA targets in water, using *Salmonella* as a model organism. *Applied and Environmental Microbiology*. 70(6):3618–3623.

Fleige, S., and Pfaffl, M.W. RNA integrity and the effect on the real-time qRT-PCR performance. 2006. *Molecular Aspects of Medicine*. 27:126–139.

Fukaba, T., Ogaw, M., Fujii, T., and Naganuma, T. 2003. Phylogenetic diversity of dissimilatory sulphide reductase genes from deep-sea cold seep sediment. *Marine Biotechnology*. 5:458–468.

Galvão, M.M., and Lutterbach, M.T.S. 2014. Application of the qPCR technique for SRB quantification in samples from the oil and gas industries. In: *Application of Molecular Microbiological Methods*. United Kingdom: Caister Academic Press, 69–75.

Gentil, V. (ed.). 1982. *Corrosão*, 3rd ed, Rio de Janeiro, RJ: Livros Técnicos e Científicos S.A.

Klein, M., Friedrich, M., Rogers, A.J., Hugenholtz, P., Fishbain, S., Abicht, H., Blackall, L.L., Stahl, D.A., and Wagner, M. 2001. Multiple lateral transfers of dissimilatory sulphite reductase genes between major lineages of sulphate reducing prokaryotes. *Journal of Bacteriology*. 183:6028–6035.

Larsen, J., Sorensen, K., Juhler, S., and Pedersen, D. The Application of Molecular Microbiological Methods for Early Warning of MIC in Pipelines, Corrosion 2013, Paper ID: 2029.

Lewandowski, Z., Stoodley, P., and Roe, F. 1995. Internal mass transport in heterogeneous biofilm: recent advances. Corrosion. Paper no. 222. NACE. Texas.

Lutterbach, M.T.S., Galvão, M.M., Araújo, M., Contador, S.C., Olivereira, V., and Gutemberg. 2011. S.P. Real-time PCR applied to quantification of SRB in bottom water from fuel tanks. Corrosion 2011, paper ID 19205 (Orlando, FL: NACE International).

Mackay, I.M. (ed.). 2007. *Real-Time PCR in Microbiology*. Norfolk, UK: Caister Academic Press.

NACE Standard Test Method TM 0194-2004 2004. Field Monitoring of Bacterial Growth in Oil and Gas Systems, Houston TX.

Nakagawa, T., Ishibashi., J., Maruyana, A., Yamanaka, T., Marimoto, Y., Kimura, H., Urabe, T., and Fukui, M. 2004. Analysis of dissimilatoryu sulphidereductase and 16S rRNA gene fragments from deep-sea hydrothermal sites of the Suiyo Seamont, Izu-Bonin Arc, Western Pacific. *Applied and Environmental Microbiology*. 70:393–403.

Neretin, L.N., Schipper, A., Pernthaler, A., Hamann, K., Amann, R., Jorgensen, B.B. 2003. Quantification of dissimilatory (bi)sulphite reductase gene expression in Desulfobacterium autotrophicum using real-time RT-PCR. *Environmental Microbiology*. 5(8):660–671.

Osburne, M.S., Griossman, T.H., August, P.R., and Macneil, I.A. 2000. Tapping into microbial diversity for natural products drug discovery. *ASM News*. 66:411–417.

Perez-Jimenez, J.R., and Kerkhof, L.J. 2005. Phylogeography of sulphate reducing bacteria among disturbed sediments, disclosed by analysis of the dissimilatory sulphite reductase genes (dsrAB). *Applied and Environmental Microbiology.* 71:1004–1011.

Price, A., Álvarez, L.A., Whitby, C., Larsen, J. 2010. Detection of SRP Activity by Quantification of mRNA for the Dissimilatory (bi) Sulfite Reductase Gene(dsrA) by Reverse Transcription Quantitative PCR. Corrosion 2010, paper ID 10253 (Houston, TX: NACE International 2010).

Qin, L.X., Beyer, R.P., Hudson, F.N., Linford, N.J., Morris, D.E., and Kerr, K.F. 2006. Evaluation of methods for oligonucletide array data via quantitative real-time PCR. *BMC Bioinformatics.* 17:7–23.

Torsvik, V., Goksoyr, J., and Daae, F. L. 1990. High diversity in DNA of soil bacteria *Applied Environmental Microbiology.* 56(3):782–787.

Videla, H.A. *Biocorrosão Biofouling e Biodeterioração de Materiais.* 1a Ed. 2003. Editora Edgard Blucher. São Paulo, SP.

Videla, H.A., and Herrera, L.K. 2005. Microbiologically Influenced Corrosion: looking to the future. *International Microbiology.* 8:169–180.

Wagner, M., Roger, A. J., Flax, J.L., Brusseau, G.A., and Stahl, D.A. 1998. Phylogeny of dissimilatory sulphide reductases supports an early origin shulphate respiration. *Journal of Bacteriology.* 180:2975–2982.

6 Rapid In-Field Collection and Ambient Temperature Preservation of Corrosion-Related Microbial Samples for Downstream Molecular Analysis

Neil Sharma and Wei Huang

CONTENTS

6.1 INTRODUCTION

Multiple families of microorganisms have been implicated in microbiologically influenced corrosion (MIC) within the oil and gas industry, as well as in other fields, resulting in billions of dollars in damage annually (Bartling 2016). The ability to detect these organisms in a timely and accurate manner can allow for the implementation of appropriate treatment regimens to mitigate costs and downtime. To address this need, we previously introduced a solution for direct in-field processing and qPCR analysis of water and biofilm samples collected from oil and gas facilities, using an easily transportable testing system packaged in a Pelican™ case (Lee et al. 2015). While this lab-in-a-box solution provides results in a matter of hours, there are still many cases where time or geographic constraints necessitate samples to be tested in a laboratory rather than directly on-site.

However, transporting unprocessed samples to a laboratory can be fraught with risk, as during transport, the microbial composition in a sample can change substantially, thereby limiting reliability of the results. Here we describe development of new user-friendly sampling kits for field-based collection and stabilization of both water and solid (biofilm) samples. These kits contain consumables and reagents allowing most samples to be collected and preserved in less than 15 minutes. Following preservation, nucleic acids within the sample are stable at ambient temperature for an extended period, allowing safe transport to a laboratory. We present initial data on developing this sampling kit with a focus on preservation buffers.

6.2 CORROSION

Corrosion costs the global economy $2.5 trillion USD annually according to the NACE International IMPACT Study (Koch et al. 2016). While virtually all industries are affected, the oil and gas industry is particularly impacted due to its large global footprint and necessity to operate in many very extreme environments. In fact, it has been estimated that corrosion is responsible for more than 25% of failures in the oil and gas industry (Kermani and Harrop 1996). Hence, effective corrosion management is essential to mitigate costs, minimize breakdowns, and prevent environmental disasters.

6.3 MIC

MIC is the term used to describe corrosion caused or accelerated by the activity of microorganisms (Little and Lee 2014). At least 10–20% of all corrosion damage is commonly attributed to MIC (Skovhus et al. 2017). Sulfate-reducing bacteria (SRB), for example, were identified as the cause of a major leak from a carbon steel pipeline in Alaska in 2006 (Rabus et al. 2006). Many groups of microorganisms have been implicated in MIC, including sulfur/sulfide-oxidizers, iron-oxidizing bacteria, acid-producing bacteria, methanogens, nitrate-reducing bacteria, and sulfur/sulfate-reducing bacteria (Skovhus et al. 2014).

Many methods are used to mitigate MIC damage, such as employing special coatings and mechanical cleaning. Using chemicals such as biocides or corrosion inhibitors on a periodic or continual basis can also be particularly effective in preventing or minimizing damage. However, using biocides and other oilfield chemicals has become more complicated over the past several decades due to: (1) increased environmental concerns, (2) stricter regulations, (3) recognition that some biocides can cause unintended damage to other parts of systems, and (4) increased understanding that different biocides can have significantly different activities in relation to different microorganisms. Therefore, the accurate diagnosis of MIC in a system is essential to designing and applying an appropriate and effective chemical mitigation that minimizes costs and maximizes impact. Having access to the right tools and the right data can mean the difference between success and failure.

6.4 DETECTION OF MIC

Molecular microbiological methods (MMM) were first broadly introduced into the oil and gas industry in 2007. MMM are microbiological analysis techniques that do not rely on the cultivation of microorganisms; rather, they commonly rely on the detection of DNA, RNA or protein-based markers (Skovhus et al. 2017). Two of the MMM regularly used in the oil and gas industry today are quantitative real-time polymerase chain reaction (qPCR) and 16S rRNA gene-based metagenomics.

6.4.1 qPCR

qPCR amplifies target nucleic acid molecules and uses fluorescently labeled dyes to detect the amplification in real-time. The technique can detect the abundance and identity of specific microorganisms by analyzing the DNA collected from the sample. Detection can be based on targeting 16S rRNA and/or functional genes such as those that have been implicated in causing corrosion damage. Additionally, the expression level of specific genes within microorganisms in the sample can also be measured using a technique called reverse transcription quantitative PCR (RT-qPCR) (Skovhus et al. 2014). In RT-qPCR, RNA is first reverse transcribed into cDNA and then analyzed in a manner similar to regular qPCR.

We previously have described an in-field qPCR system that generates results in just a few hours (Lee et al. 2015). This system is now being actively used in field situations and allows treatment decisions to be made immediately rather than having to wait for laboratory results, during which time additional damage to the system may be occurring. An example of the in-field qPCR system is shown in Figure 6.1.

FIGURE 6.1 In-field qPCR system.

6.4.2 16S rRNA Gene-Based NGS (Metagenomics)

Breakthroughs in several fields such as metagenomics have been achieved through advancements in next-generation sequencing (NGS). NGS allows hundreds of thousands to millions of pieces of DNA in a sample to be sequenced simultaneously. The field of metagenomics is defined as, "genetic analysis of genomes contained within an environmental sample without the prior need for cultivating clonal cultures" (Oulas et al. 2015). 16S ribosomal RNA (rRNA) gene-based NGS sequencing is a specific metagenomic analysis method that is commonly used to identify and compare microorganisms present within a given sample. This method is well-established for studying the phylogeny and taxonomy of samples from complex environments that are difficult or impossible to study using other techniques (Oulas et al. 2015). The NGS data is compared to publicly available DNA databases to determine the identity and relative proportion of the microorganisms present in a sample. An example of 16S metagenomic data analysis is shown in Figure 6.2.

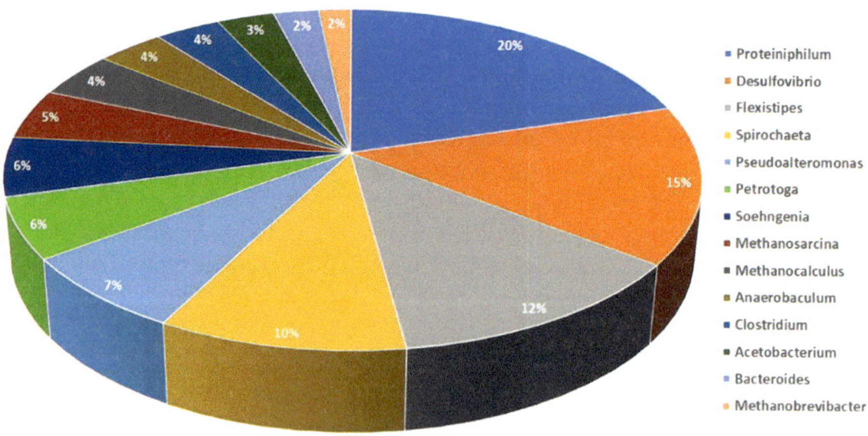

FIGURE 6.2 16S rRNA gene-based metagenomic analysis example data.

The scientific literature can subsequently be used to try to link the taxonomic information to specific metabolic activities (e.g. *Desulfovibrio* has been linked in the literature to having sulfate-reducing activity) (Heidelberg et al. 2004). This method helps users look at even more metabolic groups than qPCR and see the effects of treatments on different genera over time. NGS can be used along with qPCR; while the former provides identifying information and information on relative proportions, the latter provides absolute concentrations. While the cost of sequencing continues to decrease (Muir et al. 2016) and the technology becomes more streamlined, as of 2018 the analysis still requires time and performance off site in a laboratory.

MMM have become increasingly important in the oil and gas industry. They provide users with data that can not only be used to determine if corrosion is abiotic or biotic in nature, but also, in the case of the latter, with essential information that can be used to treat the system to mitigate damage. Knowledge of the relative proportions and types of microorganisms present in a system can allow the operator to choose more effective biocide formulations than would be possible without such data. For example, a system that has an active population of SRB and acid-producing bacteria may be better treated with different biocides than those used for a system where methanogens are the main constituent. Also, MMM provide a powerful means of precisely monitoring the effectiveness of treatments by allowing the concentrations of specific microorganisms or microorganisms with specific metabolic activities to be measured over time.

6.5 NECESSITY OF PROCESSING ON-SITE OR PRESERVATION UNTIL PROCESSING

There is no question that MMM can provide highly actionable data. Time and geography, however, often prevent samples from being processed and tested directly in the field despite the availability of in-field analysis tools (Lee et al. 2015). Instead, industry practice today is still to ship most samples for MMM analysis to an off-site laboratory for testing, which in some cases will be in another country. Ideally, these samples are stored and shipped on ice in less than 24–48 hours but these recommendations can be difficult to follow in practice due to a multitude of issues including:

1. It is easy to ship on ice in medium-to-large cities in developed countries like the US and parts of Europe. However, many oilfields and other industrial sites are in very remote locations where cold-chain transport is very difficult or impossible.
2. Many oilfields are large and can cover tens or hundreds of square miles. Thus, it can take hours or possibly days for field personnel just to collect samples, making it difficult to keep all of the samples on ice.
3. Shipping on ice also increases many associated costs, thus likely reducing the number of samples being collected for analysis.
4. Due to customs or the presence of hazardous materials, shipping between a remote site and a distant lab can take a long time. It is not uncommon for samples to take weeks or even months to arrive at a destination, especially when crossing international borders.

The problem with unprocessed samples not arriving at the laboratory very quickly (i.e., within 24–48 hours) and in a cold state is that the sample and the microorganisms within it can change.

This difficulty was demonstrated by Kilbane (2014) using culture-based analysis methods. He used a biofilm sample from a laboratory sidestream device that employed Gulf of Mexico seawater to simulate an offshore water injection system in his testing. This sample was quantified for SRB, Acid Producing Bacteria (APB), and General Heterotrophic Bacteria (GHB) and then divided into four samples for storage at different temperatures. The samples were incubated at refrigeration (4°C), room temperature (25°C), 35°C, and daily cycling between 25°C and 35°C (16 hours at 25°C, 8 hours at 35°C).

It was observed that samples stored at 4°C showed a decline in APB on day 2 while all microorganisms declined by day 7. For all the samples incubated at 25°C, 35°C, and daily cycling between 25°C and 35°C, the microbial numbers started to increase on day 1 and the values exceeded the limit of detection by day 7 (Kilbane 2014).

This suggests that changes in temperature that occur during transit and storage can influence a change in the number and ratio of microorganisms in the sample. While Kilbane's work used culture-based methods, we have used MMM including 16S rRNA gene-based metagenomics to observe this same phenomenon as shown in Figure 6.3.

In this example, two identical water samples were collected from a tributary that feeds Chesapeake Bay. One sample was immediately filtered onto a 0.2 μm filter, which was frozen at –80°C. The second sample was subjected to daily temperature cycling (12 hours at 25°C, 12 hours at 40°C) for two weeks to mimic long-distance shipping. This sample was then filtered onto a 0.2 μm filter and then DNA was extracted from both samples followed by 16S rRNA gene-based metagenomic analysis. The data show that the microbial populations (different genera represented by different colors) in the frozen sample and the temperature-cycled sample have diverged substantially at the end of the two-week testing. Hence, when samples are collected in a field and then shipped to a lab to conduct tests, the numbers of specific microorganisms in the sample may either be more or less than the original value. This depends on the time duration between sample collection and the tests, as well as on the temperature surroundings of the sample.

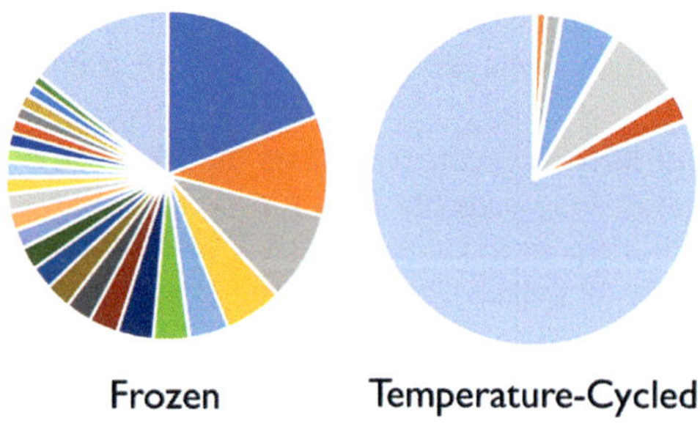

FIGURE 6.3 Change in microbial population following temperature-cycling to mimic figure extended shipment.

While it may be possible to immediately place samples on ice upon collection, the reality with oilfield operations is that the time needed to get samples from the field to the lab for testing is variable. The exact change in microbial populations that would happen in any given sample would be the result of many complex factors that are impossible to accurately calculate today. This makes much of the data being generated by the industry questionable and potentially inaccurate, thus likely resulting in the implementation of non-optimal MIC treatment plans that potentially waste money and/or put systems at undue risk.

6.6 PRESERVATION OF NUCLEIC ACIDS IN SAMPLES

The freezing of samples immediately without the use of a buffer is recommended when there are no logistical challenges (Menke et al. 2017). In the event of inability to process samples immediately, an alternative is to look at strategies to preserve samples. Storage in liquid nitrogen (–180°C) aids in preservation, but that material is impractical and dangerous to carry in the field, in addition to being costly (Doorenweerd and Beentjes 2012).

Sample preservation, especially in the absence of freezing or storage in liquid nitrogen, has seen significant research over the last decade and attempts at several approaches have been reported in the literature. A few of these approaches are summarized below.

Ethanol has been employed as a preservative for many years, as it can denature proteins that can degrade nucleic acids (Stein et al. 2013). A key reason for its use is low cost and availability, but air transport is restricted, as ethanol is deemed "dangerous" (Menke et al. 2017), and it is highly flammable making it impractical to use in many field situations. Furthermore, Song et al. (2016) reported that the concentration of ethanol makes a difference; 95% ethanol was effective but the use of 70% ethanol led to changes in microbial community composition when refrigeration was absent.

DMSO salt solution (DSS) is a dimethyl sulfoxide (DMSO) solution with saturated salt levels. It has been employed for preserving DNA when the use of ethanol is not feasible. May, Higgins and Woodley (2011) observed that this preservative was efficient in preserving DNA swab samples after four months of storage at room temperature. After storage the DNA was of sufficient purity to support subsequent PCR, but quality of the DNA was lower than that of DNA frozen using liquid nitrogen. One shortcoming of using DMSO salt solution is that it can react with ethanol, which is commonly used in downstream purification of DNA, causing irreversible damage to the DNA (Doorenweerd and Beentjes 2012). The DMSO salt solution therefore must be removed prior to beginning nucleic acid extraction procedures, which can itself lead to cell loss. Renshaw (2014) reported that the use of DMSO in a DNA preservation study also formed a precipitate during phenol–chloroform–isoamyl alcohol (PCI) DNA extraction that inhibited subsequent amplification by qPCR.

FTA cards were initially developed by Flinders Technology Associates. The typical format is small paper cards containing a matrix onto which a small aliquot of a sample can be immobilized. The FTA matrix is impregnated with protein denaturants that cause cell lysis on contact (Ahmed et al. 2011). The fixed samples can then be stored and transported in a dry state for up to 30 years while preserving nucleic acids. The primary challenges reported include high cost, the limited amount of sample that can be applied, proneness to contamination, and difficulty in processing DNA in many samples (Doorenweerd and Beentjes 2012).

RNAlater™ is a non-toxic commercial preservation reagent that is commonly used in clinical and environmental studies where freezing is not possible, such as during extended field sampling. However, Rissanen et al. (2010) reported that application of RNAlater™ resulted in a lower yield of DNA and RNA versus the unpreserved sample stored frozen at –20°C–80°C or samples stored in PCI, and also shifted the profile of bacterial species in soil sediment samples. This shift was attributed to precipitation of proteins due to ammonium sulfate in the composition of the kit. Similarly, Dominianni et al. (2014) found RNAlater™ to give lower DNA purity and that samples, to which it was applied, tended to show lower microbial diversity than identical unpreserved samples that were stored frozen at –80°C. The biggest drawback of RNAlater™ is that it must be removed or diluted prior to nucleic acid extraction, as it will interfere with commercial extraction kits. This can cause significant complications in practice, including unintentional loss of cells. Renshaw et al. (2014) tested nucleic acid preservatives in water samples and observed that insufficient removal of RNAlater™ caused a precipitate in the PCI DNA extraction that inhibited amplification by qPCR.

Renshaw et al. (2014) also evaluated use of two buffers, **CTAB and Longmire's,** to preserve environmental DNA (eDNA) collected from an aquatic environment for up to two weeks at room temperature (20°C). Filter membranes were used to collect DNA from water samples and the membranes then were stored in CTAB (cetyltrimethyl ammonium bromide) buffer or Longmire's buffer to test preservation ability. Although DNA was preserved by both the buffers at room temperature, the Longmire's buffer yielded a higher copy number of DNA than in the samples stored using CTAB. However, the very high EDTA and SDS concentrations in Longmire's buffer (100 mM and 0.5%, respectively) inhibit its use with many off-the-shelf commercial DNA extraction kits.

Menke et al. (2017) compared the preservation ability of **DNA/RNA Shield™** and RNAlater™ with homemade nucleic acid preservation (NAP) buffer on swabs of sheep fecal matter. It was found that, in the absence of any logistical constraints, the immediate freezing of forensic swabs without buffer is still the favored preservation treatment. For situations where using a preservation buffer was necessary, all three buffers were found to provide some preservation, but NAP performed the best in terms of maintaining overall microbial diversity relative to the frozen controls. However, the homemade NAP buffer and RNAlater™ have densities near that of bacterial cells, requiring the buffers to be diluted or removed prior to DNA extraction. DNA/RNA Shield™, in contrast, can be used directly in many extraction procedures but contains hazardous ingredients making it not ideal for transport to field locations.

Lastly, there is a novel technique involving an in-field sample drying system called **BioDry** to preserve aquatic samples as an alternative to freezing or preservative addition (Tuorto et al. 2015). The technique, described as inexpensive, was found to preserve microorganisms in aqueous samples for up to 30 days at room temperature. It is unclear whether such a method would be compatible with solid samples.

Some advantages and disadvantages of the methods described above are summarized in Table 6.1.

TABLE 6.1
Summary of Techniques to Preserve Nucleic Acids in Field Samples for Extended Periods Prior to Processing

Method to Preserve Nucleic Acids	Summary
Freezing	*Advantages:*
	The method recommended by most experts in the literature for short- and long-term storage
	Proven effective across virtually all biological sample types
	Disadvantages:
	Collecting and transporting samples from remote locations on ice or in liquid nitrogen is difficult or impossible in many cases
Ethanol	*Advantages:*
	Can be effective in preventing nucleic acid degradation in samples at relatively low cost
	Disadvantages:
	Tagged as dangerous for air transport; highly flammable
	Effectiveness depends on final concentration of ethanol (e.g. 95% can be more effective than 70% in many cases)
DMSO Salt Solution	*Advantages:*
	Can be effective in preventing nucleic acid degradation in samples at relatively low cost
	Disadvantages:
	Must be removed prior to beginning extraction procedures
	Formation of precipitate in chloroform-alcohol extraction of DNA
FTA Membrane	*Advantages:*
	Sample storage in dry format is possible for up to 30 years with
	color indicator to display the successful application of a sample
	Disadvantages:
	Expensive
	Limited sample volume and complicated extraction protocols
RNAlater™	*Advantages:*
	Non-toxic and commercially available
	Has been used in many studies over the past two decades
	Disadvantages:
	Effectiveness is highly mixed with several literature examples of lower yield of nucleic acids and shifts in bacterial populations
	Formation of precipitate in chloroform-alcohol extraction of DNA

(Continued)

TABLE 6.1 (Continued)
Summary of Techniques to Preserve Nucleic Acids in Field Samples for Extended Periods Prior to Processing

Method to Preserve Nucleic Acids	Summary
DNA/RNA Shield™	*Advantages:*
	Commercially available and can be used directly in many nucleic acid extraction procedures.
	Disadvantages:
	Contains hazardous ingredients making transport to and from remote locations impractical or impossible.
Home-Made Buffers:	*Advantages:*
CTAB	Low cost as these buffers can be prepared from published recipes using commonly available laboratory reagents
Longmire's	*Disadvantages:*
NAP	Published literature shows varying degrees of effectiveness for different buffers
	NAP must be removed prior to starting extraction procedures Longmire's buffer is not compatible with many commercial extraction kits due to high EDTA and SDS concentrations
BioDry	*Advantages:*
	Published results suggest preservation of water samples is comparable to freezing up to 30 days
	Disadvantages:
	Slight decrease in yield of DNA post 30 days
	Unclear if compatible with solid samples
	Little peer-reviewed literature

6.7 CASE STUDY: DEVELOPMENT OF FTA CARDS FOR QUICK ANALYSIS FOR THE OIL AND GAS INDUSTRY

Keeping in mind the change in microbial composition that occurs during transportation, as well as the hazardous nature of many common sample preservatives, Price (2012) reported a simple "Nucleic Acid Extraction Card" (NAEC) technology for the oil and gas industry. The NAEC cards were based on the FTA cards described above.

FTA cards are composed of cellulose filter paper impregnated with cell wall breaking compounds that lyse cells to release and subsequently trap nucleic acids on the paper. The presence of proteases and nuclease inactivators, along with free radical traps and UV-protective molecules, preserves the trapped nucleic acids.

NAEC cards can be used with both liquid and solid samples. Price (2012) found that efficiency of DNA extraction from samples of SRB applied on the cards is comparable to that of standard phenol-chloroform extraction with lysozyme and glass bead disruption technique. The cards were further tested in the field to investigate a group of biocide resistant general aerobic bacteria in a Middle Eastern water treatment plant (Price 2012). Following collection of the samples on the NAEC cards, extraction and denaturing gradient gel electrophoresis aided in the identification of the problematic organisms as spore forming *Acidaminobacter.* This application showed the utility of these cards in the rapid testing of MIC to devise a suitable mitigation strategy. These cards were described as being portable, void of hazardous chemicals, and preserving nucleic acids for subsequent testing (Price 2012). However, these cards have not seen widespread use in the oil and gas industry to date. This is may be due to the limited amount of sample that can be immobilized and/or issues with downstream handling and extraction procedures. Nonetheless, this case study serves as a good model for the importance of proper preservation methods in accurately diagnosing MIC issues.

6.8 OBJECTIVES OF THE KIT BEING DEVELOPED AT INSTANTLABS

MMM such as qPCR and metagenomics continue to become increasingly important in the oil and gas industry in diagnosing and treating MIC. However, different companies—and even individuals within the same company—often use different procedures to collect samples. Additionally, current collection procedures typically recommend storing and shipping samples on ice, which, as discussed previously, is not possible in many situations, so many samples are being shipped at ambient temperature either without preservation or using proprietary preservation methods. This considerable variability in procedures makes it difficult, if not impossible, to compare and gauge the accuracy of much of the MMM data being generated by the industry today. Therefore, the industry would be well-served by access to off-the-shelf kits that employ standardized protocols and include all materials necessary for collecting and preserving samples in the field, so that they can be easily transported to a laboratory for analysis without risk of the microbial population changing during collection or transport. The ideal solution would meet the following parameters:

1. The kit would include all needed sterile plasticware, reagents, and other materials for collecting water and/or solid microbiological samples.
2. The preservation reagents would be non-toxic for easy transport to and from remote sites, provide preservation of the sample for at least several weeks at ambient temperature, and would allow the preserved sample to be processed directly in a downstream nucleic acid extraction procedure without prior removal of the preservative (such as required with RNAlater™).

3. The kits would be readily available to all companies and institutions in the industry rather than being kept proprietary or as part of a contract service package.
4. The kits would be part of a portfolio of products that includes complementary nucleic acid extractions kits and MMM testing reagents (e.g. a panel of qPCR assays targeting key microorganisms and/or reagents for NGS). This would allow users across the industry to generate MMM data that are directly comparable.
5. The kits would have chain of custody features such as seals and unique sample tracking numbers and/or barcodes.
6. Materials required to ship the preserved samples from the field site to a laboratory for testing would also be provided, such as boxes with preaddressed and postage-paid labels.
7. The kits would use simple "cookbook-style" protocols for field use that are easy to learn and take as little time as necessary to conduct.

An example of where this new kit would fit into a typical MMM testing workflow is shown in Figure 6.4.

This chapter reports the initial development of a preservation reagent, as outlined in parameter 2 above, to stabilize samples before and during shipment. This reagent will be used in the future in off-the-shelf collection kits to facilitate consistent and accurate molecular testing, especially in the oil and gas industry.

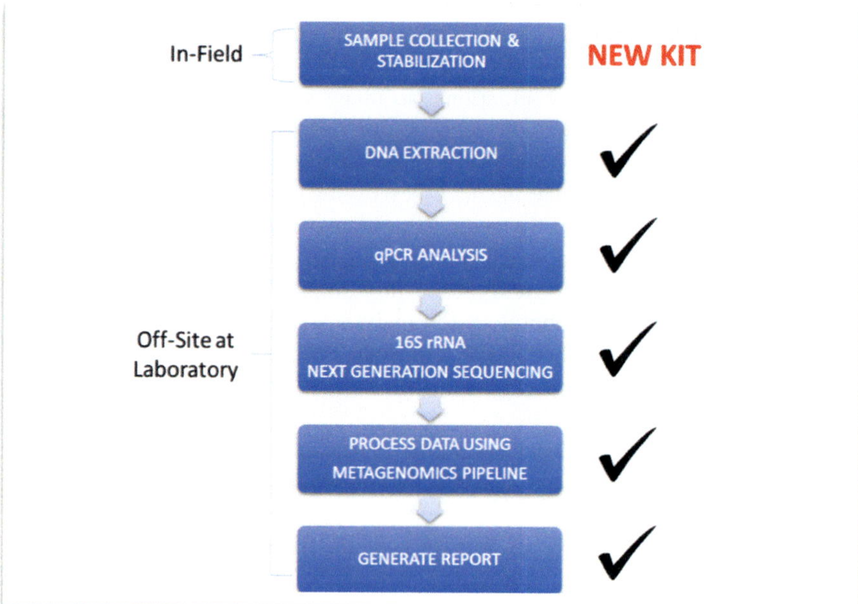

FIGURE 6.4 Proposed MMM testing workflow.

6.9 OUTLINE OF THE PROCESS/APPROACH TO DEVELOP THE PRESERVATION REAGENT

6.9.1 Initial Formulation and Screening of Potential Buffers

- A preliminary set of buffer formulations was prepared based on unpublished data from previous work in the development of freeze dried ambient stable molecular testing reagents at InstantLabs.
- Water from Chesapeake Bay (Maryland, USA) was collected and transported to a laboratory within 30 minutes. One hundred milliliters of the water were filtered through 0.2 µm pore membranes to collect microorganisms. The membranes were then folded using sterile forceps and placed into 1.5 mL microcentrifuge tubes. Replicate membranes were stored at 55°C for one week in each buffer formulation candidate. Storage at 55°C for one week mimics storage at 25°C for eight weeks based on the Arrhenius model (Anderson and Scott 1991). Unfiltered bay water was also stored under similar conditions. As zero-point controls, replicate membranes were stored at –80°C.
- Following incubation, DNA was extracted using the InstantLabs DNA extraction protocol designed for in-field use (Lee 2015).
- The extracted DNA was analyzed by 16S rRNA qPCR followed by SYBR green melt curve analysis. Buffer formulations that showed a pattern similar to the –80°C control, as analyzed by both qPCR and melt-curve analysis, were selected for subsequent analysis. Buffer C was chosen for further testing.

6.9.2 DNA-Protective Properties of Anti-Oxidant Supplements

- DNA was isolated and purified from *Bacillus coagulans* and suspended in 800 µl of Buffer C in replicate tubes.
- The tubes were supplemented with potential anti-oxidant candidates at different concentrations in duplicate along with controls lacking the anti-oxidant candidates. Household bleach (to check for protective effects of above candidates; final concentration 5%) was added to one of each duplicate tube and incubated for 30 minutes.
- DNA was then extracted using the same DNA extraction protocol described previously followed by *Bacillus coagulans* specific qPCR analysis. The supplements that were protective against the bleach were further tested to determine optimum concentrations.

6.9.3 Freeze-Drying of Preservation Buffer

- Buffer C was mixed with an anti-oxidant and excipients and then aliquoted into tubes. The tubes were subsequently freeze-dried for 48 hours in a Labconco Triad freeze-drying instrument.
- Following freeze-drying, the tubes were flushed with nitrogen gas and then sealed in gas and light impermeable foil pouches.

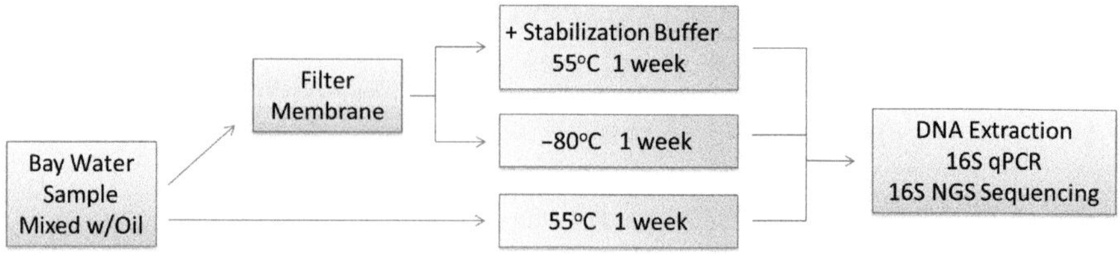

FIGURE 6.5 Flowchart for testing long-term stabilization properties using mock field samples.

- Pouches were stored at –20°C or at 55°C for 23 days. Storage at 55°C for 23 days mimics storage at 25°C for six months based on the Arrhenius model (Anderson and Scott 1991).
- Following storage, the preservation buffer was rehydrated using nuclease-free water and used in preservation experiments using environmental water samples similar to that outlined in section 6.8.1 above.

6.9.4 Testing of Long-Term Stabilization Properties Using Mock Field Samples

- Water samples from Chesapeake Bay were collected, mixed with crude oil obtained from the Gulf of Mexico, and divided into three aliquots:
  - filtered sample mixed with preservative Buffer C supplemented with anti-oxidant stored at 55°C for one week to mimic storage at 25°C for eight weeks
  - filtered –80°C control stored for one week
  - unfiltered sample without preservative buffer stored at 55°C for one week to mimic storage at 25°C for eight weeks
- Following incubation, DNA was extracted from the aliquots and subjected to 16S rRNA qPCR as well as 16S rRNA gene-based metagenomic analysis to ascertain the amount of microbial DNA and relative proportions of the microorganisms present in the samples, respectively (see Figure 6.5).

6.10 RESULTS

6.10.1 Performance of the Buffer Candidates (Eight Weeks' Accelerated Stability)

A key measure of sample preservation is the ability of a buffer to preserve the total amount and composition of amplifiable DNA prior to extraction and analysis.

As previously discussed, the –80°C control represented a zero-time point, as the recommended practice for preserving samples is freezing. When comparing the –80°C control curve with the Buffer C curve in a qPCR assay, nearly identical

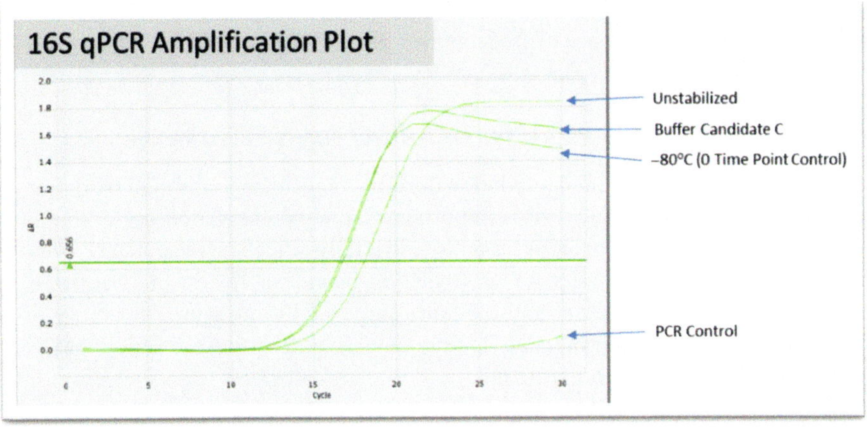

FIGURE 6.6 Example preservation qPCR amplification plot.

cycle threshold (Ct) values can be seen indicating similar amounts of microbial DNA. On the other hand, the sample that was unpreserved (to which buffer was not added) shows a two-cycle dip in Ct value, indicative of loss of amplifiable DNA (see Figure 6.6).

Analysis shows that the DNA melt curves of the –80°C control and the sample incubated with preservation Buffer C are identical. On the other hand, the melt curve of the unpreserved control without buffer is different, indicative of a change in microbial composition (see Figure 6.7).

These data show there is no significant change in the amount or composition of amplifiable DNA in the sample incubated with Buffer C compared to the control frozen at –80°C. This suggests that the preserving action of Buffer C is comparable

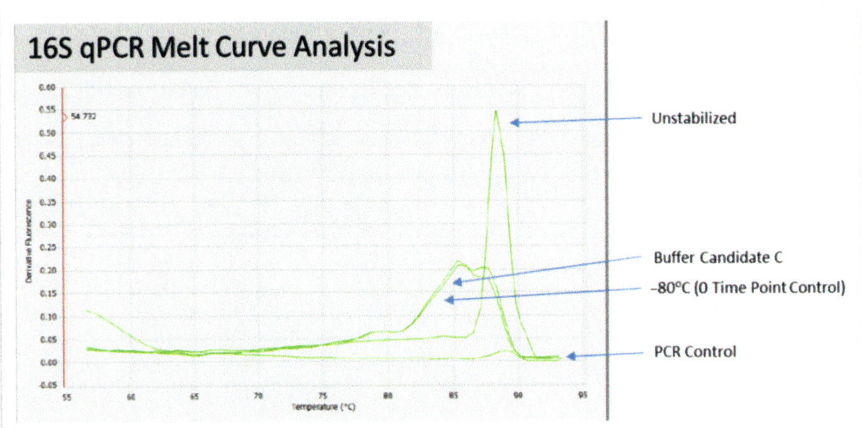

FIGURE 6.7 Example preservation qPCR melt curve plot.

to that of freezing at –80°C and supports the potential use of this buffer in a kit to preserve samples in the absence of freezing facilities.

6.10.2 DNA-Protective Properties of Anti-Oxidant Supplements

A key pathway for nucleic acid degradation during storage can be through oxidation (Cadet and Wagner 2013). Hence, it is important that any preservation buffer that is used provides resistance to oxidative agents, especially since such agents are commonly used in oil and gas systems (e.g. oxidizing biocides). To test this, Buffer C was supplemented with different anti-oxidants and tested to measure the ability of the anti-oxidants to confer protection against oxidative agents. An example qPCR amplification plot is shown in Figure 6.8.

Well 1 contains DNA in preservation buffer with a proprietary anti-oxidant and no bleach. Well 2 contains DNA in preservation buffer with bleach and no anti-oxidant,

	1	2	3	4
DNA	+	+	+	+
Antioxidant	+	-	+	-
Bleach	-	+	+	-

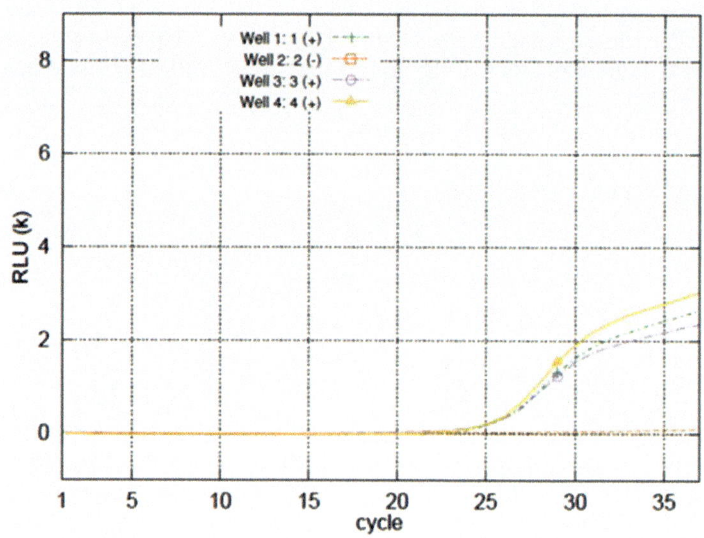

FIGURE 6.8 Example anti-oxidant qPCR amplification plot.

and Well 3 contains DNA in preservation buffer with both antioxidant and bleach. Lastly, Well 4 contains only DNA in preservation buffer.

The graph shows that in Well 2 (in red), DNA in stabilization buffer with bleach and no anti-oxidant is completely degraded as indicated by lack of amplification. However, when the buffer is supplemented with an anti-oxidant (Well 3, shown in violet), the DNA is not degraded and amplification is close to that of the control Well 4, which contains only DNA in buffer (shown in yellow). Thus, the use of an anti-oxidant in the buffer composition in a sampling kit can further aid in stabilizing nucleic acids.

6.10.3 Freeze-Dried Supplemented Buffer

In practice, it is likely that collection kits might be stored and transported for long periods before actual use. For example, a field microbiologist might carry a package of kits from site-to-site, using them as needed. Hence, it is important that any preservation buffer such as described above is *itself* stable before use (i.e. will not degrade prior to use). Freeze-drying is a commonly used method for stabilizing biological buffers for long-term storage. For example, we have used freeze-drying for many years at InstantLabs to provide clients with stable ready-to-use qPCR reagents that can be shipped at ambient temperature.

The non-toxic preservation Buffer C discussed above was mixed with excipients and an anti-oxidant and then freeze-dried to confer additional long-term stability prior to use. The freeze-dried buffer shows no difference in 16S qPCR Ct value in preservation experiments indicating no loss of sample DNA, even when stored up to six months at room temperature before use (see Figure 6.9), relative to the freeze-dried buffer stored frozen at –20°C.

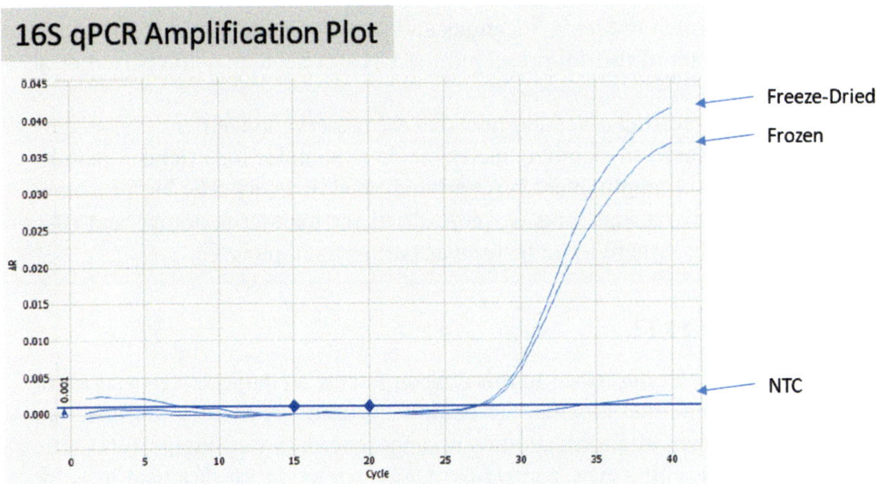

FIGURE 6.9 Example freeze-dried preservation buffer qPCR amplification plot.

Additional 16S rRNA gene-based metagenomic analysis further confirms that the freeze-dried buffer maintains preservation ability that is identical to the freshly prepared buffer (data not shown). This allows the buffer to be shipped and stored at ambient temperature for long periods and then to be rehydrated immediately prior to use.

6.10.4 Testing of Long-Term Stabilization Properties Using Mock Field Samples

A stabilization buffer needs to maintain the amount of amplifiable DNA (as described above) but also the ratios of different microbial genera present in the sample. This ensures that the data being generated using the samples is accurate and representative of the environment in which the sample was collected initially.

16S rRNA gene-based metagenomic analyses of three samples: –80°C control, sample preserved with Buffer C, and an unpreserved sample were conducted to measure the stabilization properties of Buffer C. The metagenomic data (species >1% shown) after eight weeks of accelerated storage are shown in Figure 6.10.

The microbiome content of the –80°C control and the sample preserved with buffer show nearly identical compositions of microorganisms with a major presence of *Flavobacteria* (light brown) and equivalent proportions of *Saprospirae* (yellow), *Cytophagia* (light blue), *Alphaproteobacteria* (green), *Betaproteobacteria* (dark blue), *Sphingobacteria* (gray), and chloroplast (blue).

However, the sample that was incubated without preservation buffer shows a significant change in microbial composition compared to the –80°C control. There is greater representation of *Saprospirae* (yellow) and a greater proportion of *Alphaproteobacteria* (green) and smaller ratio of *Betaproteobacteria* (dark blue) in the unpreserved sample. In addition, there is growth of *Deinococci* (shown in dark brown), which was absent in the control.

This shows that there is a drastic change in microbial composition in the sample not treated with Buffer C. This is not desirable in the oil and gas industry, where samples are often in remote locations and freezing is impractical. The use of the buffer formulation of this kit preserved the DNA collected for the period of at least eight weeks at 25°C.

Overall, these promising new data show that the preservation buffer developed by InstantLabs can be used to stabilize nucleic acids in samples for extended periods with negligible loss of amplifiable DNA and microbial diversity. The buffer is non-toxic, resists oxidative agents, and is freeze-dried for long-term storage and easy transport into the field where it can be used as part of a sampling kit.

6.11 WHAT'S NEXT?

The preservation buffer described in this chapter is now a component in a series of collection and preservation kits that are being field-tested by multiple users across different sample types to gather additional supportive data (see Figure 6.11). The kits are being used with a new compatible DNA extraction kit designed to work with diverse and highly contaminated samples such as those collected at oil and gas facilities. The work is ongoing to complete field trials.

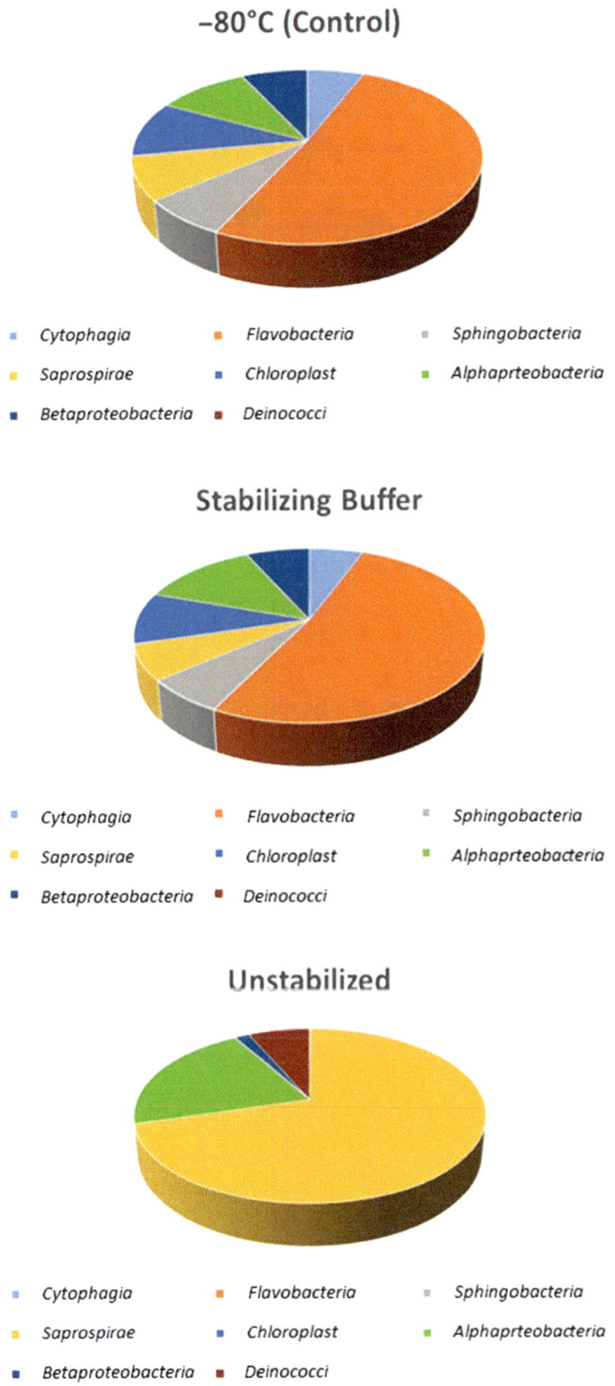

FIGURE 6.10 16S rRNA gene-based metagenomic analysis of mock field samples to assess the stabilization properties of Buffer C.

FIGURE 6.11 Samples of the kits now in field trials.

6.12 CONCLUSIONS

- This work reports development of a novel and non-toxic buffer that facilitates preservation of samples collected in field scenarios.
- The samples that were not treated with buffer showed an undesirable change in microbial community composition compared to the frozen control or the samples treated with the buffer developed in this work.
- The preservation duration of the buffer developed in this work is equivalent to a period of at least eight weeks at 25°C.
- The buffer can be part of an off-the-shelf kit that includes all necessary plasticware and standardized protocols for collecting and preserving planktonic and sessile samples from oil and gas facilities.
- Additional data are being generated across multiple sample types and this kit is being tested with real field samples to gather additional supportive data.
- A new DNA extraction kit has been developed that is optimized for samples with extensive contamination such as those coming from oil and gas facilities, and is fully compatible with the preservation method described here.

REFERENCES

Ahmed HA, MacLeod ET, Hide G, Welburn SC, Picozzi K (2011). The best practice for preparation of samples from FTA®cards for diagnosis of blood borne infections using African trypanosomes as a model system. *Parasite Vectors*. 7;4:68. doi: 10.1186/1756-3305-4-68.

Anderson G, and Scott M (1991). Determination of shelf life and activation energy for five drugs of abuse. *Clinical Chemistry*. 37(3):398–402.

Bartling C (2016). Microbially induced corrosion: silent killer of infrastructure. *Pipeline & Gas Journal*. 243(1). https://pgjonline.com/magazine/2016/january-2016-vol-243-no-1/features/microbially-induced-corrosion-silent-killer-of-infrastructure.

Cadet J, and Wagner JR (2013). DNA base damage by reactive oxygen species, oxidizing agents, and UV radiation. *Cold Spring Harbor Perspectives in Biology*. doi: 10.1101/cshperspecta012559.

Dominianni C, Wu J, Hayes RB, Ahn J (2014). Comparison of methods for fecal microbiome biospecimen collection. *BMC Microbiology*. https://doi.org/10.1186/1471-2180-14-103.

Doorenweerd C, and Beentjes K (2012). Extensive guidelines for preserving specimen or tissue for later DNA work. Version 1. https://science.naturalis.nl/media/medialibrary/2013/08/preservingdna.pdf.

Heidelberg JF, Seshadri R, Haveman SA, Hemme CL, Paulsen IT, Kolonay JF, Eisen JA, Ward N, et al. (2004). The genome sequence of the anaerobic, sulfate-reducing bacterium *Desulfovibrio vulgaris* Hildenborough. *Nature Biotechnology*. 22(5):554–559.

Kermani MB, and Harrop D (1996). The impact of corrosion on oil and gas industry. *SPE Production & Facilities*. 11(3):186–192.

Kilbane J (2014). Effect of sample storage conditions on oilfield microbiological samples. NACE. Conference Paper. March 2014. doi: 10.13140/2.1.2807.2963.

Koch, G, Varney J, Thompson N, Moghissi O, Gould M, Payer J (2016). *International measures of prevention, application, and economics of corrosion technologies study.* NACE International, Houston TX.

Lee C, Brown B, Jones D, Duet S, Cornell J, Sharma N, Lin X (2015). Field applications for on-site DNA extraction and qPCR. Corrosion 2015, 15-19 March, Dallas, Texas: IDNACE-2015-5686.

Little BJ, and Lee JS (2014). Microbiologically influenced corrosion: an update. *International Materials Reviews*. 59:7:384–393.

May LA, Higgins JL, and Woodley CM (2011). Saline-saturated DMSO-EDTA as a storage medium for microbial DNA analysis from coral mucus swab samples. NOAA Technical Memorandum NOS NCCOS 127 and CRCP 15, p. 14.

Menke S., Gillingham MAF, Wilhelm K, Sommer S (2017). Home-made cost-effective preservation buffer is a better alternative to commercial preservation methods for microbiome research. *Frontiers in Microbiology*. 8, 102. http://doi.org/10.3389/fmicb.2017.00102.

Muir P, Li S, Lou S, Wang D, Spakowicz DJ, Salichos L, Zhang J, Weinstock GM, et al. (2016). The real cost of sequencing: scaling computation to keep pace with data generation. *Genome Biology*. 17:53. doi:10.1186/s13059-016-0917-0.

Oulas A, Pavloudi C, Polymenakou P, Pavlopoulos GA, Papanikolaou N, Kotoulas G, Arvanitidis C, Iliopoulos I (2015). Metagenomics: Tools and insights for analyzing next-generation sequencing data derived from biodiversity studies. *Bioinformatics and Biology Insights*. 9:75–88.

Price A (2012) Novel DNA extraction and preservation for identification of microorganisms on-site using novel Nucleic Acid Extraction Cards for the oil and gas industry. NACE Corrosion 2012. C 2012-0001758.

Rabus R, Hansen TA, and Widdel F (2006). Dissimilatory sulfate- and sulfur-reducing prokaryotes. In *The Prokaryotes* (Dworkin M, Falkow S, Rosenberg E, Schleifer K-H, Stackebrandt E, eds). New York: Springer, 659–768.

Renshaw MA, Olds BP, Jerde CL, Mcveigh MM, Lodge DM (2014). The room temperature preservation of filtered environmental DNA samples and assimilation into a phenol–chloroform–isoamyl alcohol DNA extraction. *Molecular Ecology Resources*. 15, 1. doi: 10.1111/1755-0998.12281.

Rissanen AJ, Kurhela E, Aho T, Oittinen T, Tiirola M (2010). Storage of environmental samples for guaranteeing nucleic acid yields for molecular microbiological studies. *Applied Microbiology and Biotechnology*. 88:977–984.

Skovhus TL, Caffrey SM, Hubert CRJ (eds) (2014). *Applications of Molecular Microbiological Methods*. Poole, United Kingdom: Caister Academic Press.

Skovhus TL, and Eckert RB (2014). Practical aspects of MIC detection, monitoring and management in the Oil & Gas Industry. Corrosion 2014. Paper 3920. NACE International, Houston, TX.

Skovhus TL, Eckert RB, and Rodrigues E (2017). Management and control of microbiologically influenced corrosion (MIC) in the oil and gas industry—Overview and a North Sea case study. *Journal of Biotechnology*. 256:31–45.

Song SJ, Amir A, Metcalf JL, Amato KR, Xu ZZ, Humphrey G, Knight R (2016). Preservation methods differ in fecal microbiome stability, affecting suitability for field studies. *mSystems*. doi: 10.1128/mSystems.00021-16.

Spens J, Evans AR, Halfmaerten D, Knudsen SW, Sengupta ME, Mak SST, Sigsgaard EE, Hellstrom M (2016). Comparison of capture and storage methods for aqueous macrobial eDNA using an optimized extraction protocol: advantage of enclosed filter. *Methods in Ecology and Evolution*. doi: 10.1111/2041-210X.12683.

Stein ED, White BP, Mazor RD, Miller PE, Pilgrim EM (2013). Evaluating ethanol-based sample preservation to facilitate use of DNA barcoding in routine freshwater biomonitoring programs using benthic macroinvertebrates. *PLoS ONE*. 8(1):e51273. doi:10.1371/journal.pone.0051273.

Tatangelo V, Franzetti A, Gandolfi I, Bestetti G, Ambrosini R (2014). Effect of preservation method on the assessment of bacterial community structure in soil and water samples. *FEMS Microbiology Letters*. 356(1):32–38.

Part III

Biocides and Biofilms in the Oil and Gas Industry

7 Perchlorate and Its Application in the Oil and Gas Industry

Yi Liu, Anna Engelbrektson, Magdalena Stoeva, Tyler Barnum, Victor Reyes-Umana, and John D. Coates

CONTENTS

7.1 INTRODUCTION

Biosouring is the production of sulfide in oil reservoirs and surrounding infrastructure. It occurs as a result of the activity of sulfate-reducing microorganisms (SRM), including both archea and bacteria, which are naturally found in oil reservoirs at low abundance (L'haridon, Reysenbacht et al., 1995; Magot, Ollivier et al., 2000; Grabowski, Nercessian et al., 2005; Lybrand, Bockheim et al., 2016). These microorganisms are further introduced into reservoirs by enhanced recovery techniques, such as flooding with marine water (Patton, 1990; McInerney, Sublette et al., 1993). The combination of ample biodegradable organic carbon in oil, low oxygen concentrations in reservoirs, and high sulfate concentration in the injected seawater stimulates the growth of SRM and subsequent production of sulfide

(McInerney, Sublette et al., 1993; Hubbard, Cheng et al., 2014). Biogenic production of sulfide (as hydrogen sulfide gas—H_2S) through the activity of SRM inevitably leads to corrosion and pitting of steel infrastructure, resulting in significant damage over time and an increased risk of major oil spills (Lee et al., 1995). Furthermore, despite its ubiquity in the environment, sulfide is extremely toxic to aerobic lifeforms at levels above background (about 0.3 ppm) (Chou, Ogden et al., 2016). In humans, sulfide concentrations above 500 ppm are generally lethal, and adverse effects ranging from conjunctivitis to cyanosis can begin above 1 ppm (ATSDR, 2016). Unsurprisingly, substantial effort is placed in controlling sulfide production. Currently, most recovery operations employ a variety of chemical and physical treatments to control sulfidogenesis and its effects. These include use of corrosion resistant materials in pipelines, removal of hydrogen sulfide gas from produced fluids, and maintenance of high reservoir temperatures to preclude biological processes (80°C–90°C) (Bowker, Smith et al., 1992; Wilhelms, Larter et al., 2001). Alternatively, biological controls include introducing biocides such as tetrakis(hydroxymethyl)phosphonium sulfate (THPS), benzalkonium chloride, or formaldehyde, or specific inhibitors of microbial sulfate reduction such as nitrate (Videla and Herrera 2005; Biswas, Woodards et al., 2009; Youssef, Elshahed et al., 2009). Extensive usage of these broad-spectrum biocides could potentially lead to resistance by SRM populations due to the pervasive nature of these microorganisms in the environment. Therefore, identifying future control strategies will become increasingly important as resistance may eventually render these biocide treatments obsolete (Gieg, Jack et al., 2011).

The biocompetitive inhibitor nitrate has been used extensively in controlling biosouring in oil reservoirs (Youssef, Elshahed et al., 2009). As the most thermodynamically favorable natural electron acceptor after oxygen, nitrate addition stimulates ubiquitous nitrate-reducing bacteria (NRB) in the environment, resulting in an indirect inhibitory effect on sulfidogenesis. NRB competitively exclude SRM due to the more favorable nitrate/nitrite redox couple as compared to the sulfate/sulfide redox couple (Hubert, Voordouw et al., 2009). Additionally and probably more importantly, nitrate-reducing microbial populations produce nitrite as an intermediate of the metabolism, which also serves as a potent inhibitor of SRM (Haveman, Greene et al., 2004; He, Huang et al., 2006; Hubert and Voordouw, 2007; Korte, Saini et al., 2015). Furthermore, nitrate stimulates the growth of nitrate-reducing sulfide-oxidizing bacteria (NR-SOB) which remove sulfide by oxidation to sulfate (Gevertz et al., 2000).

In practice, nitrate has had mixed success as a biosouring inhibitor for several reasons. First, many SRM can reduce nitrate and have evolved mechanisms to cope with resultant reactive nitrogen species (Greene, Hubert et al., 2003; Carlson, Kuehl et al., 2015). These SRM have nitrite reductases, which they upregulate to alleviate nitrite inhibition and manage nitrite stress (Greene, Hubert et al., 2003). Additionally, nitrate can stimulate growth of some undesirable microorganisms such as sulfidogenic bacteria that utilize nitrate, and anammox communities that deplete nitrate in solution without inhibiting the growth of SRM (Shartau, Marcy et al., 2010). Other challenges with using nitrate as a method of biosouring control include the storage of this highly oxidizing substance, which becomes more expensive when more nitrate is needed. Additionally, nitrate and particularly nitrite are capable of causing corrosion, especially under acidic conditions (Ma, Yang et al., 2003).

Perchlorate (ClO_4^-), another sulfate analog similar to nitrate, was identified as an inhibitor of biogenic sulfate reduction in the early 1950s (Postgate, 1952). However, it is not until recently that the inhibitory potency and selectivity of this compound has been clearly demonstrated, redefining it as a promising treatment for oil reservoir biosouring (Engelbrektson, Hubbard et al., 2014; Carlson, Kuehl et al., 2015; Engelbrektson, Briseno et al., 2018; Engelbrektson, Cheng et al., 2018; Shrestha, Williamson et al., 2018). It is similar to nitrate in that it has a high redox potential ($E° = +750$ mV for the reduction of NO_3^- to N_2 and $E° = +797$ mV for the reduction of ClO_4^- to Cl) (Youngblut, Wang et al., 2016) that couples more favorably to growth compared to sulfate, and it has both *direct* inhibitory effects (i.e., specifically inhibits enzymes involved in sulfate reduction) and *indirect* inhibitory effects (i.e., effects that rely on perchlorate turnover by perchlorate reducing community members).

7.2 HISTORY OF PERCHLORATE

Perchlorate is a stable, soluble, nonvolatile ion that accumulates as salts in dry environments as a result of industrial production or natural atmospheric deposition. Due to its high oxygen content (64.6% oxygen by mass), perchlorate maintains a strong oxidizing capacity despite its stability and nonvolatility at ambient temperatures (Catling et al., 2010; Jackson, Anderson et al., 2006). Current evidence suggests perchlorate is generated abiotically from oxygen and volatilized chlorides exposed to ultraviolet radiation and ozone in the stratosphere, or from electrostatic discharge in the troposphere (Catling, Claire et al., 2010; Rao, Estrada et al., 2012). Over time, atmospheric perchlorate is deposited into the soil and the lack of rain enables it to persist in arid climates, such as the Atacama desert, parts of the western United States, and the dry valleys of the Antarctic (Catling, Claire et al., 2010; Lybrand, Bockheim et al., 2016).

In 1816 (see Figure 7.1), Prussian scientist Johann Philipp Stadion became the first person to synthesize the perchlorate ion in the form of perchloric acid (Ure, 1824). However, it was not until the 1890s that the industrial scale synthesis of potassium and ammonium perchlorate by electrolysis of sodium chlorate ($NaClO_3$) began in Masebo, Sweden, for use in munitions (Trumpolt, Crain et al., 2005). After World War II, more manufacturers in Europe and the United States arose to meet increasing demand from the munitions industry for ammonium perchlorate composite propellant (APCP), a strong oxidizer used in explosives or solid state rocket fuel (Trumpolt, Crain et al., 2005). However, by the late twentieth century, events such as the PEPCON disaster (an explosion at an ammonium perchlorate manufacturing facility in Henderson, Nevada) and widespread usage of perchlorates in military munitions, led scientists to question the environmental fate of this ion (Urbansky, 1998; Trumpolt, Crain et al., 2005). This resulted in the identification of pervasive perchlorate groundwater contamination in several regions throughout the western United States (Trumpolt, Crain et al., 2005). Due to its ability to impact thyroid function even at relatively low doses, the US Environmental Protection Agency (USEPA)

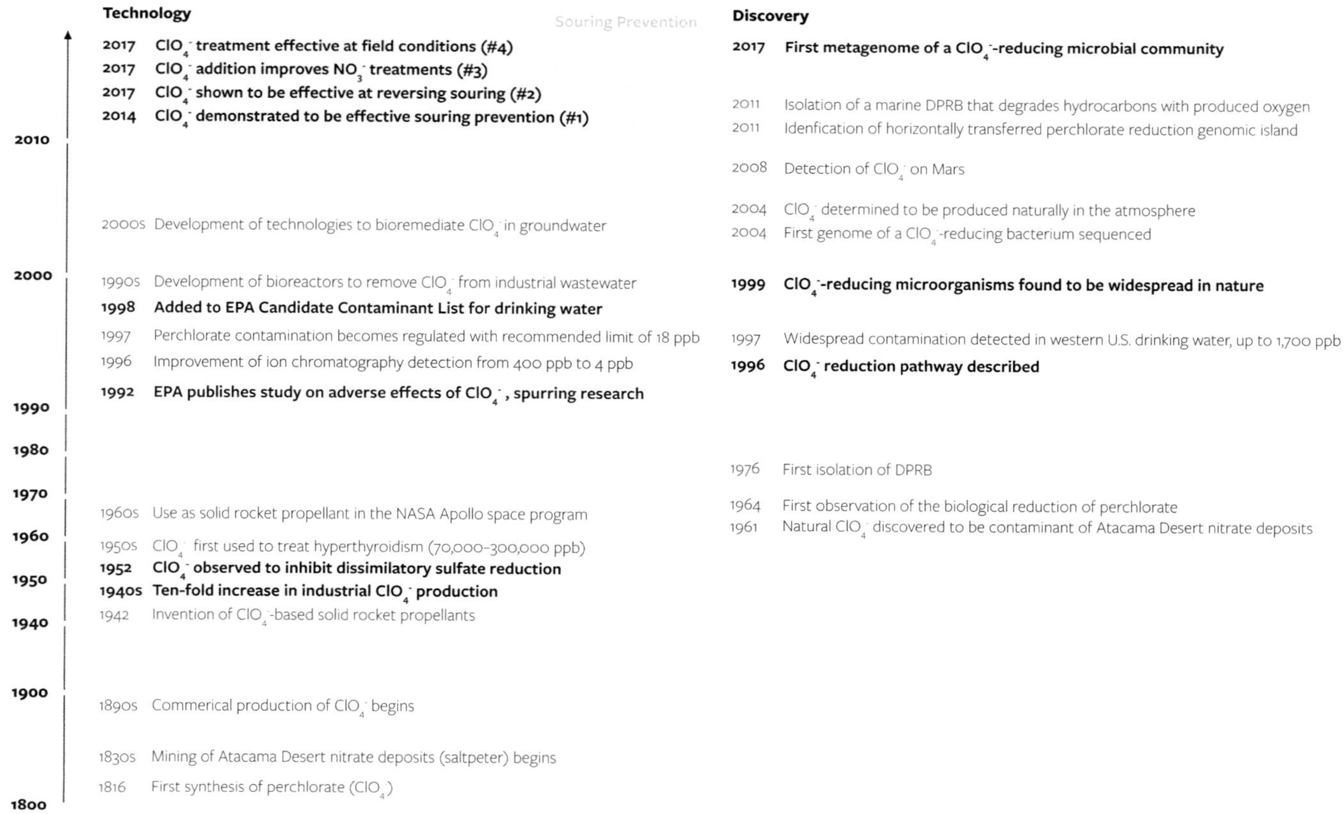

FIGURE 7.1 Timeline of perchlorate production and regulation, and discovery of DPRB.

placed perchlorate on its contaminant candidate list (CCL), and is currently working with individual states to set advisory levels for drinking water and groundwater (United States Environmental Protection Agency, 2014).

Although perchlorate is a potential groundwater contaminant, the application of perchlorate in oil reservoirs uses sodium perchlorate at concentrations less than 50 mM. Unlike ammonium perchlorate, sodium and potassium perchlorate are flammable but not explosive or combustible alone and their toxicity is comparable to that of nitrate. Nitrate causes the blue baby syndrome in infants younger than age six months; therefore, the US EPA regulates nitrate at a maximum contaminant level goal (MCLG) of 10 mg/L (United State Environmental Protection Agency 2018). Like nitrate, perchlorate is biodegradable in the environmental, and therefore, applying perchlorate in oil reservoirs should not raise concerns over its safety and persistence in the environment.

In an effort to bioremediate sites contaminated with perchlorate, past research was directed at understanding the metabolism of dissimilatory perchlorate-reducing bacteria (DPRB), which reduce perchlorate to innocuous chloride (Coates, Michaelidou et al., 1999). The first mention of biotic perchlorate reduction appeared in 1928, when researchers noticed microorganisms were capable of degrading chlorate salts added as an herbicide to control Canadian Thistle (Aslander, 1928). For many years following this observation, researchers believed that perchlorate reduction was an artifact of nitrate reductase activity; however, Romanenko et al. (1976) isolated the bacterium *Vibrio dechloraticans Cuznesove* B-1168, which grew anaerobically by degrading perchlorate and chlorate into chloride (Attaway and Smith, 1993). Studies on more recent perchlorate-reducing isolates have identified a conserved electron transport pathway encoded within a genomic unit referred to as the "perchlorate reduction genomic island" (PRI) (Melnyk, Engelbrektson et al., 2011). The PRI contains the genes *pcrABCD*, which code for the α and β subunits of the periplasmic perchlorate reductase, a c-type cytochrome, and a chaperone protein, respectively, as well as *pcrQO*, a quinone oxidase and a cytochrome, and *cld*, the chlorite dismutase (Melnyk, Engelbrektson et al., 2011; Melnyk and Coates, 2015; Wang, 2017). Collectively, this pathway takes electrons from the quinone pool, reducing perchlorate to chlorite, and subsequently dismutating chlorite into chloride and oxygen (Rikken, Kroon et al., 1996). Since the late 1990s, researchers have isolated numerous DPRB within the phyla *Proteobacteria* and, less commonly, *Firmicutes* (Coates and Achenbach, 2004; Melnyk et al., 2011; Balk et al., 2008). Although perchlorate reduction is found broadly across the *Proteobacteria*, the distribution of DPRB is mainly restricted to specific families: *Rhodospirillaceae* in the *Alphaproteobacteria*, *Rhodocyclaceae* in the *Betaproteobacteria*, *Campylobacteraceae* in the *Epsilonproteobacteria*, and *Sedimenticolaceae* in the *Gammaproteobacteria* (Carlström, Lucas et al., 2016). Of particular interest for biosouring control, several marine bacteria capable of dissimilatory perchlorate reduction were recently identified (Carlström, Wang et al., 2013; Carlstrom, Loutey et al., 2015; Carlström, Lucas et al., 2016). Other notable events and discoveries involving perchlorate and DPRB are listed in Figure 7.1.

7.3 PERCHLORATE AS AN INHIBITOR OF SRM

Perchlorate is a competitive inhibitor of key enzymes involved in sulfate reduction and as such, it constitutes a *direct* inhibitor of sulfidogenesis, meaning that its effects are enzymatic and are evident in pure cultures of SRM. Perchlorate also has a suite of *indirect* inhibitory effects mediated by other community members and not apparent in pure cultures of SRM (see Figure 7.2) but are relevant to the application of perchlorate as a biosouring inhibitor. The combination of direct and indirect inhibition enables a potent and selective suppression of sulfidogenesis in complex environmental systems. The following sections detail the direct enzymatic effects of perchlorate in model SRM, the experimental data that identified these effects, and the indirect effects of perchlorate-mediated inhibition. This serves as critical background necessary for understanding the case studies presented later in this chapter.

7.3.1 Direct Inhibition

Postgate (1952) was the first to classify perchlorate, among other sulfate analogs, as a direct inhibitor of sulfidogenesis in the model SRM *Desulfovibrio desulfuricans* Hildenborough. At 10 times the concentration of sulfate, perchlorate depressed sulfate reduction in this organism by 85% (Postgate, 1952). More recently, Carlson et al. (2015) quantified the inhibitory potency of perchlorate against *Desulfovibrio vulgaris* Hildenborough, *Desulfovibrio alaskensis* G20, *Desulfovibrio vulgaris* Miyazaki, and *Desulfovibrio desulfuricans* 27774. The parameter IC_{50} is defined here as the concentration of compound needed to inhibit 50% of the growth or sulfide production of an SRM strain over 48h. Depending on the *Desulfovibrio* species, the IC_{50} for perchlorate is between 14–50 mM (Carlson, Kuehl et al., 2015). Further, the concentration of perchlorate required to inhibit sulfidogenesis in a marine enrichment is approximately nine times lower than that required to inhibit overall growth, demonstrating that perchlorate is not only a potent but also a *selective* inhibitor of sulfidogenesis.

Carlson et al. (2015) gained insight into the mechanism of this inhibition via profiling of perchlorate stress in *D. alaskensis* G20 using a barcoded transposon insertion library (BarSeq) (Carlson et al., 2014). BarSeq is a powerful tool for outlining the molecular effects of a stressor. The approach generates thousands of barcoded transposon insertion mutants, creating a pool of mutants, each with an inactivation in a single gene. Collectively the library (pool) hits every non-essential gene in the genome of an organism multiple times. Subsequent growth of this pool in the presence of a stressor is used to profile the fitness of all mutants in parallel. Mutants with transposon insertions in a gene detrimental to growth under the stressor will grow more than average or wildtype control. Conversely, insertions that disrupt the function of a gene, which is beneficial under the selective pressure, will grow poorly. Comparing the frequency of mutants, via their barcodes using next-generation sequencing technology, pre- and post-selection can thereby inform on important and detrimental genes for growth under the condition in question (Van Opijnen, Bodi et al., 2009; Smith, Heisler et al., 2010; Korte, Fels et al., 2014; Wetmore, Price et al., 2015). The *D. alaskensis* G20 insertion library highlights the *Dde_2702* gene: the

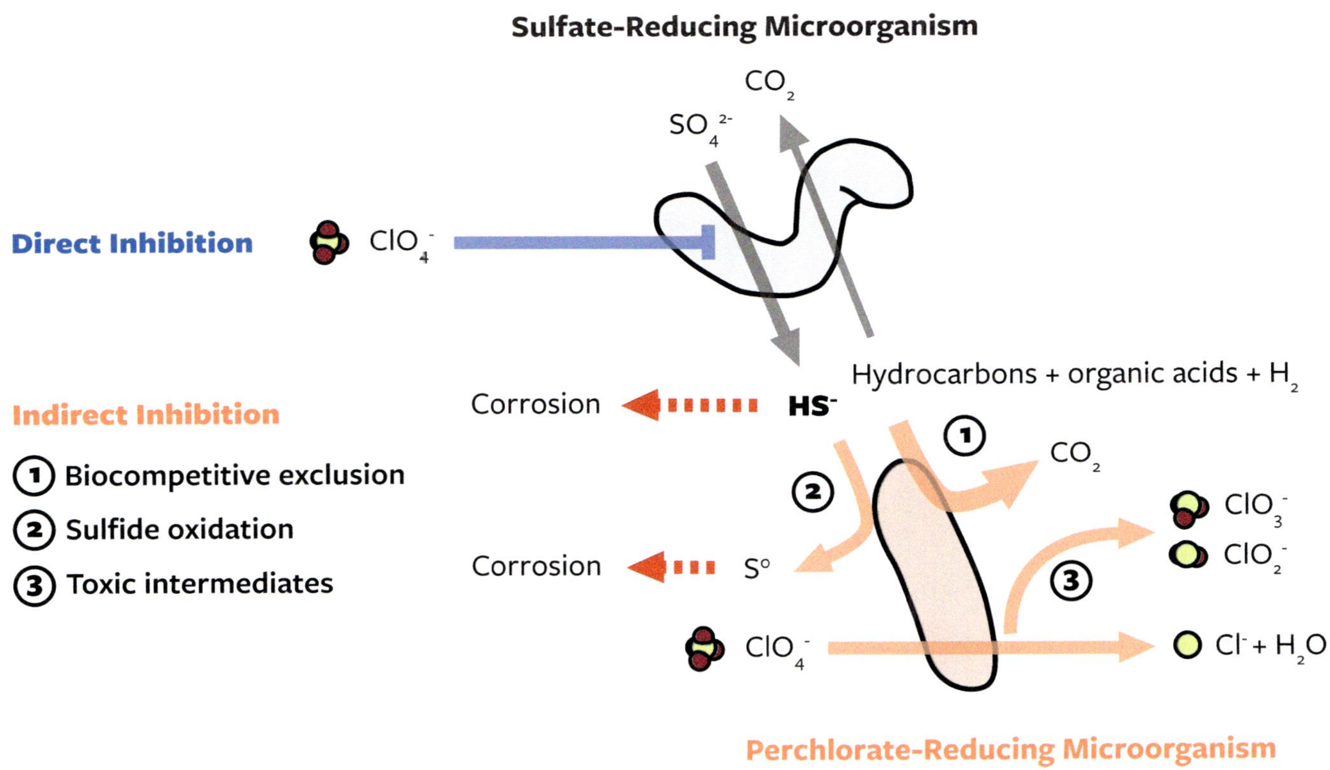

FIGURE 7.2 Mechanisms of perchlorate inhibition of SRM. Circled numbers are the action pathways of perchlorate reducers. 1 is biocompetitive exclusion, 2 is sulfide oxidation, and 3 is the toxic intermediates of perchlorate reduction.

Tn insertion mutant in this gene increased fitness over other insertion mutants when grown in perchlorate, defining this gene as detrimental to perchlorate stress (Carlson, Kuehl et al., 2015). Dde_2702 is an NADH:NAD+ sensitive Rex repressor and its regulon, the Rex regulon, includes core respiratory enzymes of sulfate reduction. An insertion in, and thereby inactivation of *rex*, leads to de-repression of the regulon (i.e., increased transcription), and a concurrent twofold increase in resistance to perchlorate. This suggests that perchlorate targets the central pathway for sulfate reduction, and de-repression of this pathway can alleviate inhibition (see Figure 7.2).

Interestingly, within the Rex regulon only sulfate transporters and the sulfate adenylyltransferase (*sat*, catalyzing the first step in the SR pathway) interact directly with sulfate, making these proteins primary targets for sulfate analogs such as perchlorate (Kuehl, Price et al., 2014; Christensen, Zane et al., 2015). Indeed, sulfate turnover in eukaryotic *sat* enzymes is competitively inhibited by perchlorate (Renosto, Patel et al., 1993; Hanna, MacRae et al., 2002; Hanna, Ng et al., 2004). Hence, the likely mechanism of perchlorate inhibition is competitive binding at the *sat*, leading to a reduction in sulfate respiration. Consequential to this is the growth defect seen in pure cultures of SRM exposed to perchlorate.

7.3.2 Indirect Inhibition

In a mixed sulfidogenic community, perchlorate can have indirect effects, which are effects mediated through various community members. It functions as an alternative electron acceptor, enriching for DPRB. These organisms can enhance the efficacy of the inhibitor in several ways. First, dissimilatory perchlorate reduction is a more energetically favorable respiratory metabolism than sulfate reduction. Thus, when competing for the same electron donor, DPRB will outcompete SRM through biocompetitive exclusion (see Figure 7.2) (Engelbrektson, Hubbard et al., 2014; Engelbrektson, Briseno et al., 2018; Engelbrektson, Cheng et al., 2018). DPRB also reduce perchlorate to chloride and water (see Figure 7.2) through the intermediates chlorate, chlorite, and oxygen, all of which are toxic to SRM (Carlson, Stoeva et al., 2015). Finally, the innate ability of all DPRB to oxidize sulfide to elemental sulfur (see Figure 7.2) can reduce the human health and economic costs associated with SRM-mediated sulfide production as discussed below (Gregoire, Engelbrektson et al., 2014; Mehta-Kolte, Loutey et al., 2017).

7.3.3 Sulfur Oxidation

The oxidation of sulfide to sulfur by perchlorate-reducing microorganisms inhibits growth and may significantly alter the geochemistry of reservoir systems. While sulfide oxidation by other respiratory organisms conserves energy from the oxidation of sulfide to sulfate, all mesophilic DPRB tested oxidized sulfide partially, to elemental sulfur (S^0) and polysulfide species, without concomitant growth (Gregoire, Engelbrektson et al., 2014). Biochemical, genetic, and transcriptional studies on this pathway demonstrate that sulfide is oxidized both by the perchlorate reductase enzyme and the reactive intermediates chlorate, chlorite, and oxygen (Mehta-Kolte, Loutey et al., 2017). Because sulfide oxidation bypasses the electron transport system and removes energy-yielded pathway intermediates, sulfide inhibits the growth of DPRB until most sulfide is oxidized to elemental sulfur (Mehta-Kolte, Loutey et al., 2017). Elemental sulfur then can be reduced, oxidized, or disproportionated into sulfide and sulfate as part of the sulfur cycle (Jørgensen and Kasten, 2006; Barton, Fardeau et al., 2014). Accordingly, when sulfide is present with DPRB in a flow-through system, the elemental sulfur concentration is initially elevated in the short term until the sulfide is completely removed, after which, the elemental sulfur levels also return to baseline levels (Engelbrektson, Briseno et al., 2018). Further sulfide production and thus elemental sulfur production is prevented by the inhibitory effects of the perchlorate on SRM activity. Based on phylogenetic analysis of active microbial communities, a number of studies hypothesized that elemental sulfur production supported the growth of elemental sulfur-reducing microorganisms in communities (Engelbrektson, Hubbard et al., 2014; Carlström, Lucas et al., 2016; Engelbrektson, Briseno et al., 2018; Engelbrektson, Cheng et al., 2018). Genome-resolved metagenomics has confirmed the presence of genes for sulfur reduction in most of these suspected elemental sulfur-reducing microorganisms (Barnum, Figueroa et al., 2018). By reducing sulfur to sulfide, sulfur-reducing bacteria may slow DPRB growth in batch cultures (Barnum, Figueroa et al., 2018) and may similarly limit their growth in flow-through systems. However, this is currently speculative. Alternatively, in the thermophile *Archaeoglobus fulgidus*, any oxidized sulfur species are co-reduced to sulfide during perchlorate reduction, which reacts abiotically with the chlorite, formed from perchlorate reduction by the AF0174-0176 enzyme, to form chloride (Liebensteiner, Pinkse et al., 2013). Therefore, perchlorate-dependent sulfur cycling could be expected across a range of temperatures, although the effect on DPRB growth and SRM inhibition may differ depending on the DPRB organisms present.

7.4 BENCH SCALE COLUMN STUDIES

Bench scale continuous flow-through sediment systems (*columns*) inoculated with mixed sulfidogenic communities can be used to gain insight into perchlorate inhibition in a complex environment. These systems are designed to better mimic the conditions found in oil reservoirs than is possible in batch systems and/or provide greater insight into the geochemistry and community interactions in similar environments. Here, four column studies pertaining to the effects of perchlorate, chlorate, and nitrate on sulfidogenesis are summarized in the following sections.

7.4.1 Case Study 1: Prevention of Biosouring

This was the first proof of concept study examining the efficacy of perchlorate and the related oxyanion chlorate compared to equimolar nitrate as an inhibitor of SRM activity (Engelbrektson, Hubbard et al., 2014). Small (50 mL) up-flow sand-packed columns were inoculated with un-soured San Francisco Bay sediment and provided with yeast extract as a multivariate labile electron donor. Effluent geochemistry and microbial community were monitored throughout the experiment. Immediately after setup, inhibitor treatments (perchlorate, nitrate, and chlorate) were added in a pulsed fashion (10 mM for 2 days followed by no treatment for 2 days) in triplicate throughout

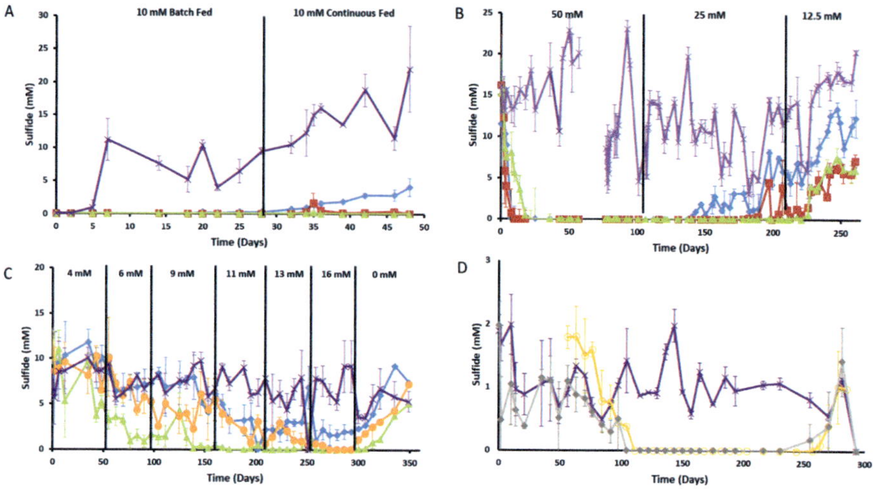

FIGURE 7.3 Sulfide production over time in the four case studies. Purple stars represent untreated columns, blue diamonds represent nitrate treated columns, green triangles represent perchlorate treated columns, red squares represent chlorate treated columns, orange circles represent mixed perchlorate, and nitrate treated columns, yellow hallow circles represent perchlorate continuous treated columns, and gray diamonds represent perchlorate pulsed column. Panel A: Case study 1 (Engelbrektson, Hubbard et al., 2014). Panel B: Case study 2 (Engelbrektson, Cheng et al., 2018). Panel C: Case study 3 (Engelbrektson, Briseno et al., 2018). Panel D: Case study 4 (Shrestha, Williamson et al., 2018). Black bars represent treatment changes with approximate treatment concentration. Error bars are calculated standard deviations.

the first 28 days of the experiment. In the second phase, treatments were continuously administered at 10 mM. Perchlorate and chlorate treatments completely inhibited biosouring during both treatment phases, except when the concentration of treatment was lowered to 5 mM mid-experiment and a small amount of sulfide was observed (see Figure 7.3A). However, sulfide immediately disappeared when the treatment was returned to 10 mM suggesting that concentration is important in perchlorate inhibition of sulfate reduction and suggesting that perchlorate may be able to reverse biosouring in a soured system. With nitrate treatment sulfide was observed after 23 days of treatment and the levels continuously increased to about 4 mM by the cessation of the experiment. In all cases inhibition was coupled to microbial reduction of the inhibitor. Perchlorate reduction in the columns took time to initiate (10 days), whereas nitrate was immediately and completely consumed. This study demonstrated that at equimolar concentrations, perchlorate is more effective than nitrate at preventing biosouring and hinted that perchlorate may be effective at reversing biosouring.

7.4.2 Case Study 2: Reversal of Biosouring

This study explored the potential of perchlorate to reverse biosouring (*sweetening*) and investigated the amount of inhibitor (perchlorate, chlorate, or nitrate) necessary to achieve complete *sweetening* of the system (Engelbrektson, Cheng et al., 2018).

Similar to case study 1, triplicate up-flow columns were monitored and compared to untreated control columns. However, the treatments were not initiated until after the columns were pre-soured (a period of flow with electron donor resulting in maximum sulfidogenesis in the columns). The columns used in this experiment were larger (1 L) and treatments were fed in continuously, with three treatment concentrations tested across three phases of the experiment (50 mM for phase 1; 25 mM for phase 2; and 12.5 mM for phase 3).

During phase 1 biosouring was completely reversed by day 10 for both nitrate and chlorate treatments versus day 18 for perchlorate. Since inhibition at these concentrations is thought to be largely biocompetitive, perchlorate took the longest to become effective compared to chlorate and then nitrate likely due to the low abundance of DPRB in the environment. Biosouring rebounded in both nitrate and chlorate treated columns during phase 2 (25 mM treatment), but perchlorate treated columns didn't begin to rebound until phase 3 (12.5 mM treatment) (see Figure 7.3B). This is likely due to the relative electron accepting capacities of these different anions (NO_3^-/N_2—5 electrons; ClO_3^-/Cl^-—6 electrons; ClO_4^-—8 electrons). In support of this difference in electron-accepting capacities, nitrate was undetected in the effluent for all phases, whereas both chlorate and perchlorate were detected until phase 3 (12.5 mM), indicating that both NRB and DPRB are present and active. Furthermore, the rate of sulfide production after rebound was much lower for perchlorate-treated columns than nitrate-treated columns. Interestingly, chlorate appeared to react abiotically with both the sulfide and iron present in the system resulting in an immediate 40% loss after entering the column when treatment was initiated. Because of this reactivity and the corrosive nature of chlorate, further study of this potential inhibitor was abandoned. This study concluded that perchlorate, although slower to inhibit biosouring than nitrate, maintained sweetening at a lower concentration.

7.4.3 Case Study 3: Synergism of Perchlorate and Nitrate

Research indicates that a combination of synergistic treatments can lead to a larger effect than treatment with one inhibitor alone (Greene, Brunelle et al., 2006; Carlson, Kuehl et al., 2015). As such Engelbrektson et al. (2018) used continuous flow packed columns to explore the potential synergistic effects of combined perchlorate and nitrate treatment in a soured system with crude oil as the electron donor. Crude oil was chosen as the electron donor instead of yeast extract to more closely replicate an oil reservoir system. Triplicate lateral flow sand packed 1 L columns were inoculated with pre-soured bay sediment to compare untreated, perchlorate treated, nitrate treated, and combined nitrate and perchlorate treatment (50/50). In order to also explore effective treatment concentrations, treatments were increased over six phases, from 4 mM to 16 mM. From this range of concentrations, the IC_{50} of each treatment was calculated: 9.8–12.0 mM for nitrate, 4.4–5.2 mM for perchlorate, and 2.2–9.9 mM for the combined treatment. Interestingly, sulfide production was never completely inhibited in the nitrate-treated columns (see Figure 7.3C). This study again shows that perchlorate alone is more effective than nitrate at equimolar concentrations. When combined, a perchlorate/nitrate mix is more inhibitory than nitrate alone, but not more effective than perchlorate alone. This study also indicates that perchlorate is an effective addition to a reservoir already undergoing nitrate treatment.

7.4.4 Case Study 4: Pulsed vs. Continuous Perchlorate Treatment

In an attempt to more closely simulate oil reservoir conditions than the previous case studies, Shrestha et al. (2018) used 3.8 L columns packed with sand and crude oil and inoculated with both North Seawater and produced water from a North Sea oil reservoir. These columns were controlled at a temperature of 40°C, the predicted temperature of the biozone in the source material oil reservoir. Also, the retention time of these columns were designed to closely represent that of the simulated oil reservoir (14 days). The seawater was amended with a 1 mM total concentration of volatile fatty acids (acetate, formate, propionate, and butyrate), phosphate, ammonium and trace elements to mimic concentrations measured in field produced water injected back into the reservoir. Two treatment regimens were compared in terms of their inhibitory effect: pulsed (brief high concentration injections of inhibitor) and continuous (low but constant injection of inhibitor). The perchlorate concentration was set at a concentration deemed feasible for real field operations: continuous treatment at 3.5 mM, and pulsed treatment initially applied for 24 hours weekly at 16.8 mM and then at 8.4 mM biweekly. The weekly pulsed regime was changed to biweekly when the retention time of the column was determined to be less than half (three–five days) of that expected due to short circuiting of flowpaths in the packed beds. Treatment addition in the pulsed columns was found to initially be unable to completely control biosouring, probably due to the short contact time of the perchlorate in the pulsed column matrices before it was washed out. However, once the pulsed regiment was changed to biweekly addition, complete biosouring inhibition was observed. The continuous treatment showed complete inhibition of sulfidogenesis. Interestingly, the biosouring rebound time for continuous treatment was much longer than pulsed, giving a longer lasting effect of inhibition. The mode of action of perchlorate in both continuous and pulsed was likely at least partially biocompetitive exclusion of SRM by DPRB as ~0.5 mM of perchlorate was consumed in the system (see Figure 7.3D). This case study, like case study 1, suggests that continuous perchlorate treatment is an effective treatment for oil reservoirs.

7.4.5 Summary of Column Case Studies

The column studies all strongly suggest that perchlorate is a much more effective, predictable and reliable inhibitor of SRM compared to nitrate. Case study 1 showed that perchlorate is more effective than nitrate at prevention of biosouring. Case study 2 concluded that perchlorate maintained sweetening at a lower concentration than nitrate. Case study 3 demonstrated that perchlorate can inhibit biosouring by itself and also added to nitrate treated oil reservoirs to increase effectiveness. Case study 3 also showed that when treatment stopped perchlorate treated column re-soured more slowly. Finally, case study 4 compared the treatment regimens that could be performed on an oil reservoir using perchlorate and concluded that continuous treatment is the more effective and logical solution to inhibiting biosouring. Therefore, taking into account all the case studies, using perchlorate as either a treatment or addition to treatment of biosouring is more effective and predictable compared to nitrate.

7.4.6 Community Patterns in Columns

Perchlorate amendment fundamentally alters the microbial ecology of seawater-based systems. Cultivation-independent methods have been used to characterize the composition of perchlorate-treated marine microbial communities in acetate-fed batch cultures (Carlström, Lucas et al., 2016), yeast extract-fed batch cultures (Carlson, Kuehl et al., 2015), yeast extract-fed column systems (Engelbrektson, Briseno et al., 2018; Engelbrektson, Cheng et al., 2018), and moderately thermophilic oil and VFA fed columns (Engelbrektson, Briseno et al., 2018; Shrestha, Williamson et al., 2018). 16S rRNA amplicon sequencing can be used to identify taxa with previously cultivated SRM (Müller, Kjeldsen et al., 2015). Additionally, the reconstruction of genomes from the environment can identify genes for dissimilatory sulfate and perchlorate reduction in uncultivated taxa never before associated with the metabolism (Anantharaman, Jungbluth et al., 2017), confirming or refuting 16S rRNA amplicon sequencing data when available. These measurements, in support of geochemical data, demonstrate that perchlorate treatment leads to a decline in SRM populations across all communities. Notably, in case study 2, SRM populations rebounded once perchlorate treatment ceased and columns began to re-sour (Engelbrektson, Cheng et al., 2018), whereas in case study 3, the cessation of treatment did not significantly change the community composition, suggesting the persistence of SRM throughout treatment although sulfidogenesis was inhibited (Engelbrektson, Briseno et al., 2018).

Another consistent effect of perchlorate treatment is the growth of DPRB. The perchlorate-reducing taxa most frequently enriched during treatment are *Arcobacter*, *Sedimenticola*, *Marinobacter*, and *Denitromonas* (Engelbrektson, Hubbard et al., 2014; Carlström, Lucas et al., 2016). Additionally, evidence from genomics and genome-resolved metagenomics has identified the ability to reduce perchlorate in strains of *Sulfurospirillum* spp., *Sulfurimonas* spp., unclassified *Rhodocyclaceae* spp., and unclassified *Rhodospirillales* spp. (Barnum, Figueroa et al., 2018; Shrestha, Williamson et al., 2018). The correlation between SRM inhibition and the growth of diverse DPRB supports biocompetitive exclusion of SRM by DPRB as a major mechanism of inhibition.

7.5 REACTIVE TRANSPORT MODELING

Modeling provides a vital tool to identify which mechanism of inhibition is occurring. As previously discussed, the mechanism of perchlorate inhibition can be direct or indirect and modeling allows the researcher to assess the contribution of each mechanism in complex systems. Models can also integrate data sets on column experiments performed under different conditions (i.e., different electron donor, different temperature, or different source of microorganisms), and merge geochemical, isotope, and microbial community data. Two different types of models have been used to investigate oil reservoir biosouring and treatment. Empirical modeling, constrained by experimental data, has been used to understand sulfide generation in oil reservoirs (Eden, Laycock et al., 1993; Ligthelm, de Boer et al., 1991). Whereas, mechanistic-based models, structured to represent a causal relationship to the underlying biological and chemical processes occurring, have been used to model both biosouring and treatment of biosouring (Coombe, Jack et al., 2010; Farhadinia, Bryant et al., 2010; Haghshenas, Sepehrnoori et al., 2012; Cheng, Hubbard et al., 2016). Cheng et al. (2016) were the first to model perchlorate treatment using data from case study 1. The goal of this model was to investigate the roles of the three different potential mechanisms of perchlorate treatment; (i) direct inhibitory effect of perchlorate; (ii) microbial sulfide oxidation coupled to perchlorate reduction; and (iii) heterotrophic biocompetitive perchlorate reduction (Engelbrektson, Hubbard et al., 2014). The model also integrated sulfur isotopic data and concluded that biocompetitive exclusion was the dominant effect in this particular yeast extract–based column system. Case study 2, a similar column-based perchlorate inhibition system, was also modeled (Engelbrektson, Cheng et al., 2018). In this system biocompetitive exclusion was again the main mechanism of perchlorate inhibition. However direct inhibition was vital during the initial phase of treatment in order to bring down the sulfide levels enough for heterotrophic perchlorate reduction to begin. Sulfur oxidation coupled to perchlorate reduction was important during treatment phase where perchlorate concentrations were lowest. These studies illustrate how important modeling can be for gaining insight into the various mechanisms of inhibition that would otherwise be obscured by the complexity and heterogeneity of these systems.

7.6 CONCLUSIONS

Biosouring in oil reservoirs is of major concern for maintaining industrial sustainability and safety and for preventing ecological disasters. Nitrate, a sulfate analog, is commonly used in the field to prevent biosouring, but it is unpredictable in its inhibitory effects. Perchlorate is a promising alternative to nitrate or biocide use. Perchlorate acts through both *direct* and *indirect* inhibition of SRM (see Figure 7.2). It also rapidly decreases residual sulfide levels through the ability of DPRB to enzymatically oxidize sulfide to elemental sulfur. In the column case studies presented, perchlorate is a more effective and predictable inhibitor of SRM than either nitrate or chlorate at equimolar concentration with a longer inhibitory effect after cessation of treatment. Also, perchlorate can be added with nitrate treatment to increase the effectiveness. While perchlorate directly inhibits SRM in pure culture, multiple community-based analyses and computer modeling suggest that indirect inhibition by biocompetitive exclusion is the primary mechanism observed in treated communities. Collectively, the studies presented have led to the conclusion that perchlorate is a specific, predictable, versatile, and effective inhibitor of sulfidogenesis and aids in sulfide removal from soured systems. Therefore, perchlorate shows great promise as a novel inhibitor of sulfidogenesis for the oil and gas industry.

REFERENCES

Anantharaman, K., S.P. Jungbluth, R.S. Kantor, A. Lavy, L.A. Warren, M.S. Rappé, B.C. Thomas, and J.F. Banfield. 2017. "Dramatic expansion of microbial groups that shape the global sulfur cycle." *bioRxiv*. doi: https://doi.org/10.1101/166447.

Aslander, A. 1928. "Experiments on the eradication of Canada thistle, Cirsium arvense, with chlorates and other herbicides." *Journal of Agricultural Research*. 36 (11): 915–935.

Attaway, H., and M. Smith. 1993. "Reduction of perchlorate by an anaerobic enrichment culture." *Journal of Industrial Microbiology Society for Industrial Microbiology*. 1293 (6): 408–412.

Balk, M., T. van Gelder, S.A. Weelink, and A.J. Stams, 2008. "(Per) chlorate reduction by the thermophilic bacterium Moorella perchloratireducens sp. nov., isolated from underground gas storage." *Applied and environmental microbiology,* 74 (2): 403–409.

Barnum, T.P., I.A. Figueroa, C.I., Carlström, L.N., Lucas, A.L., Engelbrektson, and J.D. Coates. 2018. "Genome-resolved metagenomics identifies genetic mobility, metabolic interactions, and unexpected diversity in perchlorate-reducing communities." *The ISME Journal*. 12(6): 1568–1581.

Barton, L.L., M.-L. Fardeau, and G.D. Fauque. 2014. Hydrogen sulfide: a toxic gas produced by dissimilatory sulfate and sulfur reduction and consumed by microbial oxidation. *The Metal-Driven Biogeochemistry of Gaseous Compounds in the Environment*. P. M. H. Kroneck and M. E. S. Torres. Dordrecht, Springer Dordrecht: 237–277.

Biswas, K.C., N.A. Woodards, H. Xu, and L.L. Barton. 2009. "Reduction of molybdate by sulfate-reducing bacteria." *BioMetals*. 22(1): 131–139.

Bowker R. P. G., G. A. Audibert, H. J. Shah, and N. A. Webster. "Detection, control, and correction of hydrogen sulfide corrosion in existing wastewater systems." EPA 832-R-92-001, U.S. EPA Office of Water. Washington, D C, USA, 1992.

Carlson, H.K., J.V. Kuehl, A.B. Hazra, N.B. Justice, M.K. Stoeva, A. Sczesnak, M.R. Mullan, A.T. Iavarone, A. Engelbrektson, and M.N. Price. 2015. "Mechanisms of direct inhibition of the respiratory sulfate-reduction pathway by (per) chlorate and nitrate." *The ISME Journal*. 9(6): 1295.

Carlson, H.K., M.K. Stoeva, N.B. Justice, A. Sczesnak, M.R. Mullan, L.A. Mosqueda, J.V. Kuehl, A.M. Deutschbauer, A.P. Arkin, and J.D. Coates. 2015. "Monofluorophosphate is a selective inhibitor of respiratory sulfate-reducing microorganisms." *Environmental Science & Technology*. 49(6): 3727–3736.

Carlström, C.I., D.E. Loutey, O. Wang, A. Engelbrektson, I. Clark, L.N. Lucas, P.Y. Somasekhar, and J.D. Coates. 2015. "Phenotypic and genotypic description of Sedimenticola selenatireducens strain CUZ, a marine (per)chlorate-respiring gammaproteobacterium, and its close relative the chlorate-respiring Sedimenticola strain NSS." *Appl Environ Microbiol*. 81(8): 2717–2726.

Carlström, C.I., L.N. Lucas, R.A. Rohde, A. Haratian, A.L. Engelbrektson, and J.D. Coates. 2016. "Characterization of an anaerobic marine microbial community exposed to combined fluxes of perchlorate and salinity." *Applied Microbiology and Biotechnology*. 100(22): 9719–9732.

Carlström, C.I., O. Wang, R.A. Melnyk, S. Bauer, J. Lee, A. Engelbrektson, and J.D. Coates. 2013. "Physiological and genetic description of dissimilatory perchlorate reduction by the novel marine bacterium Arcobacter sp. strain CAB." *MBio*. 4(3): e00217–00213.

Catling, D., M. Claire, K. Zahnle, R. Quinn, B. Clark, M. Hecht, and S. Kounaves. 2010. "Atmospheric origins of perchlorate on Mars and in the Atacama." *Journal of Geophysical Research: Planets*. 115(E1).

Cheng, Y., C.G. Hubbard, L. Li, N. Bouskill, S. Molins, L. Zheng, E. Sonnenthal, M.E. Conrad, A. Engelbrektson, and J.D. Coates. 2016. "Reactive transport model of sulfur cycling as impacted by perchlorate and nitrate treatments." *Environmental Science & Technology*. 50(13): 7010–7018.

Chou, C.H., J.M. Ogden, H.R. Pohl, F. Scinicariello, L. Ingerman, L. Barber, and M.J. Citra, 2016. Toxicological profile for hydrogen sulfide and carbonyl sulfide. Agency for Toxic Substances and Disease Registry: Atlanta, GA.

Christensen, G., G. Zane, A. Kazakov, X. Li, D. Rodionov, P. Novichkov, I. Dubchak, A. Arkin, and J. Wall. 2015. "Rex (encoded by DVU_0916) in Desulfovibrio vulgaris Hildenborough is a repressor of sulfate adenylyl transferase and is Regulated by NADH." *Journal of Bacteriology*. 197(1): 29–39.

Coates, J.D., U. Michaelidou, R.A. Bruce, S.M. O'Connor, J.N. Crespi, and L.A. Achenbach. 1999. "Ubiquity and diversity of dissimilatory (per) chlorate-reducing bacteria." *Applied and Environmental Microbiology*. 65(12): 5234–5241.

Coombe, D.A., T. Jack, G. Voordouw, F. Zhang, B. Clay, and K. Miner. 2010. "Simulation of Bacterial Souring Control in an Alberta Heavy-Oil Reservoir." *Journal of Canadian Petroleum Technology*. 49(05): 19–26.

Eden, B., P. J. Laycock, and M. Fielder. 1993. *Oilfeild Reservior Souring*. United Kingdom.

Engelbrektson, A., V. Briseno, I. Figueroa, M. Yee, G.L. Shao, H. Carlson, and J.D. Coates. 2018. "Mitigating Sulfidogenesis with Simultaneous Perchlorate and Nitrate Treatment." *Frontiers in Microbiology*. Vol. 9: 2305.

Engelbrektson, A., Y. Cheng, C.G. Hubbard, Y.T. Jun, B. Arora, L. Tom, P. Hu, A.-L. Grauel, M.E. Conrad, G. Andersen, J. Ajo-Franklin, and J.D. Coates. 2018. "Attenuating Sulfidogenesis in a Soured Continuous Flow Column System with Perchlorate Treatment." *Fronties in Microbiology*. Vol. 9: 1575.

Engelbrektson, A., C. Hubbard, L. Tom, A. Boussina, Y.T. Jin, H. Wong, Y.M. Piceno, H.K. Carlson, M. Conrad, G.L. Andersen, and J.D. Coates. 2014. "Inhibition of microbial sulfate reduction in a flow-through column system by (per)chlorate treatment." *Frontiers in Microbiology*. 5: 315.

Farhadinia, M., S. Bryant, K. Sepehrnoori, and M. Delshad. 2010. "Development and implementation of a multidimensional reservoir souring module in a chemical flooding simulator." *Petroleum Science and Technology*. 28(6): 535–546.

Gieg, L.M., T.R. Jack, and J.M. Foght. 2011. "Biological souring and mitigation in oil reservoirs." *Applied Microbiology and Biotechnology*. 92(2): 263.

Grabowski, A., O. Nercessian, F. Fayolle, D. Blanchet, and C. Jeanthon. 2005. "Microbial diversity in production waters of a low-temperature biodegraded oil reservoir." *FEMS Microbiology Ecology*. 54(3): 427–443.

Greene, E., C. Hubert, M. Nemati, G. Jenneman, and G. Voordouw. 2003. "Nitrite reductase activity of sulphate-reducing bacteria prevents their inhibition by nitrate-reducing, sulphide-oxidizing bacteria." *Environmental Microbiology*. 5(7): 607–617.

Greene, E.A., V. Brunelle, G.E. Jenneman, and G. Voordouw. 2006. "Synergistic inhibition of microbial sulfide production by combinations of the metabolic inhibitor nitrite and biocides." *Applied and Environmental Microbiology*. 72(12): 7897–7901.

Gregoire, P., A. Engelbrektson, C.G. Hubbard, Z. Metlagel, R. Csencsits, M. Auer, M.E. Conrad, J. Thieme, P. Northrup, and J.D. Coates. 2014. "Control of sulfidogenesis through bio-oxidation of H2S coupled to (per) chlorate reduction." *Environmental Microbiology Reports*. 6(6): 558–564.

Haghshenas, M., K. Sepehrnoori, S.L. Bryant, and M. Farhadinia. 2012. "Modeling and Simulation of Nitrate Injection for Reservoir-Souring Remediation." *SPE Journal*. 17(03): 817–827.

Hanna, E., I.J. MacRae, D.C. Medina, A.J. Fisher, and I.H. Segel. 2002. "ATP sulfurylase from the hyperthermophilic chemolithotroph Aquifex aeolicus." *Archives of Biochemistry and Biophysics*. 406(2): 275–288.

Hanna, E., K.F. Ng, I.J. MacRae, C.J. Bley, A.J. Fisher, and I.H. Segel. 2004. "Kinetic and stability properties of Penicillium chrysogenum ATP sulfurylase missing the C-terminal regulatory domain." *Journal of Biological Chemistry*. 279(6): 4415–4424.

Haveman, S.A., E.A. Greene, C.P. Stilwell, J.K. Voordouw, and G. Voordouw. 2004. "Physiological and Gene Expression Analysis of Inhibition of Desulfovibrio vulgaris Hildenborough by Nitrite." *Journal of Bacteriology* 186: 7944–7950.

He, Q., K.H. Huang, Z. He, E.J. Alm, M.W. Fields, T.C. Hazen, A.P. Arkin, J.D. Wall, and J. Zhou. 2006. "Energetic consequences of nitrite stress in Desulfovibrio vulgaris Hildenborough, inferred from global transcriptional analysis." *Applied and Environmental Microbiology*. 72: 4370–4381.

Hubbard, C.G., Y. Cheng, A. Engelbrekston, J.L. Druhan, L. Li, J.B. Ajo-Franklin, J.D. Coates, and M.E. Conrad. 2014. "Isotopic insights into microbial sulfur cycling in oil reservoirs." *Front Microbiol*. 5: 480.

Hubert, C., and G. Voordouw. 2007. "Oilfield souring control by nitrate-reducing Sulfurospirillum spp. that outcompete sulfate-reducing bacteria for organic electron donors." *Applied and Environmental Microbiology*. 73(8): 2644–2652.

Hubert, C., G. Voordouw, and B. Mayer. 2009. "Elucidating microbial processes in nitrate- and sulfate-reducing systems using sulfur and oxygen isotope ratios: the example of oil reservoir souring control." *Geochimica et Cosmochimica Acta*. 73(13): 3864–3879.

Jackson, W. A., Anderson, T., Harvey, G., Orris, G., Rajagopalan, S., and Kang, N. (2006). Occurrence and Formation of Non-Anthropogenic Perchlorate. *Perchlorate: Environmental Occurrence, Interactions and Treatment*. B. Gu and J. D. Coates. Boston, MA, Springer US: 49–69.

Jørgensen, B.B., and S. Kasten. 2006. Sulfur cycling and methane oxidation. *Marine Geochemistry*. Springer: 271–309.

Korte, H.L., S.R. Fels, G.A. Christensen, M.N. Price, J.V. Kuehl, G.M. Zane, A.M. Deutschbauer, A.P. Arkin, and J.D. Wall. 2014. "Genetic basis for nitrate resistance in Desulfovibrio strains." *Frontiers in Microbiology*. 5: 153.

Korte, H.L., A. Saini, V.V. Trotter, G.P. Butland, A.P. Arkin, and J.D. Wall. 2015. "Independence of Nitrate and Nitrite Inhibition of Desulfovibrio vulgaris Hildenborough and Use of Nitrite as a Substrate for Growth." *Environmental Science & Technology*. 49(2): 924–931.

Kuehl, J.V., M.N. Price, J. Ray, K.M. Wetmore, Z. Esquivel, A.E. Kazakov, M. Nguyen, R. Kuehn, R.W. Davis, and T.C. Hazen. 2014. "Functional genomics with a comprehensive library of transposon mutants for the sulfate-reducing bacterium Desulfovibrio alaskensis G20." *MBio*. 5(3): e01041–14.

L'haridon, S., A.-L. Reysenbacht, P. Glenat, D. Prieur, and C. Jeanthon. 1995. "Hot subterranean biosphere in a continental oil reservoir." *Nature*. 377(6546): 223–224.

Liebensteiner, M.G., M.W. Pinkse, P.J. Schaap, A.J. Stams, and B.P. Lomans. 2013. "Archaeal (per) chlorate reduction at high temperature: an interplay of biotic and abiotic reactions." *Science*. 340(6128): 85–87.

Ligthelm, D.J., R.B. de Boer, J.F. Brint, and W.M. Schulte. 1991. Reservoir Souring: An Analytical Model for H2S Generation and Transportation in an Oil Reservoir Owing to Bacterial Activity, In *Offshore Europe*. Society of Petroleum Engineers.

Lybrand, R.A., J.G. Bockheim, W. Ge, R.C. Graham, S.R. Hlohowskyj, G. Michalski, J.S. Prellwitz, J.A. Rech, F. Wang, and D.R. Parker. 2016. "Nitrate, perchlorate, and iodate co-occur in coastal and inland deserts on Earth." *Chemical Geology*. 442: 174–186.

Ma, H., C. Yang, G. Li, W. Guo, S. Chen, and J. Luo. 2003. "Influence of nitrate and chloride ions on the corrosion of iron." *Corrosion*. 59(12): 1112–1119.

Magot, M., B. Ollivier, and B.K. Patel. 2000. "Microbiology of petroleum reservoirs." *Antonie van Leeuwenhoek*. 77(2): 103–116.

Mehta-Kolte, M.G., D. Loutey, O. Wang, M.D. Youngblut, C.G. Hubbard, K.M. Wetmore, M.E. Conrad, and J.D. Coates. 2017. "Mechanism of H2S Oxidation by the Dissimilatory Perchlorate-Reducing Microorganism Azospira suillum PS." *MBio*. 8(1): e02023–16.

Melnyk, R.A., and J.D. Coates. 2015. "The Perchlorate Reduction Genomic Island: Mechanisms and Pathways of Evolution by Horizontal Gene Transfer." *Bmc Genomics.* 16: 1–11.

Melnyk, R.A., A. Engelbrektson, I.C. Clark, H.K. Carlson, K. Byrne-Bailey, and J.D. Coates. 2011. "Identification of a perchlorate reduction genomic island with novel regulatory and metabolic genes." *Applied and Environmental Microbiology.* 77(20): 7401–7404.

McInerney, M.J., K.L. Sublette, V.K. Bhupathiraju, J.D. Coates, and R.M. Knapp, 1993. "Causes and control of microbially induced souring." *Developments in petroleum science.* 39: 363–371.

Müller, A.L., K.U. Kjeldsen, T. Rattei, M. Pester, and A. Loy. 2015. "Phylogenetic and environmental diversity of DsrAB-type dissimilatory (bi) sulfite reductases." *The ISME Journal.* 9(5): 1152.

Patton, C.C. 1990, "Injection Water Quality." *Journal of Petroleum Engineers.* 42(10).

Postgate, J. 1952. "Competitive and non-competitive inhibitors of bacterial sulphate reduction." *Microbiology.* 6(1–2): 128–142.

Rao, B., N. Estrada, S. McGee, J. Mangold, B. Gu, and W. A. Jackson. 2012. "Perchlorate production by photodecomposition of aqueous chlorine solutions." *Environmental Science & Technology.* 46(21): 11635–11643.

Renosto, F., H.C. Patel, R.L. Martin, C. Thomassian, G. Zimmerman, and I. Segel. 1993. "ATP sulfurylase from higher plants: kinetic and structural characterization of the chloroplast and cytosol enzymes from spinach leaf." *Archives of Biochemistry and Biophysics.* 307(2): 272–285.

Rikken, G., A. Kroon, and C. Van Ginkel. 1996. "Transformation of (per) chlorate into chloride by a newly isolated bacterium: reduction and dismutation." *Applied Microbiology and Biotechnology.* 45(3): 420–426.

Romanenko, V., V. Koren'kov, and S. Kuznetsov, 1976. "Bacterial decomposition of ammonium perchlorate." *Mikrobiologiia.* 45(2): 204–209.

Shartau, S. L. C., Y. Marcy, S. Lin, A. A. Grigoryan, L. Adewal, H. Park, B. P. Lomans, E. Van Der Biezen, M. S. M. Jetten, and. Voordouw. "Ammonium concentrations in produced waters from a mesothermic oilfield subjected to nitrate injection decrease through formation of denitrifying biomass and anammox activity." *Applied and environmental microbiology* 76, no. 15 (2010): 4977–4987.

Shrestha, P.M., A.J. Williamson, T.P. Barnum, Y. Liu, J. Nowak, D. Loutey, W. Louie, H. Kim, W.S. Chin, K.S. Mayilvahanan, T. Shabbir, A. Goldstein, and J.D. Coates. 2018. "Microbial Biosouring Control Using Perchlorate Treatment in Crude Oil Reservoir Simulated Condidtion." *In Prep.*

Smith, A.M., L.E. Heisler, R.P. St. Onge, E. Farias-Hesson, I.M. Wallace, J. Bodeau, A.N. Harris, K.M. Perry, G. Giaever, and N. Pourmand. 2010. "Highly-multiplexed barcode sequencing: an efficient method for parallel analysis of pooled samples." *Nucleic Acids Research.* 38(13): e142–e142.

Trumpolt, C.W., M. Crain, G.D. Cullison, S.J. Flanagan, L. Siegel, and S. Lathrop. 2005. "Perchlorate: sources, uses, and occurrences in the environment." *Remediation Journal.* 16(1): 65–89.

United State Environmental Protection Agency. 2018. "National Primary Drinking Water Regulations." Retrieved April 26, 2018, from https://www.epa.gov/ground-water-and-drinking-water/national-primary-drinking-water-regulations.

United State Environmental Protection Agency. 2014. "Technical Fact Sheet - Perchlorate." https://www.epa.gov/sites/production/files/2014-03/documents/ffrrofactsheet_contaminant_perchlorate_january2014_final.pdf

Urbansky, E.T. 1998. "Perchlorate chemistry: implications for analysis and remediation." *Bioremediation Journal.* 2(2): 81–95.

Ure, A. 1824. *A dictionary of chemistry: on the basis of Mr. Nicholson's* London, Printed for Tegg.

Van Opijnen, T., K.L. Bodi, and A. Camilli. 2009. "Tn-seq: high-throughput parallel sequencing for fitness and genetic interaction studies in microorganisms." *Nature Methods*. 6(10): 767–772.

Videla, H. A. and L. K. Herrera 2005. "Microbiologically influenced corrosion: looking to the future." *International Microbiology*. 8: 169–180.

Wang, O. 2017. "No Title."

Wetmore, K.M., M.N. Price, R.J. Waters, J.S. Lamson, J. He, C.A. Hoover, M.J. Blow, J. Bristow, G. Butland, and A.P. Arkin. 2015. "Rapid quantification of mutant fitness in diverse bacteria by sequencing randomly bar-coded transposons." *MBio*. 6(3): e00306–00315.

Wilhelms, A., S.R. Larter, I. Head, P. Farrimond, R. di-Primio, and C. Zwach. 2001. "Biodegradation of oil in uplifted basins prevented by deep-burial sterilization." *Nature*. 411: 1034.

Youngblut, M.D., O. Wang, T.P. Barnum, and J.D. Coates. 2016. "(Per) chlorate in Biology on Earth and Beyond." *Annual Review of Microbiology*. 70: 435–457.

Youssef, N., M.S. Elshahed, and M.J. McInerney. 2009. Microbial Processes in Oilfields. *Advances in Applied Microbiology*. 66: 141–251.

8 Lab-on-a-Chip Model for Investigating the Effect of Biocides on Co-Culture Biofilms

Susmitha Purnima Kotu, Song-I Han, Arum Han, Sam Mannan, and Arul Jayaraman

CONTENTS

8.1 INTRODUCTION

8.1.1 MIC and Biofilms

Microbiologically influenced corrosion (MIC), commonly referred to as microbial corrosion or biocorrosion, is an electrochemical process where the metabolic activity of microorganisms results in corrosion (Videla, 1996). Based on the global costs of corrosion estimated by the NACE IMPACT (Koch G, 2016) and the 10–20% contribution of MIC to total corrosion, MIC amounted to about \$250–\$500 billion in 2013. It is now widely accepted that surface attached biofilms of microbial communities, and not floating planktonic microorganisms, causes MIC on metal surfaces (Costerton et al., 1995).

Biofilms are microbial communities embedded in a matrix of extracellular polymeric substance (EPS) and have been associated with tissue surfaces, natural aquatic environments and industrial systems (Costerton et al., 1987). Eradication of biofilms is a significant problem because the effective concentration of biocides needed for complete killing of biofilm microorganisms is orders of magnitude higher than that needed for planktonic microorganisms (Lock et al., 1984). An important characteristic of MIC is the difference in the composition of biofilms at different field locations impacted by MIC (Geissler et al., 2014; Sun et al., 2014). Therefore, an effective microbial control program requires selecting the right type and dosage of biocide that can eradicate the specific MIC causing microorganisms present at the field locations. Identification of such a biocide (or combination of biocides) requires testing multiple experimental conditions (type of biocide, concentration, duration of exposure) before finalizing the biocide treatment strategy. Thus, there is a need for systems that can be effectively used for determining the appropriate biocide type and dosage for controlling MIC.

8.1.2 Flow Systems to Investigate Biofilms and MIC

Most MIC studies in the laboratory are conducted using batch reactors, where metal coupons are immersed into bacterial cultures and incubated for extended periods of time (Jayaraman, Sun, and Wood, 1998; Enning et al., 2012; Castaneda and Benetton, 2008). The difference in mass of the coupons due to exposure is used as the measure of corrosion. While these systems are simple to use, they have several disadvantages. First, static systems are not always representative of the environment where MIC occurs (i.e., pipelines where liquid is constantly flowing). Second, the metabolism of the bacterial suspension is also affected by nutrient limitation and accumulation of waste products. As an alternative, continuous flow systems (Jayaraman et al., 1997b), circulating flow loops (Enning, Smith, and Stolle, 2016), and modified continuous stirred tank reactors (Tidwell et al., 2016) have been developed. While these models address some of the drawbacks of batch systems, they typically require large volumes and are not amenable for testing where the amount of sample is limited. The model described in this paper overcomes these drawbacks of both batch and recirculating systems by using minimal reagents and mimicking once–flow–through system conditions.

8.1.3 Lab-on-a-Chip Model to Investigate Biofilms and MIC

Lab-on-a-chip technology refers to the miniaturization of several laboratory tasks onto a chip that measures a few millimeters to few centimeters (Dittrich and Manz, 2006). Microfluidics is a subset of the lab-on-a-chip technology and has been used extensively for several applications like microarray analysis (Situma, Hashimoto, and Soper, 2006), genomics and sequencing (Williams, 2002), disease diagnostics (Chin et al., 2011) etc. Microfluidics has been used to investigate biofilm growth dynamics in several applications (Hong et al., 2012; Kim, Hegde, and Jayaraman, 2010; Kim et al., 2012; Lee, Kaplan, and Lee, 2008). However, the use of lab-on-a-chip technology for investigating biofilms and MIC has not been reported before. We have previously developed a lab-on-a-chip model that integrated electrochemical impedance spectroscopy and confocal microscopy to investigate biofilm growth dynamics and MIC (Kotu et al., 2016). We recently modified this system and developed a lab-on-a-chip model using carbon steel as the corroding metal for investigating MIC using either corrosive process waters or laboratory grown biofilms (Kotu et al., 2018). This model is used in the current study to determine biocide treatment based on the laboratory grown biofilms. This model allows real–time monitoring of biofilm growth dynamics with microscopy, with less reagent usage, and can be further modified for simultaneous monitoring of biofilms and corrosion in a high–throughput manner.

8.1.4 Model System Based Approach for Investigating Single-Species and Co-Culture Biofilms

Different types of microbial cultures, from single organisms to mock communities, have been used for laboratory MIC experiments. For example, inoculum from the field such as liquid or biofilm scrapings from pipelines with MIC–causing communities were grown in the different batch or continuous flow systems to investigate their impact on metal coupons and corrosion (Stipanicev et al., 2014). In addition to field samples, several sulfate–reducing bacteria (SRB) have been used to understand MIC mechanisms in the laboratory (Lopes et al., 2006). In this study, two model microorganisms that have been detected at locations impacted with MIC, and representing metabolic functions characteristic to locations impacted with MIC, were selected. *V. natriegens* is a biofilm former identified to be present in the locations impacted with MIC (Nivens et al., 1986; Benbouzid-Rollet et al., 1991; Yin et al., 2008; Cheng et al., 2009) while *S. oneidensis* is a marine biofilm forming iron–reducing bacterium that has also been isolated from MIC impacted industrial locations like cooling water towers (McLeod, MacDonald, and Brozel, 2002). However, *S. oneidensis* has been found to inhibit corrosion in several microbial communities resulting in its dual role in MIC (De Windt et al., 2003; Kotu et al., 2016; Herrera and Videla, 2009).

8.1.5 Use of Biocides for Controlling MIC

Various physical, electrochemical, biological, and chemical methods have been implemented for the prevention and mitigation of MIC. Cleaning pigs are typically employed to prevent solid accumulation in pipelines (Quarini and Shire, 2007). These cleaning pigs are also helpful to physically remove the surface attached biofilms and collect samples from pipelines. Frequent use of cleaning pigs can reduce microbial attachment and further prevent the sustenance of MIC (Bhola et al., 2010). Electrochemical methods like cathodic protection have been used in systems susceptible to MIC by using a sacrificial anode that provides electrons and hence prevents the metal from corroding (Wilson and Jack, 2017). However, this method is ineffective once a biofilm has been formed (de Romero, De Rincon, and Ocando, 2009). Corrosion inhibition by aerobic biofilms in the laboratory has been reported on various metals (Jayaraman et al., 1997b; Jayaraman et al., 1997a; Jayaraman, Sun, and Wood, 1998). In addition, engineered biofilms were used to generate antimicrobials *in situ* and eliminate SRB from biofilms (Jayaraman et al., 1999). However, industrial applications of these biological methods have not been successful (Arps et al., 2003).

One of the most widely used MIC treatment methods is treating with different chemicals that show microbiocidal activity. Based on the physical conditions at the field locations impacted by MIC, oxidizing or non–oxidizing biocides are used to kill microorganisms and control MIC (Javaherdashti, 2008). However, identification of the correct biocide for the microorganisms present and the appropriate dosage is important for maximum effectiveness. Laboratory methods for testing the performance of biocides include traditional bioassays and antimicrobial tests, while corrosion coupons are commonly used for field testing (Skovhus, 2014). The two most commonly used biocides for microbial control are glutaraldehyde and tetrakis–hydroxymethyl–phosphonium sulfate (THPS), as they are broad–spectrum biocides with quicker kill time compared to the other commonly used biocides such as 2–hydroxymethyl–2–nitropropane–1,3–diol (THNM), 1–(3–chloroallyl) –3,5,7–triaza–1–azoniaadamantane chloride (CTAC), and 4,4–dimethyl oxazolidine (DMO) (Yin et al., 2018; Williams and Schultz, 2015) in addition to being biodegradable (Videla, 2002). The biocidal action of glutaraldehyde is due to the cross–linking of the proteins in the cell wall disrupting routine functions (Maillard, 2002). The phosphine group in THPS reduces and disrupts the disulfide bonds to thiols in cell proteins and enzymes resulting in the death of cells (Ballantyne and Jordan, 2004). In this study, a lab-on-a-chip model was used to investigate the efficacy of short-term (4 h) glutaraldehyde and THPS treatment on laboratory grown single-species biofilms and co-culture biofilms.

8.2 METHODS

8.2.1 Lab-on-a-Chip Model Design and Fabrication

Glass slides were acid–cleaned and then coated with a 200 nm layer of carbon steel 1018 alloy using carbon steel 1018 target discs in a Lesker DC Sputter (Jefferson Hills, PA). The coated metal was then patterned into rectangular bands (see Figure 8.1). The transparent polymer polydimethyl siloxane (PDMS) was poured into 3D printed molds to obtain rectangular micro channels. These channels measure 500 μm in height, 1000 μm in width, and 10000 μm in length. PDMS with rectangular micro channels is placed on top of the patterned rectangular metal band to obtain closed channels. Syringes connected to Tygon tubing were used to introduce

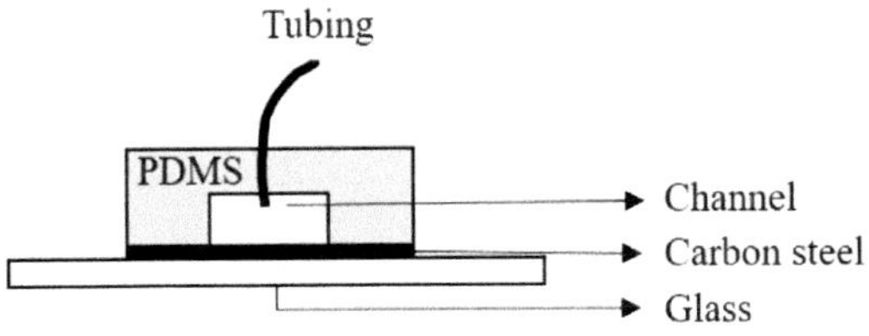

FIGURE 8.1 Schematic of the lab-on-a-chip model for investigating MIC.

fluids through the system. Flow was maintained in the micro channels by mounting syringes onto a Chemyx Fusion 200 syringe pump (Stafford, TX).

8.2.2 Biofilm Culture in Microfluidic Devices

For single-species biofilms, overnight cultures of *V. natriegens* and *S. oneidensis* grown in LB media were diluted to an optical density of ~1.0 before seeding into the microfluidic flow cell. For co-culture biofilms, overnight cultures of *V. natriegens* and *S. oneidensis* diluted to an optical density of ~1.0 were mixed in equal ratio before seeding into the microfluidic channel. After seeding, these cultures were kept for 2 h without flow to initiate attachment to the metal. After 2 h, LB media was continuously flowed through the system at 0.25 mL/h to remove unattached cells as well as provide nutrients for the growth of biofilms throughout the duration of the experiment. Biofilms were grown under aerobic environment for either 12 h or 36 h at 30°C prior to treatment with 400 ppm of glutaraldehyde or THPS for 4 h in batch mode (i.e., without flow). Staining of the live and dead cells in the biofilms was performed by diluting 3 μL of SYTO 9 and propidium iodide dyes from the Filmtracer biofilm Live/Dead stain kit (Carlsbad, CA) in 1 mL autoclaved DI water and flowing through the microfluidic device for 10 minutes at 0.25 mL/h, followed by 15 minutes incubation before imaging. All experiments were done in triplicates with three independent cultures.

8.2.3 Confocal Microscopy and Image Analysis

A Leica SP8 confocal laser scanning microscope (Wetzlar, Germany) was used to acquire z–stacks of the stained biofilms every 2.5 μm with a 20× objective. Z–stacks were taken at five random locations in all of the devices. Argon laser (488 nm) and Helium–Neon laser (561 nm) corresponding to excitation wavelengths of the dyes were used to image the biofilms over the metal surface. At the end of the experiment, the metal surface was also examined using a 20× objective for macroscopic indicators of corrosion after removing biofilm. Live/dead staining was used to visualize as well as differentiate the living and dead biofilm biomass on the metal surface. Live cells stain green while dead cells stain red in this assay. The fold changes in the percentage of dead cells before and after the biocide treatment was used as a measure of the effectiveness of the biocide. To account for variability in the percentage of dead biomass before the biocide treatment across different experiments, the dead cell percentage after the treatment was normalized to the percentage of dead cells before the biocide treatment.

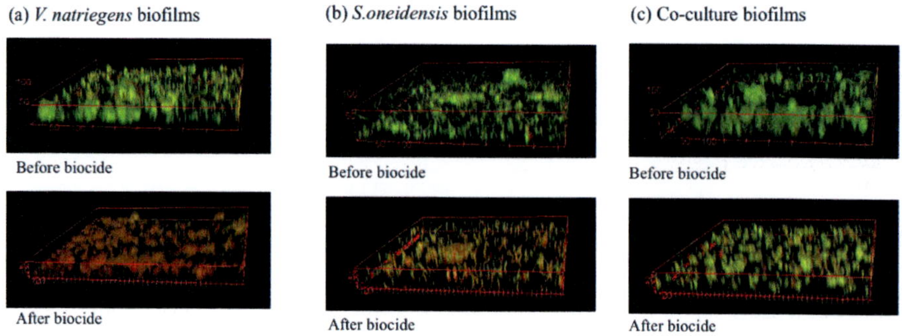

FIGURE 8.2 3D view of *V. natriegens* biofilms (a), *S. oneidensis* biofilms (b), and co-culture biofilms (c) grown for 12 h before and after glutaraldehyde treatment.

8.3 RESULTS AND DISCUSSION

8.3.1 Treatment of Short-Term Biofilms with Glutaraldehyde and THPS

Single-species biofilms grown for 12 h and treated with 400 ppm of glutaraldehyde of *V. natriegens* showed a 3.6–fold increase in the percentage of dead cells while single-species biofilms of *S. oneidensis* showed a 2.7–fold increase in the percentage of dead cells (see Figure 8.2). For co-culture biofilms of *V. natriegens* and *S. oneidensis* grown for 12 h, 2.3–fold increase in dead cells was observed (see Figure 8.2).

Single-species biofilms of *V. natriegens* grown for 12 h and treated with 400 ppm of THPS showed a 4.9–fold increase in the percentage of dead cells while single-species biofilms of *S. oneidensis* showed a 3.1–fold increase in the percentage of dead dells (see Figure 8.3). With co-culture biofilms of *V. natriegens* and *S. oneidensis* grown for 12 h, a 2.9–fold increase in the percentage of dead cells was observed (see Figure 8.3).

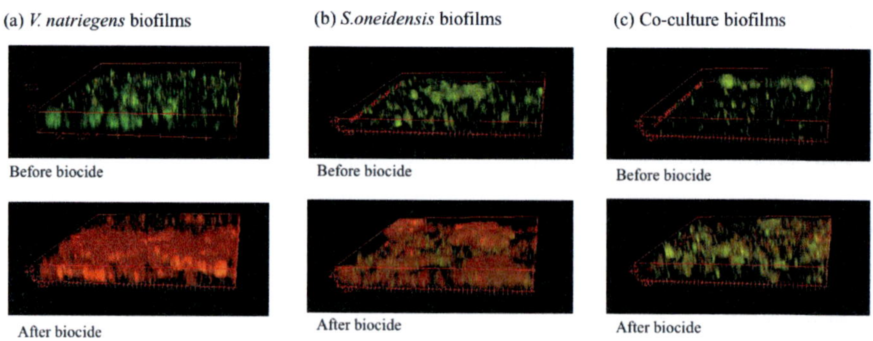

FIGURE 8.3 3D view of *V. natriegens* biofilms (a), *S. oneidensis* biofilms (b), and co-culture biofilms (c) grown for 12 h before and after THPS treatment.

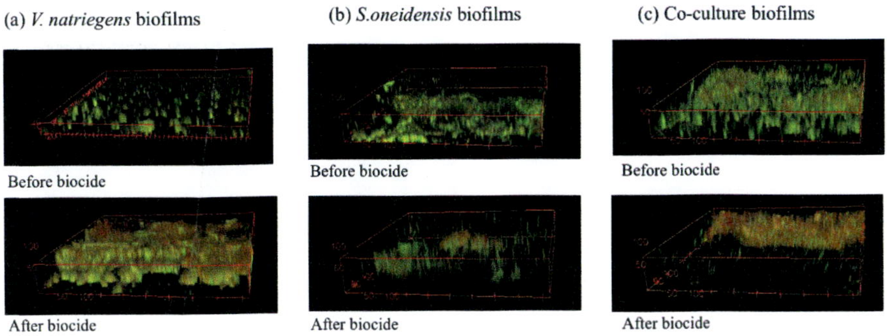

FIGURE 8.4 3D view of *V. natriegens* biofilms (a), *S. oneidensis* biofilms (b), and co-culture biofilms (c) grown for 36 h before and after glutaraldehyde treatment.

8.3.2 Treatment of Long-Term Biofilms with Glutaraldehyde and THPS

Single-species biofilms grown for 36 h and treated with 400 ppm of glutaraldehyde of *V. natriegens* showed a 2.9–fold increase in the percentage of dead cells while single-species biofilms of *S. oneidensis* showed a 2.0–fold increase in the percentage of dead cells (see Figure 8.4). With 36 h co-culture biofilms of *V. natriegens* and *S. oneidensis*, a 1.9–fold increase in the percentage of dead cells was observed (see Figure 8.4).

Single-species biofilms of *V. natriegens* grown for 36 h and treated with 400 ppm of THPS showed a 3.0–fold increase in the percentage of dead cells while single-species biofilms of *S. oneidensis* showed a 1.9–fold increase in the percentage of dead cells (see Figure 8.5). With co-culture biofilms of *V. natriegens* and *S. oneidensis* grown for 36 h, 1.5–fold increase in the percentage of dead cells was observed (see Figure 8.5).

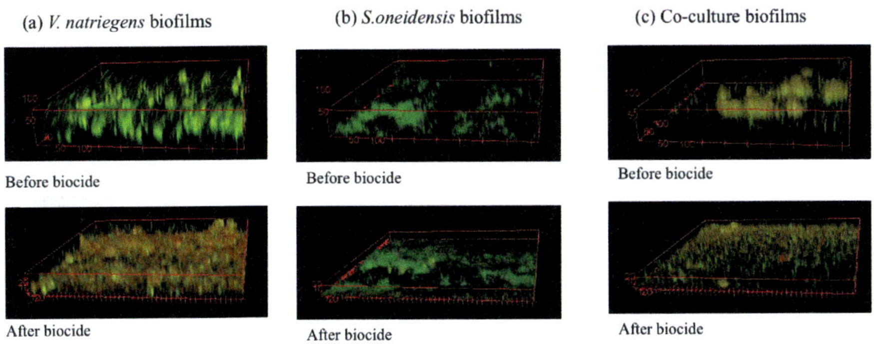

FIGURE 8.5 3D view of *V. natriegens* biofilms (a), *S. oneidensis* biofilms (b), and co-culture biofilms (c) grown for 36 h before and after THPS treatment.

Unexposed Metal Metal Exposed to 12 or 36 h Biofilms

FIGURE 8.6 Images of the metal surface in the microfluidic flow system unexposed (left) and after removal of biofilms grown for 12 h or 36 h (right).

8.3.3 Carbon Steel Corrosion

The metal surface in the lab-on-chip model did not show any indicators of corrosion after single-species biofilms or co-culture biofilms were cultured for 12 h or 36 h. Representative micrographs of metal surfaces are shown (Figure 8.6). Control experiments were done with LB growth media and no signs of corrosion was seen after 12 h and 36 h exposure.

8.3.4 Comparison of Single-Species Biofilms to Co-Culture Biofilms

Comparing the fold changes in percentage of dead biomass within each of the Tables 8.1–8.4, showed that the susceptibility of biofilms to biocide treatment varied with the biofilm species. The highest susceptibility to biocide treatment was seen with *V. natriegens* biofilms, while *S. oneidensis* biofilms were ~1.5–fold more resistant to biocide treatment than *V. natriegens* biofilms. Co-culture biofilms and single-species biofilms of *S. oneidensis* were more resistant to biocide treatment than the single-species biofilms of *V. natriegens* (p–value<0.05). Previous work (Berry, Xi, and Raskin, 2006;, Adam, Baillie, and Douglas, 2002) comparing single-species biofilms to multi-species biofilms also reported multi-species biofilms to be more resistant to antimicrobial treatments than single-species biofilms corroborating the results. The differences in the composition of extracellular polymeric substance (EPS) in these biofilms was suggested as a reason to justify the increased resistance (Adam, Baillie, and Douglas, 2002). However, the exact mechanisms behind the observed increased resistance of co-culture biofilms to biocides are unknown. Studying the complex biochemical interactions between *V. natriegens* and *S. oneidensis* with the extracellular polymeric substance in the co-culture biofilms is likely to provide insights on the increased resistance.

TABLE 8.1
Effect of Glutaraldehyde Treatment on Percentage of Dead Biomass in Single-Species and Co-Culture Biofilms Grown for 12 Hours

Biofilms	% Dead Biomass before Biocide	% Dead Biomass after Biocide	Fold Change in Dead Biomass
Vibrio natriegens	25.1±3.1	89.1±9.7	3.6
Shewanella oneidensis	31.8±9.7	85.8±14.4	2.7
Co-culture	32.2±7.5	73.1±11.9	2.3

TABLE 8.2
Effect of THPS Treatment on Percentage of Dead Biomass in Single-Species and Co-Culture Biofilms Grown for 12 Hours

Biofilms	% Dead Biomass before Biocide	% Dead Biomass after Biocide	Fold Change in Dead Biomass
Vibrio natriegens	18.1±0.6	88.7±3.3	4.9
Shewanella oneidensis	27.5±5.3	86.0±3.8	3.1
Co-culture	24.4±2.3	70.0±12.0	2.9

TABLE 8.3
Effect of Glutaraldehyde Treatment on Percentage of Dead Biomass in Single-Species and Co-Culture Biofilms Grown for 36 Hours

Biofilms	% Dead Biomass before Biocide	% Dead Biomass after Biocide	Fold Change in Dead Biomass
Vibrio natriegens	26.5±1.4	76.8±8.9	2.9
Shewanella oneidensis	26.7±6.2	53.7±10.7	2.0
Co-culture	33.1±5.6	63.6±5.9	1.9

TABLE 8.4
Effect of THPS Treatment on Percentage of Dead Biomass in Single-Species and Co-Culture Biofilms Grown for 36 Hours

Biofilms	% Dead Biomass before Biocide	% Dead Biomass after Biocide	Fold Change in Dead Biomass
Vibrio natriegens	22.8±8.2	67.9±2.8	3.0
Shewanella oneidensis	13.4±7.8	24.9±8.0	1.9
Co-culture	32.7±1.5	50.2±6.7	1.5

8.3.5 Comparison of Biocide Susceptibility of Short-Term and Long-Term Biofilms

We compared biofilms grown for 12 h and 36 h (see Tables 8.1–8.4) and inferred that biofilms grown for 12 h were more susceptible (p–value<0.05) to biocide treatment than 36 h. This clearly points out the significance of selecting a representative time for biofilm growth for determining the efficacy of biocide treatment. Further analysis was conducted to see if this difference was due to the difference in thickness of the biofilms or the biofilm biomass. No significant differences in terms of the biofilm thickness or biofilm biomass was seen for short-term biofilms compared to long-term biofilms (see Figure 8.7). Hence, the difference in the susceptibility of short-term and long-term biofilms to the biocide treatment cannot be attributed to diffusion limitations. The observed differences in the susceptibility are likely due to the differences in the extracellular matrix composition or phenotypic mutations of the cells or persister cells or efflux pumps.

From biofilm susceptibility data, it was inferred that short-term biofilms were more susceptible to THPS than glutaraldehyde (p–value<0.05) while with long-term biofilms the increased susceptibility to biocide treatment was only observed for co-culture biofilms (p–value<0.05). Since it is well established that biocides exhibit different degree of antimicrobial effects on different classes of microorganisms and are not as specific in their action as compared to antibiotics (Russell, 2003b; Williams, 1990), some variation in the susceptibility of microorganisms to biocides is expected. The different susceptibility to biocides can also be attributed to the differences in the composition of extracellular polymeric

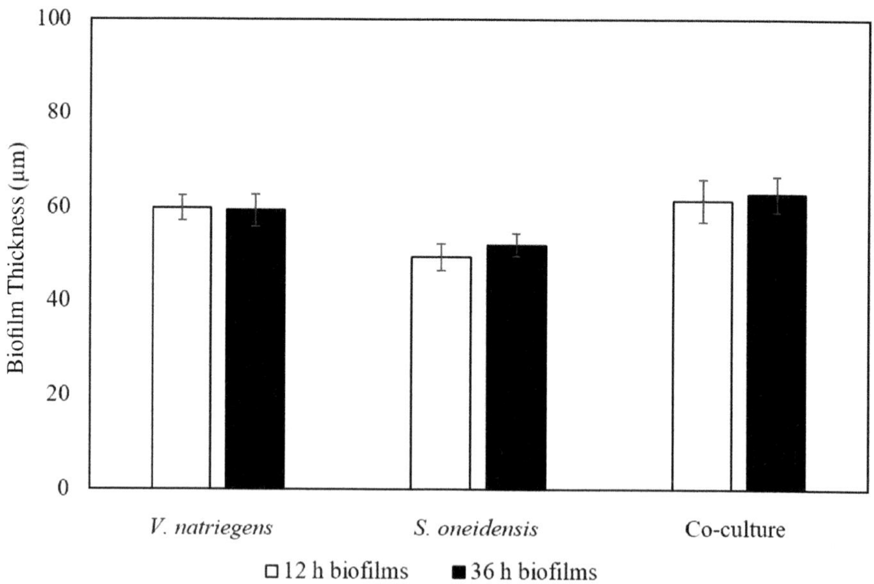

FIGURE 8.7 Comparison of the thickness of *V. natriegens*, *S. oneidensis*, and co-culture biofilms grown for 12 h and 36 h.

substance and/or the differences in the biocide adsorption mechanisms and subsequent uptake of biocides (Russell, 2003a). It is also possible that using Live/Dead staining underrepresents the biocidal effectiveness of glutaraldehyde as it is not clear if membranes that are crosslinked (upon glutaraldehyde exposure) are comparably permeable to the SYTO9 and propidium iodide dyes as normal (untreated) membranes.

8.4 CONCLUSIONS

This chapter demonstrates the application of a lab-on-a-chip flow system for screening biocides against MIC causing biofilms. This microfluidic system offers a platform for simulating pipeline flow conditions that overcomes the drawbacks of static systems and large–scale flow systems. Using minimal reagents for the establishment of biofilms and corrosive conditions in the system is advantageous particularly when using field waters to determine appropriate biocide type and dosage. While it is not possible to replicate field conditions in the laboratory, this lab-on-a-chip model can be used to test a wide range of relevant process conditions (e.g., pressure, flowrate, temperature).

From the confocal microscopy images and quantitative analysis of the biofilm viability, it was concluded that co-culture biofilms were more resistant to biocide treatment than single-species biofilms. This reiterates the importance of investigating co-culture laboratory grown biofilms for understanding MIC mechanisms rather than solely investigating the obvious corroding species like sulfate reducers. Differences in the susceptibility of the biofilms to the two biocides was seen only for short-term biofilms, with THPS being more effective in killing biofilm bacteria than glutaraldehyde under these experimental conditions. Further, co-culture biofilms grown for longer time were more resistant to biocide treatment than single-species biofilms indicating the importance of biofilm growth time as an important parameter in biocide efficacy studies. Further investigation of the extracellular matrix composition, biocide adsorption, cellular uptake, and efflux is essential for elucidating the differences in the biocide susceptibility of biofilms with species, time of growth, and type of biocide.

The variability of biocide resistance seen across different biofilms highlights the importance of executing biocide efficacy screening and qualification studies for the specific microorganisms present in the locations impacted by MIC than performing a random selection of biocide. While the current work is limited to 36 h biofilm culture and no noticeable metal corrosion was observed, this lab-on-a-chip model can be adapted for longer term biocide efficacy studies to allow for metal corrosion and hence investigate MIC using model organisms or using inoculum from MIC impacted field locations.

ACKNOWLEDGMENTS

The authors would like to thank the Mary Kay O' Connor Process Safety Center at Texas A&M University for the administrative and financial support. This work was partially supported by the Ray Nesbitt Chair Endowment to Arul Jayaraman.

REFERENCES

Adam, B., G.S. Baillie, and L.J. Douglas. 2002. "Mixed species biofilms of Candida albicans and Staphylococcus epidermidis." *Journal of Medical Microbiology*. 51 (4):344–349.

Arps, P.J., J.C. Earthman, L.–c. Xu, B.C. Syrett, R. Green, T. Wood, and F.B. Mansfeld. 2003. "Field evaluation of corrosion control using regenerative biofilms (CCURB)." CORROSION 2003, San Diego, California.

Ballantyne, B., and S.L. Jordan. 2004. "Biocides." *Pesticide Toxicology and International Regulation*. 1:365.

Benbouzid-Rollet, N.D., M. Conte, J. Guezennec, and D. Prieur. 1991. "Monitoring of a Vibrio natriegens and Desulfovibrio vulgaris marine aerobic biofilm on a stainless steel surface in a laboratory tubular flow system." *Journal of Applied Bacteriology*. 71 (3):244–251.

Berry, D., C. Xi, and L. Raskin. 2006. "Microbial ecology of drinking water distribution systems." *Current Opinion in Biotechnology*. 17 (3):297–302.

Bhola, R., S.M. Bhola, B. Mishra, and D.L. Olson. 2010. "Microbiologically influenced corrosion and its mitigation: A review." *Material Science Research India*. 7 (2):407–412.

Castaneda, H., and X.D. Benetton. 2008. "SRB–biofilm influence in active corrosion sites formed at the steel–electrolyte interface when exposed to artificial seawater conditions." *Corrosion Science*. 50 (4):1169–1183.

Cheng, S., J. Tian, S. Chen, Y. Lei, X. Chang, T. Liu, and Y. Yin. 2009. "Microbially influenced corrosion of stainless steel by marine bacterium Vibrio natriegens:Corrosion behavior." *Materials Science and Engineering: C*. 29 (3):751–755.

Chin, C.D., T. Laksanasopin, Y.K. Cheung, D. Steinmiller, V. Linder, H. Parsa, J. Wang, H. Moore, R. Rouse, and G. Umviligihozo. 2011. "Microfluidics–based diagnostics of infectious diseases in the developing world." *Nature Medicine*. 17 (8):1015–1019.

Costerton, J.W., K.J. Cheng, G.G. Geesey, T.I. Ladd, J.C. Nickel, M. Dasgupta, and T.J. Marrie. 1987. "Bacterial biofilms in nature and disease." *Annual Reviews in Microbiology*. 41:435–64.

Costerton, J.W., Z. Lewandowski, D.E. Caldwell, D.R. Korber, and H.M. Lappin–Scott. 1995. "Microbial biofilms." *Annual Reviews in Microbiology*. 49 (1):711–745.

de Romero, M., O. De Rincon, and L. Ocando. 2009. "Cathodic protection efficiency in the presence of SRB: State of the art." CORROSION 2009, Atlanta, Georgia.

De Windt, W., N. Boon, S.D. Siciliano, and W. Verstraete. 2003. "Cell density related H2 consumption in relation to anoxic Fe (0) corrosion and precipitation of corrosion products by Shewanella oneidensis MR-1." *Environmental Microbiology*. 5 (11):1192–1202.

Dittrich, P.S., and A. Manz. 2006. "Lab-on-a-chip: microfluidics in drug discovery." *Nature Reviews Drug Discovery*. 5 (3):210–218.

Enning, D., R. Smith, and J. Stolle. 2016. "Evaluating the efficacy of weekly THPS and glutaraldehyde batch treatment to control severe microbial corrosion in a simulated seawater injection system." CORROSION 2016, Vancouver, Canada

Enning, D., H. Venzlaff, J. Garrelfs, H.T. Dinh, V. Meyer, K. Mayrhofer, A.W. Hassel, M. Stratmann, and F. Widdel. 2012. "Marine sulfate-reducing bacteria cause serious corrosion of iron under electroconductive biogenic mineral crust." *Environmental Microbiology*. 14 (7):1772–1787.

Geissler, B., R. De Paula, C. Keller–Schultz, J. Lilley, and V. Keasler. 2014. "Data mining to prevent microbiologically influenced corrosion?" CORROSION 2014, San Antonio, Texas.

Herrera, L.K., and H.A. Videla. 2009. "Role of iron–reducing bacteria in corrosion and protection of carbon steel." *International Biodeterioration & Biodegradation*. 63 (7):891–895.

Hong, S.H., M. Hegde, J. Kim, X. Wang, A. Jayaraman, and T.K. Wood. 2012. "Synthetic quorum–sensing circuit to control consortial biofilm formation and dispersal in a microfluidic device." *Nature Communications*. 3:613.

Javaherdashti, R. 2008. "Microbiologically influenced corrosion (MIC)." In *Microbiologically Influenced Corrosion: An Engineering Insight*, 29–71. London: Springer

Jayaraman, A., E. Cheng, J. Earthman, and T. Wood. 1997a. "Importance of biofilm formation for corrosion inhibition of SAE 1018 steel by axenic aerobic biofilms." *Journal of Industrial Microbiology and Biotechnology*. 18 (6):396–401.

Jayaraman, A., E.T. Cheng, J.C. Earthman, and T.K. Wood. 1997b. "Axenic aerobic biofilms inhibit corrosion of SAE 1018 steel through oxygen depletion." *Applied Microbiology and Biotechnology*. 48 (1):11–17.

Jayaraman, A., P. Hallock, R. Carson, C.-C. Lee, F. Mansfeld, and T. Wood. 1999. "Inhibiting sulfate–reducing bacteria in biofilms on steel with antimicrobial peptides generated *in situ*." *Applied Microbiology and Biotechnology*. 52 (2):267–275.

Jayaraman, A., A.K. Sun, and T.K. Wood. 1998. "Characterization of axenic Pseudomonas fragi and Escherichia coli biofilms that inhibit corrosion of SAE 1018 steel." *Journal of Applied Microbiology*. 84 (4):485–492.

Kim, J., M. Hegde, and A. Jayaraman. 2010. "Co-culture of epithelial cells and bacteria for investigating host–pathogen interactions." *Lab Chip*. 10 (1):43–50.

Kim, J., M. Hegde, S. H. Kim, T. K. Wood, and A. Jayaraman. 2012. "A microfluidic device for high throughput bacterial biofilm studies." *Lab Chip*. 12 (6):1157–63.

Koch G, J. Varney, N. Thompson, O. Moghissi, M. Gould, and J. Payer. 2016. "NACE International IMPACT (International Measures of Prevention, Application, and Economics of Corrosion Technologies) Study." accessed 4/8/2017. http://impact.nace.org/economic–impact.aspx.

Kotu, S.P., C. Erbay, N. Sobahi, A. Han, S. Mannan, and A. Jayaraman. 2016. "Integration of electrochemical impedance spectroscopy and microfluidics for investigating microbially influenced corrosion using co-culture biofilms." CORROSION 2016, Vancouver, Canada, March 8, 2016.

Kotu, S.P., A. Jayaraman, M.S. Mannan, and A. Han. 2018. M–MIC: Microfluidic microbiologically influenced corrosion model. U.S. Patent App No. PCT/US2018/030198.

Lee, J.–H., J.B. Kaplan, and W.Y. Lee. 2008. "Microfluidic devices for studying growth and detachment of Staphylococcus epidermidis biofilms." *Biomedical Microdevices*. 10 (4):489–498.

Lock, M.A., R.R. Wallace, J.W. Costerton, R.M. Ventullo, and S.E. Charlton. 1984. "River epilithon: Toward a structural–functional model." *Oikos*. 42 (1):10–22.

Lopes, F.A., P. Morin, R. Oliveira, and L.F. Melo. 2006. "Interaction of Desulfovibrio desulfuricans biofilms with stainless steel surface and its impact on bacterial metabolism." *Journal of Applied Microbiology*. 101 (5):1087–95.

Maillard, J.Y. 2002. "Bacterial target sites for biocide action." *Journal of Applied Microbiology*. 92:16S–27S.

McLeod, E.S., R. MacDonald, and V.S. Brozel. 2002. "Distribution of Shewanella putrefaciens and Desulfovibrio vulgaris in sulphidogenic biofilms of industrial cooling water systems determined by fluorescent *in situ* hybridisation." *Water Sa*. 28 (2):123–128.

Nivens, D., P. Nichols, J. Henson, G. Geesey, and D. White. 1986. "Reversible acceleration of the corrosion of AISI 304 stainless steel exposed to seawater induced by growth and secretions of the marine bacterium Vibrio natriegens." *Corrosion*. 42 (4):204–210.

Quarini, J., and S. Shire. 2007. "A review of fluid–driven pipeline pigs and their applications." *Proceedings of the Institution of Mechanical Engineers, Part E: Journal of Process Mechanical Engineering*. 221 (1):1–10.

Russell, A.D. 2003a. "Similarities and differences in the responses of microorganisms to biocides." *Journal of Antimicrobial Chemotherapy*. 52 (5):750–763.

Russell, A.D. 2003b. "Biocide use and antibiotic resistance: the relevance of laboratory findings to clinical and environmental situations." *The Lancet Infectious Diseases*. 3 (12):794–803.

Situma, C., M. Hashimoto, and S.A. Soper. 2006. "Merging microfluidics with microarray–based bioassays." *Biomolecular Engineering*. 23 (5):213–231.
Skovhus, T.L. 2014. "Biocide testing using single and multispecies biofilms." In *Biofouling Methods*, S. Dobretsov, D.N. Williams. and J.C. Thomason (eds.). Hoboken, NJ: John Wiley & Sons.
Stipanicev, M., F. Turcu, L. Esnault, O. Rosas, R. Basseguy, M. Sztyler, and I.B. Beech. 2014. "Corrosion of carbon steel by bacteria from North Sea offshore seawater injection systems: laboratory investigation." *Bioelectrochemistry*. 97:76–88.
Sun, H., B. Shi, Y. Bai, and D. Wang. 2014. "Bacterial community of biofilms developed under different water supply conditions in a distribution system." *Science of The Total Environment*. 472 (Supplement C):99–107.
Tidwell, T.J., R. De Paula, Z. Broussard, and V. Keasler. 2016. "A comprehensive approach to diagnosing a solution for a South Texas production system with severe MIC." CORROSION 2016, Vancouver, Canada.
Videla, H.A. 1996. *Manual of Biocorrosion*. Boca Raton, FL: CRC Press.
Videla, H.A. 2002. "Prevention and control of biocorrosion." *International Biodeterioration & Biodegradation*. 49 (4):259–270.
Williams, J. 2002. Microfluidics system for single molecule DNA sequencing. U.S. Patent Application 10/228,729.
Williams, J.D. 1990. "Understanding antibacterial action and resistance." *Journal of Hospital Infection*. 16 (2):188.
Williams, T.M., and C.M. Schultz. 2015. "Effect of environmental parameters on the stability and performance of oil and gas biocides." SPE International Symposium on Oilfield Chemistry.
Wilson, S.L., and T.R. Jack. 2017. "MIC mitigation: Coatings and cathodic protection." In *Microbiologically Influenced Corrosion in the Upstream Oil and Gas Industry*, T. Lund Skovhus, D. Enning, and J.S. Lee (eds.). Boca Raton, FL: CRC Press.
Yin, B., T. Williams, T. Koehler, B. Morris, and K. Manna. 2018. "Targeted microbial control for hydrocarbon reservoir: Identify new biocide offerings for souring control using thermophile testing capabilities." *International Biodeterioration & Biodegradation*. 126 (Supplement C):204–207.
Yin, Y.S., S. Cheng, K.T. Lau, S.G. Chen, and T. Liu. 2008. "The influence of the marine bacterium Vibrio natriegens growth on metallic corrosion as studied by microscope technologies." *Advanced Materials Research*. 47:169–172.

9 Considerations for Evaluating Biocidal Efficacy on Biofilm Formation of Oilfield Relevant Microorganisms

Rikke Markfoged, Steffen Larsen Aggerholm, Lea Holsten, Lisa Struve, Philipp Trénel, Morten Poulsen, and Katrin Steinhauer

CONTENTS

9.1 INTRODUCTION

Microbial growth in offshore oil production facilities results in complications such as biofouling, MIC, and souring. All of those are of severe economic impact to the industry. Several microbial groups have been shown to influence corrosion, including sulfate-reducing prokaryotes (SRP), thiosulfate-reducing bacteria, nitrate-reducing bacteria, and methanogenic archaea (Sørensen et al., 2012; Magot et al., 1997; Vik et al., 2007; Daniels et al., 1987; Larsen et al., 2010). SRPs, are one of the most common and thoroughly described MIC-related organisms. These organisms degrade organic compounds (e.g., acetate) to produce energy. Within this process, sulfate is used as the terminal electron acceptor and hydrogen sulfide (H_2S) is formed (see Equation 9.1). The production of hydrogen sulfide results in souring of the reservoir and is highly toxic (Lundgaard et al., 2010). Furthermore, hydrogen sulfide is a powerful corrosive reactant that rapidly reacts with metallic iron to form iron sulfides (see Equation 9.2).

$$SO_4^{2-} + CH_3COO^- => 2HCO_3^- + HS^- \quad \Delta G^0 = -47.7 \text{ kJ}\left(\text{mol } SO_4^{2-}\right)^{-1} \tag{9.1}$$

$$Fe^0 + H_2S \rightarrow H_2 + FeS \quad \Delta G^0 = -72.5 \text{ kJ}\left(\text{mol } Fe^0\right)^{-1} \tag{9.2}$$

(ΔG^0: Gibbs Free energy at standard conditions (1 atm, 25°C = 298 Kelvin, 1 mol/L of all reactants))

Microorganisms can settle on surfaces and, when growing, form a biofilm composed of cells from different types of microorganisms tightly connected and stabilized by extracellular polymeric substances (EPS). A thick biofilm with the presence of corrosion products protects the microorganisms against chemical mitigating strategies.

Control of microbial growth is currently handled by use of antimicrobial chemicals (biocides) as well as physical removal of internal debris (pigging) for some environments. Aldehyde-based biocides such as formaldehyde and glutaraldehyde are favored in the oil industry (NACE-31205-2006). Aldehydes are broad-spectrum biocides that kill the organisms by cross-linkage of proteins in the bacterial cell membrane. Aldehydes also react with hydrogen sulfide abundant in most production systems and may function as a sulfide scavenger rather than an antimicrobial chemical, if for example, high amounts of sulfide are present.

The lack of compatibility for chemical cocktails used in the oil and gas industry such as scale inhibitors, oxygen scavengers, biocides, hydrogen sulfide scavengers, and EOR chemicals can potentially downregulate the efficacy of these chemicals (NACE-31205-2006). This lack of compatibility is exemplified by sulfites (e.g., sodium bisulfite, ammonium bisulfite, or hydrosulfite), the most commonly used oxygen scavengers for injection water treatment, which react with a variety of biocides (NACE-31205-2006). Therefore, evaluation of a biocide efficacy and compatibility with on-site chemicals and conditions (as e.g., NACE-TM0194-2014) is highly cost-efficient when compared to daily savings of correct dosing (Larsen et al., 2015).

In this study, the importance of choosing appropriate experimental conditions for a given biocide was investigated by testing three biocides with respect to their efficacy towards inhibiting biofilm development in the presence of different oxygen scavengers. Additionally, the impact of two different oxygen scavengers on a biocide was analyzed as an example, as well as the effect of hydrogen sulfide on the efficacy of an aldehyde-based biocide.

9.2 METHODS

9.2.1 Experimental Overview

Three chemicals with known biocidal effect (biocides) were evaluated with different oxygen scavengers commonly used in media for growing sulfate-reducing bacteria (SRB). Efficacy of the biocides regarding inhibition of biofilm formation was evaluated in a 96-well plate-based setup enabling simultaneous evaluation of a high number of different biocide concentrations. The chemical stability of the biocides in the growth media was evaluated with one biocide (Biocide B, bisoxazolidine, formaldehyde donor) by analytical determination. Additionally, the impact hydrogen sulfide has on the biocidal efficacy of biofilm inhibition was studied for an aldehyde-based biocide (Biocide A).

9.2.2 Biocides

Three selected biocides were used for evaluating inhibition of biofilm formation: Biocide A (glutaraldehyde), Biocide B (bisoxazolidine; 3,3′-methylenebis[5-methyloxazolidine]) and Biocide C (a mixture of 1,2-benzisothiazol-3(2H)-one and N-(3-aminopropyl)-N-dodecylpropane-1,3-diamine). Compatibility with growth media was evaluated as an example for Biocide B whereas impact of sulfide was studied with Biocide A.

9.2.3 Growth Media

Two modified low iron Postgate growth media were used for this study: one growth medium contained sodium-dithionite as the only oxygen scavenger (Dith-medium), and the other growth medium contained ascorbic acid and sodium thioglycolate as oxygen scavengers (As+Th-medium). Both media were based on lactate as the carbon source and were phosphate buffered with a pH of 6.94 and 7.16, respectively. The redox indicator dye, resazurin, was added for a visual indication of the redox potential.

Dithionite, also known as sodium hydrosulfite ($S_2O_4^{2-}$), is an oxygen scavenger ($E^0 = -290$ mV, $2SO_3^{2-}/S_2O_4^{2-}$, E^0: Standard electron potential of the oxidation/reduction reaction) that slowly decomposes in water to form bisulfite (HSO^{3-}), which also serves as an excellent oxygen scavenger, both of which are often used in the oil and gas industry (Spencer 1967; Kelland 2014). Ascorbic acid, also known as synthetic vitamin C ($C_6H_8O_6$), is an antioxidant ($E^0 = +58$ mV, dehydroascorbate/ascorbate) used in the food industry and various other industries. Thioglycolate, also known as TGA ($C_2H_3NaO_2S$), is a well-known oxygen scavenger (E^0 = −140 mV, disulfide/thioglycollate) in microbial media for growth of anaerobic organisms.

9.2.4 Pure Culture

A pure culture of *Desulfovibrio alaskensis* (DSM16109) was selected as a model organism, due to its capability to form biofilms and for its relevance with respect to sulfide production (souring) and effect on MIC in the oil and gas industry (Feio et al., 2004). The strict anaerobe *D. alaskensis* was originally isolated from a soured oil well in Alaska, USA. Handling of *D. alaskensis* was conducted under strictly anoxic conditions and the organism was incubated at 37°C. *D. alaskensis* was cultivated by incubating for three days and was transferred three successive times prior to biocide evaluation test, ensuring the organism being in exponential growth state and ensuring similar stock cultures before exposure to biocide treatment.

9.2.5 Chemical Stability of Biocide B in Growth Media

The chemical stability of Biocide B (bisoxazolidine) in Dith- and As+Th-medium was determined indirectly via formaldehyde and steam distillation by adding approximately 200 ppm Biocide B to the respective medium under anoxic conditions. Measurement of Biocide B (bisoxazolidine) concentration was performed directly after addition of the biocide (two replicates) and three days after incubation at 36°C under anoxic conditions. Media without biocide served as negative controls. During steam distillation, all formaldehyde molecules bound to Biocide B were released. Subsequently, acetylacetone was added in presence of ammonium acetate, which reacts with formaldehyde to 3,5-diacetyl-1,4-dihydrolutidine. The absorption of this compound was measured at 412 nm with a UV/VIS photometer (Kakac and Vejdelek, 1974). The concentration of formaldehyde was determined by a formaldehyde calibration curve, which then was used to calculate the amount of Biocide B (bisoxazolidine) in the samples.

9.2.6 Biocide Efficacy Screening

A *Multiwell Plate System* enables evaluation of any selected water-soluble biocide for the biocidal activity against either biofilm formation or sessile microorganisms of any biofilm forming microbial culture. The method has been proven for fast screening and for ranking many biocides and concentrations as well as for testing any synergistic (or antagonistic) effect of combinations of chemical treatments (Peeters et al., 2008).

Sterile growth medium was inoculated with actively growing *D. alaskensis* culture to a final inoculum concentration of 1–10% (specified in results). Medium with cells was transferred into 96-well plates and exposed to a series of different biocide treatments (see Table 9.1). The 96-well plates with culture and biocides were incubated for 72 hours at 37°C. Anoxic conditions were maintained throughout the experiment. After 72 hours, medium with suspended and loosely attached cells was

TABLE 9.1
Treatment Overview Performed with Dith-Medium and/or As+Th-Medium and 1%, 5% and/or 10% *D. alaskensis* Active Culture as Inoculum

Treatment	Concentration (ppm)	Abbreviation	Concentration of Inoculum in Dith-Medium in %	Concentration of Inoculum in As+Th-Medium in %
Control	0	Control	10%, 5%	1%
Biocide A	10	A10	10%	1%
Biocide A	50	A50	10%	1%
Biocide A	100	A100	10%	1%
Biocide A	150	A150	10%	1%
Biocide A	200	A200	10%	1%
Biocide A	1000	A1000	10%, 5%	1%
Biocide B	10	B10	10%	1%
Biocide B	50	B50	10%	1%
Biocide B	100	B100	10%	1%
Biocide B	150	B150	10%	1%
Biocide B	200	B200	10%	1%
Biocide C	50	C50	10%	1%
Biocide C	100	C100	10%	1%

removed from each well, leaving only the surface-attached biofilm in the wells. Biofilm attached to the well surface was stained using crystal violet (CV) (Peeters et al., 2008), and subsequently washed three times with distilled water to remove surplus CV, leaving only CV bound to biofilm. The remaining CV was dissolved by adding absolute ethanol and quantified spectrophotometrically (570 nm). Absorbance average readings of wells exposed to sterile growth medium with individual biocide concentrations (As+Th-medium experiments), or wells exposed to sterile growth medium without biocides (Dith-medium experiments) were used as biocide-specific background and subtracted from equivalent readings.

Different concentrations of the three selected biocides were tested for the inhibitory effect on biofilm formation of *D. alaskensis*. An overview of the evaluated concentrations and conditions are listed in Table 9.1.

9.2.7 Statistics for Comparison of Inhibitory Biocide Concentrations

Statistics were applied to identify which biocides and biocide concentration caused a significant alteration of the amount of biofilm compared to untreated controls. To ensure comparability, data were treated and transformed by the following procedures: Biofilm density was calculated by subtracting corresponding average background reading, and complete inhibition of biofilm formation was defined as average treatment readings being equal to the corresponding average background reading (i.e., density = 0). For normalizing of residuals, density measurements obtained from all replicates (n = 128 for control, n = 16 for all individual treatments; n: number of replicates) were boxcox-transformed (see specifications in figures) before any statistical analysis. For each analysis, the boxcox-transformed density was then modeled as a function of treatment, i.e., biocide/dose combination, using a generalized least squares model, in order to account for nonconstant variances within treatments. Least squares means and 95% confidence bounds for each treatment were calculated. The significance of treatment differences was assessed by means of post-hoc pairwise t-tests using the Sidak method for multiple test correction. Finally, results were back-transformed using the inverse boxcox and divided by the control least squares estimate to obtain estimates of relative biofilm of control. All statistical analyses were carried out using the R-environment (R-Core Team, 2016).

9.3 RESULTS AND DISCUSSION

9.3.1 Biofilm Inhibition Studies in Dith-Medium

When grown in 96-well plates under anoxic conditions, *D. alaskensis* formed a dense biofilm after three days. Wells containing Dith-medium and different biocide concentrations (see Table 9.1) were inoculated with 10% actively growing *D. alaskensis* culture. Under these conditions, no biofilm inhibition was observed at any of the biocide concentrations applied (see Figure 9.1, ≤200 ppm Biocide A, ≤200 ppm Biocide B, or ≤100 ppm Biocide C). Additionally, a signal increase in the crystal violet staining was observed for Biocide B (bisoxazolidine) at higher concentrations (100–200 ppm), which seemed to be due to precipitations.

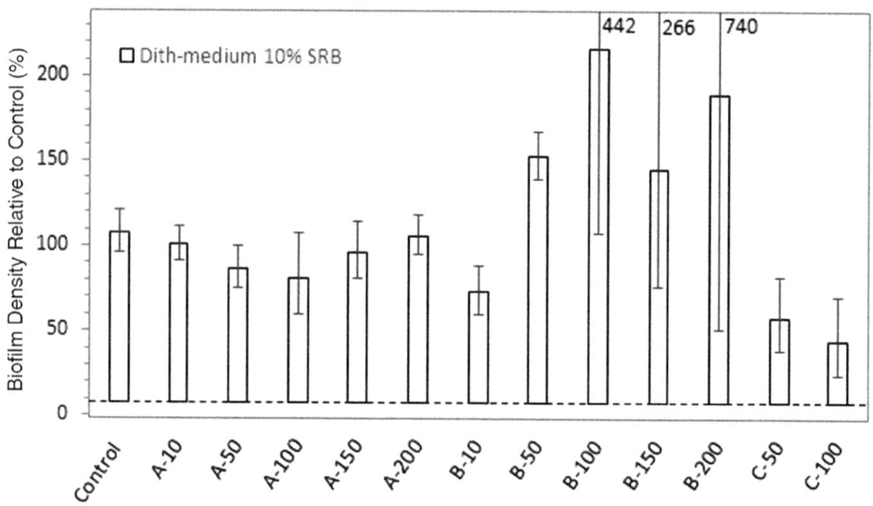

FIGURE 9.1 Biofilm density of *D. alaskensis* relative to a control with no treatment, after treatment with different biocides (A, B, or C) and different biocide concentrations (ppm). Biofilm growth was determined in Dith-medium and 10% SRB inoculum. $n = 16$ used for least squares means and standard bars (95% confidence). Boxcox-transformation with $\lambda = \{-0.4\}$. Data labels at B-100, B-150, and B-200 are upper standard bars + least squares mean.

Possible reasons for these results were chemical reactions between individual biocides and growth media components (e.g., dithionite) as well as metabolites (i.e., sulfide produced by SRB), which could lead to deactivation of the biocides or precipitations. These conclusions were investigated in the following experiments.

9.3.2 Biocide Compatibility with Growth Media

Biocide B (bisoxazolidine) was not only ineffective in inhibiting *D. alaskensis* biofilm formation but also showed precipitations in Dith-medium, which seemed to interfere with the biofilm density evaluation. Therefore, Biocide B was used as an example to determine chemical compatibility with the Dith-medium. Furthermore, As+Th-medium was used to evaluate chemical compatibility with a medium without dithionite to determine whether the oxygen scavenger interfered with the biocide.

Biocide B (bisoxazolidine) exposed to growth medium with either dithionite or ascorbate and thioglycolate was analyzed analytically and by visual inspection for precipitation. When dithionite was present as an oxygen scavenger in the growth medium (Dith-medium), the concentration of Biocide B was equal to background measurements already five minutes after addition (see Table 9.2). In contrast, no decrease of Biocide B was observed in growth medium with ascorbate and thioglycolate as oxygen scavengers (As+Th-medium) (see Table 9.2). These results showed clearly that Biocide B was inactivated in Dith-medium whereas it was stable for several days in As+Th-medium.

It can be assumed that Biocide A (aldehyde) will also be unstable in Dith-medium because it is inactivated by bisulfite which is a decomposition product of dithionite in aqueous solution (Jordan et al., 1996; Spencer, 1967).

Interestingly, the redox indicator dye, resazurin, shifted from colorless to pink, hence from a redox potential of ≤–110 mV to >–51 mV, when high biocide concentrations (200 ppm in Dith-medium and 1000 ppm in As+Th-medium) were added to the medium (see Figure 9.2). This clearly indicated a reaction within the growth media after the addition of Biocide B. Precipitations and markedly increased redox potential already occurred at 200 ppm Biocide B (bisoxazolidine) in Dith-medium compared to 1000 ppm in As+Th-medium.

TABLE 9.2
Analytical Determination of Residual Biocide B (Bisoxazolidine) after Incubation of Approximately 200 ppm Biocide B in Dith-Medium or As+Th-Medium for 5 Minutes or 3 Days at 36°C, Control Media without Biocide Were below Detection Limit (BD, 11 ppm)

Time in Media	5 Minutes	3 Days
Dith-medium	13 ppm; BD	13 ppm
As+Th-medium	311 ppm; 245 ppm	181 ppm

BD: Below Detection Limit

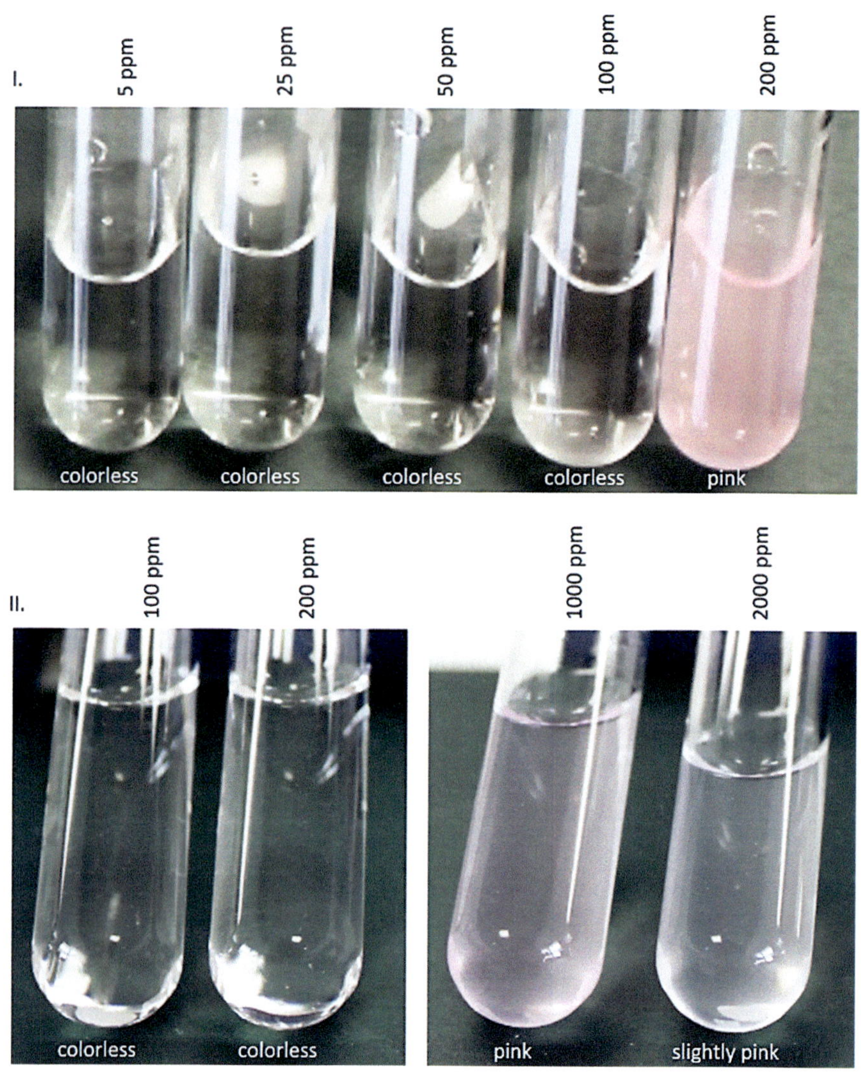

FIGURE 9.2 Chemical stability of Biocide B (bisoxazolidine) with oxygen scavengers in growth media. Visual precipitation in a concentration series of Biocide B (I. 5–200 ppm or II. 100–2000 ppm) 24 hours after addition to Dith-medium (I) or As+Th-medium (II), respectively. Pink coloring reflects redox potential >–51 mV.

This showed that the redox potential was more stable in As+Th-medium than in Dith-medium after addition of Biocide B and that less precipitations occurred in As+Th-medium.

Taking these results into account, As+Th-medium was clearly more suitable to test the efficacy of Biocide B (bisoxazolidine) on *D. alaskensis* biofilm inhibition.

9.3.3 Impact of Sulfide on Biocide Efficacy

Similar to most SRPs, hydrogen sulfide is a primary metabolite end-product of *D. alaskensis*. Consequently, hydrogen sulfide is abundant in an actively growing population. Within this study the impact of hydrogen sulfide on the biocide efficacy of Biocide A was studied by reducing the initially transferred hydrogen sulfide amount. When inoculum of *D. alaskensis*, and thereby the transferred amount of hydrogen sulfide, was decreased by 50% (from 10% to 5% inoculum), a significant increase of the biocidal effect of Biocide A (aldehyde) was observed (see Figure 9.3). This effect can be explained by the hydrogen sulfide scavenging abilities of Biocide A (aldehyde), which leads to a decrease in the biocidal capabilities (Amosa et al., 2010).

A sulfide scavenging capability of Biocide B (bisoxazolidine) has also been described previously (Patent: EP 0 882 112 B1). Hence, it can be presumed that the

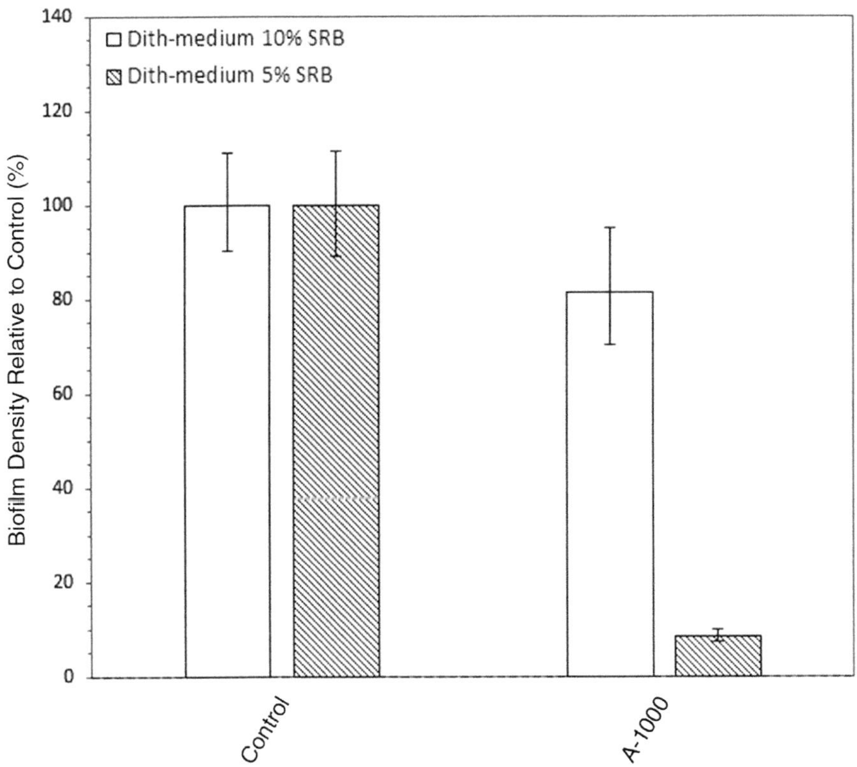

FIGURE 9.3 Biofilm density of *D. alaskensis* after treatment with 1000 ppm Biocide A (aldehyde) relative to control with no biocide treatment. Increasing H_2S concentration with increasing inoculum during transfer (5 and 10% SRB culture, respectively). $n = 14–16$, used for least squares means and standard bars (95% confidence). Boxcox-transformation with $\lambda = \{-0.1, 0.5\}$ for Dith-medium 10% and 5%, respectively.

efficacy of Biocide B (bisoxazolidine) would also be affected by hydrogen sulfide in a similar way as demonstrated for Biocide A (aldehyde).

Overall, minimizing the initial sulfide concentration should be considered especially when sulfide scavenging biocides are compared to nonsulfide scavenging biocides. Although it has to be considered that the size of the inoculum can have an impact on the inhibitory concentration (Lambert 2000). When using sulfide scavenging biocides in production facilities with high hydrogen sulfide present, the use of higher biocide concentrations has to be considered to ensure microbial control.

9.3.4 Ranking of Biocide Efficacy against Biofilm Formation with Optimized Conditions

Biocidal efficacy against *D. alaskensis* biofilm formation was evaluated in 96-well plates under anoxic conditions. Instead of Dith-medium As+Th-medium was used and the initial sulfide concentration was minimized by reducing the inoculum from 10% to 1% *D. alaskensis* culture. Under these optimized conditions significant ($p < 0.001$; p: the probability of obtaining a value at least as extreme, if the null hypothesis is true (Grafen and Hails, 2002)) inhibition of biofilm formation could be detected for all three tested biocides (see Figure 9.4).

Specifically, Biocide A (aldehyde) and B (bisoxazolidine) exhibited complete biofilm inhibition at concentrations ≥50 ppm, whereas Biocide C (isothiazolinone and amine) exhibited complete biofilm inhibition at concentrations ≥100 ppm. At concentrations of 10 ppm, both Biocide A and B exhibited incomplete biofilm inhibition,

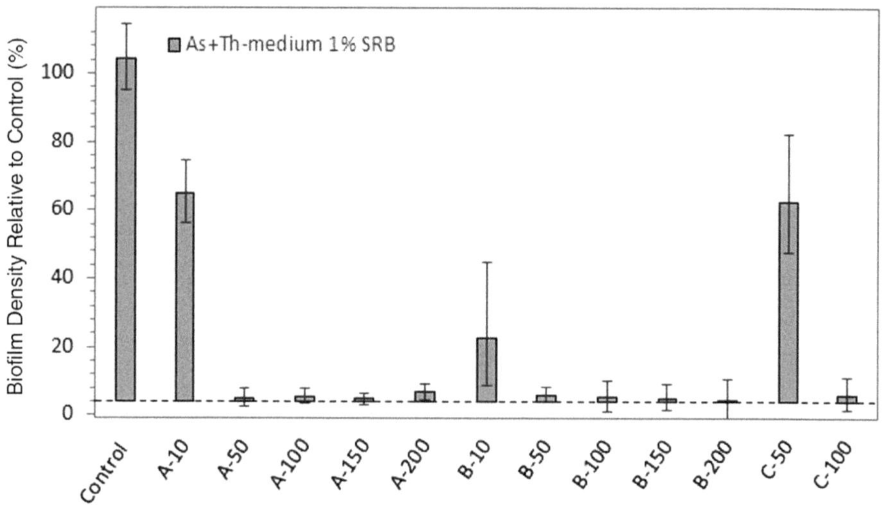

FIGURE 9.4 Biofilm density of *D. alaskensis* relative to a control with no treatment, after treatment with different biocides (A, B, or C) and different biocide concentrations (ppm). Biofilm growth was determined in As+Th-medium and 1% SRB inoculum. $n = 16$ used for least squares means and standard bars (95% confidence). Boxcox-transformation with $\lambda = \{-0.7\}$.

with Biocide B displaying a significantly better ($p < 0.001$) biofilm inhibition as compared to Biocide A (see Figure 9.4). Under these optimized conditions the performances of the biocides were distinguishable.

9.4 CONCLUSION AND PERSPECTIVES

This study demonstrated how the selection of various types of oxygen scavengers and inoculum levels during biocide evaluation can result in two very different findings on the efficacy of biocides. Data obtained in this study indicates that carry over of sulfide from the stock culture as well as the choice of oxygen scavenger can impact biocidal efficacy significantly. Thus, reflection and application of adequate experimental conditions are of paramount importance when evaluating biocidal efficacy under laboratory conditions in anaerobic complex systems.

Based on the experiences from this study, the three tested biocides were only comparable if suitable experimental conditions were applied. Hence, valid data could be obtained resulting in an inhibitory effect of the three tested biocides on biofilm formation of *D. alaskensis* with Biocide B (bisoxazolidine) performing better than Biocide A (aldehyde) and Biocide C (isothiazolinone and amine).

The efficacy of a biocide is highly dependent on the conditions to which the biocide is applied. Thus, a chemical compatibility test with production chemicals should be conducted prior to the use of new biocides or chemicals—a procedure that agrees with the recommendation given in the NACE standard TM0194-2014. Specifically, it is recommended to perform compatibility and biocide efficacy evaluation when a new biocide is considered for use, when new chemicals are considered for use together with already applied biocides, and if the effect of the biocides is lowered in the system or consumption is increased over time. The business case for conducting these chemical compatibility tests is high cost efficiency compared to the daily savings of correct chemical dosing (Larsen et al., 2015).

REFERENCES

Amosa, M.K., Mohammed, I.A., and Yaro, S.A. 2010. Sulphide Scavengers in Oil and Gas Industry—A Review. *NAFTA*. 61:85–92.

Daniels, L., Belay, N., Rajagopal, B.S., and Weimer, P.J. 1987. Bacterial methanogenesis and growth from CO_2 with elemental iron as the sole source of electrons. *Science*. 237:509.

Feio, M.J., Zinkevich, V., Beech, I.B., et al. 2004. Desulfovibrio alaskensis sp. nov., a sulphate-reducing bacterium from a soured oil reservoir." *International Journal of Systematic and Evolutionary Microbiology*. 54:1747.

Grafen, A., and Hails, R. 2002. *Modern Statistics for the Life Science*. New York: Oxford University Press.

Jordan, S.L., Russo, M.R., Blessing, R.L., and Theis, A.B. 1996. Inactivation of Glutaraldehyde by Reaction with Sodium Bisulfite. *J. Tox. Envir. Health*. 47:299–309.

Kakac, B., and Vejdelek, Z.J., 1974. Handbuch der photometrischen Analyse organischer Verbindungen. Verlag Chemie. *Band* 1, 257.

Kelland, M.A. 2014. *Production Chemicals for the Oil and Gas Industry*. Boca Raton, FL: CRC Press.

Lambert, R.J.W. 2000. Susceptibility testing: inoculum size dependency of inhibition using the Colworth MIC technique. *Journal of Applied Microbiology*. 89:275–279.

Larsen, J., Rasmussen, K., Pedersen, H., Sørensen, K., Lundgaard, T., and Skovhus, T.L. 2010. Consortia of mic bacteria and archaea causing pitting corrosion in top side oil production facilities. NACE_C 2012-10252, NACE Corrosion 2010, San Antonio (TX), NACE International.

Larsen, J., Tang, L., Sørensen, K.B., and Thomsen, U.S. 2015. Co-effect of production chemicals for biocide optimization in different types of injection water. NACE_C 2015-5457, NACE Corrosion 2015, Dallas (TX), NACE International.

Lundgaard, T., Sorensen, K., Holmkvist, L., et al. 2010. Investigation of Significant Microorganisms Responsible for Souring in a Hot Oil Reservoir in the North Sea. TEKNA Oilfield Chemistry Symposium, Geilo, Norway.

Magot, M., Ravot, G, Campaignolle, X., et al. 1997. Dethiosulfovibrio peptidovorans gen. nov., sp. nov., a new anaerobic slightly halophilic, thiosulfatereducing bacterium from corroding offshore oil wells. *Int. J. Syst. Bacteriol.* 47:818–824.

NACE 31205-2006. Selection, Application, and Evaluation of Biocides in the Oil and Gas Industry. NACE Item No. 24227.

NACE TM0194-2014. Standard test method. Field monitoring of bacterial growth in oil and gas systems. NACE Item No. 21224.

Patent: EP 0 882 112 B1. Baker-Hughes Incorporated, 1998. Bisoxazolidine hydrogen sulfide scavenger.

Peeters, E., Nelis, H.J., and Coenye, T. 2008. Comparison of multiple methods for quantification of microbial biofilms grown in microtiter plates. *Journal of Microbiological Methods*. 72:157.

R-Core Team. 2016. R: A language and environment for statistical computing. R Foundation for Statistical Computing, Vienna, Austria. https://www.R-project.org/.

Sørensen, K.B., Thomasen, U.S., Juhler, S., and Larsen, L. 2012. Cost Efficient MIC Management System based on Molecular Microbiological Methods. NACE_C 2012-1111.

Spencer, M.S. 1967. Chemistry of dithionite; Part 1 Kinetics of decompositions in aqueous bisulfate solution. *Trans. Faraday Soc.* 63, 2510–2515.

Vik, E.A., Janbu, A.O., Garshol, F., et al. 2007. Nitrate-based souring mitigation of produced water—side effects and challenges from the Draugen produced water re-injection pilot. Paper SPE 106178. SPE International Symposium on Oilfield Chemistry, Houston, TX.

Part IV

Hydrocarbon Biodegradation

10 Insights into the Mechanisms and Microorganisms Catalyzing Methanogenic Hydrocarbon Biodegradation in Petroleum Reservoirs

Courtney R.A. Toth and Lisa M. Gieg

CONTENTS

10.1 INTRODUCTION

Hydrocarbon-bearing reservoirs are carbon-rich deep subsurface environments known to harbour diverse microbial life. Several lines of geological and physiological evidence collected over the past 30 years, coupled with the more recent adoption of microbiological evidence, have proven that saturated and aromatic hydrocarbons present within crude oils can drive carbon transformations to methane (Jiménez et al., 2016). Head et al. (2003) and Larter et al. (2006) have estimated that the microbial conversion of crude oil hydrocarbons to methane at oil water contacts across all biodegrading reservoirs approximates 100,000,000 kg/year. Methanogenic hydrocarbon biodegradation in reservoirs has therefore not only been significant on a geological timescale, but also occurs in our modern age. As the leading model to explain the widespread occurrence of biodegraded oils and gas formation in oxidant-free reservoirs that are <80°C (Head et al., 2003; 2014;

Jones et al., 2008), extensive efforts have been made to understand the pathways, environmental factors, and microorganisms contributing to *in situ* methanogenic hydrocarbon biodegradation. There is also considerable interest in exploiting hydrocarbon transformation to natural gas for enhanced energy recovery purposes (e.g., of trapped oil from mature oilfields; Gieg et al., 2008; Head et al., 2014), as well as for bioremediation applications (Gieg et al., 1999; Gieg et al., 2009; Rabus et al., 2016). Jiménez et al. (2016) recently summarized the extent of our current understanding of hydrocarbon biodegradation by methanogenic microbial communities; a brief overview of anaerobic hydrocarbon metabolic pathways and common methanogenic hydrocarbon-degrading taxa is presented here as an introduction to the case study presented in this chapter.

This case study, based on the work reported by Toth and Gieg (2018), was conducted to gain a better understanding of the mechanisms and microorganisms catalyzing the methanogenic degradation of hydrocarbons in crude oil as they may occur during the formation of heavy oil in petroleum reservoirs. Research consisted of a multi-pronged approach to assess the chemical, functional, and microbial community responses of a produced water consortium (from a heavy oil-producing field in Alberta, Canada) to incubation under methanogenic conditions on different crude oils, building on preliminary evidence for methanogenic crude oil degradation occurring in this oilfield (Agrawal et al., 2012; Berdugo-Clavijo and Gieg, 2014). Based on the widespread evidence that fumarate addition catalyzes the activation of model hydrocarbon substrates (e.g., saturated and alkyl substituted aromatic hydrocarbons) under all anoxic conditions, we hypothesized that fumarate addition is a prevalent process used by diverse methanogenic reservoir-associated microorganisms when susceptible hydrocarbons are present in crude oil.

10.2 METHANOGENIC HYDROCARBON METABOLISM

The process of methanogenic crude oil biodegradation prevails in hydrocarbon-rich environments depleted of electron acceptors (e.g., NO_3^-, SO_4^{2-}, Fe^{3+}), a condition often encountered in oil reservoirs as well as in coal deposits, oil-contaminated groundwater aquifers, and deep sediments (Foght, 2008; Musat, 2015; Rabus et al., 2016). Many pure compounds including monoaromatic hydrocarbons, alkanes (linear and cyclic), alkenes, and polycyclic aromatic hydrocarbons have been shown to be susceptible to methanogenic degradation based on oil decay studies and enrichment culture experiments (e.g., Zengler et al., 1999; Anderson and Lovley, 2000; Beller and Edwards, 2000; Jones et al., 2008; Berdugo-Clavijo et al., 2012; and references therein); hydrocarbons can also be consumed in crude oil mixtures (see Table 10.1). Though the mechanisms underpinning their transformation to methane remain largely unknown, the detection of (signature) anaerobic hydrocarbon metabolites and corresponding biodegradative genes can offer clues as to what processes may be involved (Gieg and Suflita, 2002; Gieg and Toth, 2017). A common strategy is to search for metabolites diagnostic of anaerobic hydrocarbon activation mechanisms, which include (1) fumarate addition, (2) oxygen-independent hydroxylation, (3) carboxylation, and (4) methylation (Foght, 2008; Gieg and Toth, 2017).

TABLE 10.1
Summary of Studies Reporting Fumarate Addition Genes and/or Fumarate Addition Products from Crude Oil or Hydrocarbon Mixtures under Methanogenic Conditions

Inoculum Source	Experimental Details	Fumarate Addition Genes Detected	Fumarate Addition Metabolites Detected	Number of Time Points Sampled	References
Produced water from the Medicine Hat Glauconitic C oilfield (30°C), Canada	Fluids immediately processed	-	C_6-C_8 alkylsuccinates (with and without saturation)	Single	Agrawal et al., 2012
	Fluids enriched and incubated with light oil (°API = 37)	*assA*, *bssA*	Up to C_6 alkylsuccinates	Single (day 302)	Berdugo-Clavijo and Gieg, 2014
	Fluids incubated with light (°API = 32) or heavy oil (°API = 16); no nutrients added	*assA*, *bssA*	C_1-C_9 alkylsuccinates (with and without saturation) from both oils	7 over 525 days	Toth and Gieg, 2018
Mature fine tailings (MFT) from the Mildred Lake Settling Basin, Canada	MFT enriched and maintained with C_6-C_{10} alkanes (linear and cyclic)	*assA*, *bssA*, *nmsA*	n.d.	1 or 2 over 714 days	Tan et al., 2013; Tan et al., 2015b
	MFT enriched and maintained with naphtha	*assA*, *bssA*, *nmsA*	n.d.	Single (day 103)	Tan et al., 2015a
	MFT enriched and maintained with C_7-C_8 *iso*-alkanes	*assA*	C_7 and C_8 *iso*-alkylsuccinates	Single (day 173)	Abu Laban et al., 2015b
River Tyne sediments, England	Sediment slurry enriched and incubated with light oil (°API ~ 38)	*assA*	C_{19}-C_{25} alkylsuccinates at T_0, no change in abundance during incubation period	5 over 686 days	Aitken et al., 2013

(Continued)

TABLE 10.1 (Continued)
Summary of Studies Reporting Fumarate Addition Genes and/or Fumarate Addition Products from Crude Oil or Hydrocarbon Mixtures under Methanogenic Conditions

Inoculum Source	Experimental Details	Fumarate Addition Genes Detected	Fumarate Addition Metabolites Detected	Number of Time Points Sampled	References
Produced water from the Jiangsu (80–90°C), Xinjiang (32°C) and Huabei oilfields (37–45°C), China	Fluids immediately processed	*assA*, *bssA*, *nmsA*	C_1-C_8 alkylsuccinates, benzylsuccinate	Single	Bian et al., 2015
Produced water from the Alaska North Slope oilfield (~68°C), USA	Fluids immediately processed	-	C_1-C_4 alkylsuccinates	Single	Duncan et al., 2009
	Fluids immediately processed	-	C_1-C_3 alkylsuccinates, benzylsuccinate	Single	Gieg et al., 2010
Produced water from the Huabei oilfield (37°C), China	Fluids enriched and incubated with C_{15}-C_{20} *n*-alkanes	*assA*	-	Single (day 356)	Li et al., 2012
Produced water from the Shengli oilfield (55°C), China	Fluids enriched and incubated with C_{15}-C_{20} *n*-alkanes	*assA*	n.d.	Single (day 528)	Mbadinga et al., 2012
	Fluids enriched and incubated with C_{15}-C_{20} *n*-alkanes	*assA*	n.d.	6 over 749 days	Zhou et al., 2012
	Fluids enriched and incubated with C_{15}-C_{20} *n*-alkanes	*assA*	n.d.	Single (day 1323)	Liang et al., 2015; Wang et al., 2012
Production water from the San Juan Basin, USA	Fluids immediately processed	*assA* (*bssA* not evaluated)	C_1-C_{16} alkylsuccinates, (alkyl)benzylsuccinates	Single	Wawrik et al., 2012

Fields are left blank (-) if assay was not evaluated; n.d. = not detected.

Of these, fumarate addition (also referred to as "addition to fumarate") is the most widely reported mechanism used by diverse anaerobic taxa to activate linear alkanes (e.g., Zengler et al., 1999; Anderson and Lovley, 2000; Gieg and Suflita, 2002; Gieg et al., 2008; 2010; Duncan et al., 2009; Agrawal et al., 2012; Mbadinga et al., 2012; Zhou et al., 2012; Wawrik et al., 2012; 2016; Cheng et al., 2013; Berdugo-Clavijo and Gieg, 2014; Bian et al., 2015; Gründger et al., 2015; Liang et al., 2015; Tan et al., 2013; 2015a; 2015b; Abu Laban et al., 2015b; Oberding and Gieg, 2017; Toth and Gieg, 2018) and alkyl-substituted aromatic hydrocarbons (e.g., Beller and Edwards, 2000; Gieg and Suflita, 2002; Washer and Edwards, 2007; Gieg et al., 2010; Fowler et al., 2012; 2014; Wawrik et al., 2012; Berdugo-Clavijo and Gieg, 2014; Abu Laban et al., 2015a; Bian et al., 2015; Tan et al., 2015a; Toth and Gieg, 2018). For a methyl-substituted aromatic hydrocarbon such as toluene, the methyl group is added via a glycyl-radical-driven C–C addition reaction to the double bond of fumarate, forming benzylsuccinate (Foght, 2008) (see Figure 10.1a). In the case of *n*-alkanes, fumarate addition typically occurs at the subterminal C_2 position yielding the corresponding (1-methylalkyl)succinate, more commonly known as an alkylsuccinate (Musat 2015) (see Figure 10.1b). Due to their transient nature, benzylsuccinates and alkylsuccinates are only sometimes detected in oilfield fluids (Duncan et al., 2009; Gieg et al., 2010; Agrawal and Gieg, 2012; Bian et al., 2015), whereas genes encoding for the catalytic subunits of fumarate addition enzymes (e.g., alkylsuccinate synthase, *assA/masD*; benzylsuccinate synthase, *bssA*; naphthylsuccinate synthase, *nmsA*) are more commonly found (see Table 10.1).

Garnering evidence for aromatic/alkane addition to fumarate in crude oil mixtures has been challenging under methanogenic conditions, even though this mechanism has clearly been demonstrated under other electron-accepting conditions (Foght, 2008; Musat, 2015; Rabus et al., 2016). Aitken et al. (2013) performed a two-year time course experiment comparing the extent of crude oil alkane degradation under sulfate-reducing and methanogenic conditions. Though the consumption of alkanes was confirmed in both sets of microcosms (corresponding with a decrease in sulfate and formation of methane, respectively), the authors only detected alkylsuccinate metabolites and an increase in *assA/masD* gene abundance in the sulfate-reducing incubations. From these results, Aitken and colleagues (2013) postulated that an alternate pathway may be responsible for alkane activation under methanogenic conditions. If the enzymes downstream of alkylsuccinate synthase are more efficient in the methanogenic consortia than under sulfate-reducing conditions, then

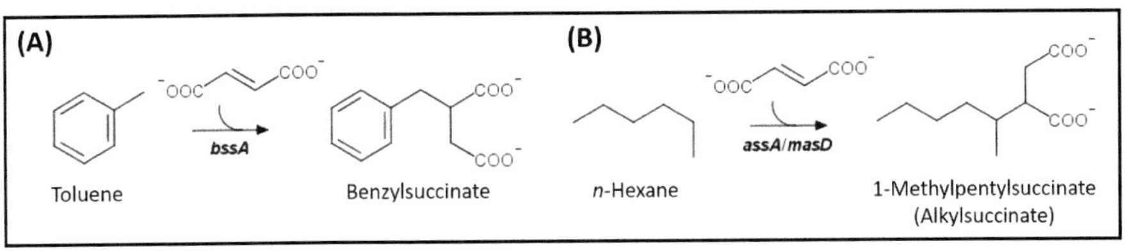

FIGURE 10.1 Fumarate addition reactions catalyzed during the initial anaerobic degradation of (A) aromatic hydrocarbons and (B) *n*-alkanes, using toluene and *n*-hexane as model hydrocarbons, respectively.

the alkylsuccinates may not accumulate/be detectable. It is also likely that other (undiscovered) mechanisms of hydrocarbon activation exist (Callaghan, 2013). Thus, while the potential for fumarate addition appears to be near-ubiquitous among reservoir environments, its overall importance to crude oil degradation under methanogenic conditions remains unclear. To address this, multiple lines of evidence are required to provide the strongest case supporting the metabolic processes that may be occurring.

10.3 OILFIELD-ASSOCIATED BACTERIA CATALYZING HYDROCARBON ACTIVATION

Microbial diversity analyses have indicated that similar types of prokaryotes are present in methanogenic hydrocarbon-degrading communities, including members of the Firmicutes, Proteobacteria, Bacteroidetes, and Spirochaetes (Foght, 2008; Musat, 2015; Rabus et al., 2016; Li et al., 2017). Most of these organisms have largely speculative roles within these microbial communities, but some are known to carry hydrocarbon biodegradative gene sequences within their genetic material and/or have been definitively shown to participate in hydrocarbon degradation. For example, Gray et al. (2011) and Gründger et al. (2015) each reported the enrichment of a *Smithella* sp. in parallel with significant alkane degradation and methane production in oil-amended enrichment cultures. Cheng et al. (2013) demonstrated that carbon from ^{13}C-labelled hexadecane was incorporated into *Smithella* during DNA-based stable isotope probing of a methanogenic consortium, while Wawrik et al. (2016) and Oberding and Gieg (2017) also demonstrated increased *assA* gene expression (assigned to *Smithella* spp.) in methanogenic paraffin-degrading enrichment cultures. Thus, *Smithella* is likely the chief alkane degrader in each of these communities and in fact *Smithella* is also commonly detected in reservoir fluid samples, oil sands tailings, and crude oil/alkane-degrading enrichment cultures (e.g., Agrawal et al., 2012; Siddique et al., 2012; Tan et al., 2013; 2015a; 2015b; Berdugo-Clavijo and Gieg, 2014; Bian et al., 2015; Toth and Gieg, 2018).

Phylotypes belonging to the Firmicutes, specifically within the *Clostridiales* order, have been shown to participate in the activation of aromatic hydrocarbons via fumarate addition (e.g., Fowler et al., 2012; 2014; Abu Laban et al., 2015a) or are thought to facilitate secondary degradation steps (e.g., *Clostridium* spp.; Imachi et al., 2006; Gieg et al., 2010). The genus *Desulfosporosinus* is perhaps the most widely reported putative aromatic hydrocarbon degrader in enrichment culture investigations (e.g., Berdugo-Clavijo et al., 2012; Siddique et al., 2012; Tan et al., 2013; Tan et al., 2015a; 2015b; Abu Laban 2015a; 2015b), but new aromatic hydrocarbon-degrading phylotypes have recently been brought to light. In a metagenomic survey of oilfield production fluids, Hu et al. (2016) detected sequences orthologous to fumarate addition enzyme subunits in more than a dozen genomes belonging mostly to *Desulfotomaculum*, a metabolically versatile organism. Some members of this genus (cluster I) are obligate syntrophs that grow in concert with methanogens (Imachi et al., 2006), and have been detected in or enriched from an increasing number of oilfield-related fluids (e.g., Gieg et al., 2008; Siddique et al., 2012; Berdugo-Clavijo and Gieg, 2014; Abu Laban et al., 2015b; Tan et al., 2015b; Gründger et al., 2015; Toth and Gieg, 2018). Thus, *Desulfotomaculum* may also serve as a widespread contributor to the methanogenic degradation of aromatic hydrocarbons in oil systems, but experimental verification is required.

10.4 CASE STUDY: TIME-COURSE ANALYSIS OF METHANOGENIC CRUDE OIL BIODEGRADATION

The anaerobic consortium used to study methanogenic crude oil biodegradation in this study was enriched from a mixture of five production water (PW) samples collected from the Medicine Hat Glauconitic C (MHGC) field in southern Alberta, Canada (Voordouw et al., 2009; Toth and Gieg, 2018). Though this oilfield has undergone more than 10 years of sulfide remediation via nitrate injection, anoxic "zones" free of exogenous electron acceptors (e.g., nitrate and sulfate) have been identified within the field and shown to harbour high abundances of methanogenic archaea (Agrawal et al., 2012). Further, a previous study also confirmed that consortia enriched from this field degraded crude oil hydrocarbons coupled to methane production (Berdugo-Clavijo and Gieg, 2014). In this case study, we thus sought to assess the prevalence of fumarate addition activity by methanogenic MHGC reservoir-associated microorganisms.

Oil-free produced water containing no nitrate or sulfate (verified using methods described by Voordouw et al., 2009) was amended with 20 mL of either a light oil (°API = 32) or heavy oil (MHGC oil, °API = 16) and incubated at MHGC reservoir temperature (30°C) under dark and static conditions for 17–20 months. No supplemental nutrients or reducing agents were added to microcosms to assess the extent of methanogenic activity under natural, low nutrient conditions. Due to limited available PW fluids collected during the sampling trip, only one live replicate was prepared per oil type so as to ensure that adequate sterile (autoclaved) replicates (one for each oil type) and an oil-free control could also be established, for a total of five microcosms. We monitored methanogenic activity during the incubation period by routinely monitoring methane production (gas chromatography, GC), hydrocarbon metabolite formation (GC equipped with a mass spectrometer, GC–MS), anaerobic hydrocarbon activation genes (*assA*, *bssA* by PCR/qPCR), and microbial community composition dynamics (16S rRNA gene sequencing); an endpoint crude oil loss measurement was also performed (GC–MS) to determine hydrocarbon loss.

Throughout the experiment, PW-containing microcosms incubated with light or heavy oil produced near identical amounts of CH_4 at uniform rates over time (0.14 µmol CH_4/day/g oil vs. 0.13 µmol CH_4/day/g oil, respectively) (see Figure 10.2). Less than a tenth of the amount of methane was formed in the microcosm headspace when the inoculum was incubated in the absence of crude oil, and no detectable CH_4 was formed at any point in either sterile control. Endpoint GC–MS crude oil analyses (methods described in Berdugo-Clavijo and Gieg, 2014) showed measurable losses of *n*-alkanes $\leq C_8$ (47–79%), cyclohexane (22–71%), and some alkyl-substituted aromatic hydrocarbons (toluene and xylenes) in oil-amended PW cultures after 17 months of incubation (as compared to sterile controls), suggesting

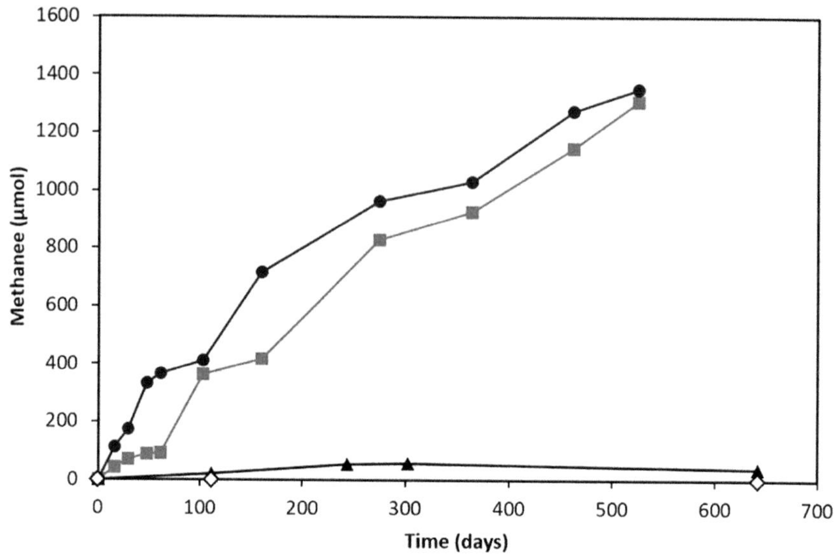

FIGURE 10.2 Methane production from produced water-containing incubations enriched on light (squares) and heavy (circles) oil relative to an oil-free control (triangles). No methane was produced from sterile control replicates (white diamonds).

that hydrocarbon biodegradation within the crude oil material was driving methane production, and that exogenous nutrients were not required to stimulate methanogenic activity.

To assess the metabolic mechanism(s) involved in hydrocarbon biodegradation, cultures were subsampled in duplicate (2 × 25 mL) after 0, 1, 2, 4, 8, 12, and 17 months for chemical and functional gene analyses. Following centrifugation, supernatants were acidified with 6 M HCl (pH < 2) and putative hydrocarbon metabolites were extracted (Berdugo-Clavijo and Gieg, 2014), while the resulting pellets were used for genomic DNA extraction (Toth and Gieg, 2018). Putative hydrocarbon metabolites from silylated PW organic extracts were positively matched to authentic standards and quantified by GC–MS (Gieg and Toth, 2018). DNA extracts were probed for the presence of anaerobic hydrocarbon activation genes using a series of established primer sets and thermocycling conditions (Toth and Gieg, 2018). Purified amplicons of the expected size were cloned, sequenced by Sanger sequencing, and queried against the NCBI non-redundant nucleotide database using BLASTn to identify homology to known sequences. The total abundance of recovered hydrocarbon biodegradative genes was estimated by qPCR (Aitken et al., 2013; Toth and Gieg, 2018).

Extracted components with mass spectral profiles diagnostic of C_1–C_9 alkylsuccinates (with and without saturation), aromatic acids (e.g., benzoate, toluate isomers, and 2-methylnaphthoate), and cyclohexane carboxylate were detected exclusively in the oil-amended PW cultures over time, coinciding with losses of their corresponding parent hydrocarbons (GC–MS). It should be noted here that the formation of benzylsuccinates could not be detected at any time point. We found that the average concentration of hydrocarbon metabolites fluctuated over time, with an initial accumulation during the first 2–4 months of incubation. Greater amounts of C_1–C_4 alkylsuccinates were detected in the PW culture incubated with light oil (0.07–0.44 μM) than with heavy oil (≤0.07 μM). Additionally, four peaks were detected in the light oil culture with MS fragment ions corresponding to propane and butane fumarate addition products (two each), suggesting that hydrocarbon activation was occurring at both the C_2 and terminal carbon position (Musat, 2015). At T_8 and T_{12}, the concentrations of most identifiable alkylsuccinates fell below detectable limits (<10 nM), and the cumulative concentration aromatic/other hydrocarbon metabolites decreased up to 50-fold (2.6–7.6 μM). This observation was assumed to be due to an increase in hydrocarbon metabolism activity (occurring sometime between two and eight months of incubation) leading to the consumption of transient hydrocarbon intermediates. Further, this effect was strongly supported by qPCR analyses (Toth and Gieg, 2018), where a $>10^5$ increase in fumarate addition gene copies (*bssA* and *assA*) was reported in DNA extracts for both oil-amended cultures by T_8. By the end of the incubation period (T_{17}), some unsaturated alkylsuccinates and aromatic acids were found to reaccumulate in each culture, but most other positively identified metabolites of interest remained below detectable limits (<10 nM).

From the extracted genomic DNA, a total of three unique *assA* and 14 *bssA* gene fragments were retrieved across both oil-degrading microcosms (NCBI accession numbers MG460804–MG460820); no biodegradative genes were detected in the T_0 inoculum or in control replicates. All three *assA* gene sequences shared 82–99% homology to uncultured prokaryotic clones of *assA* recovered from several Chinese oilfield production fluids (Zhou et al., 2012; Bian et al., 2015). Maximum likelihood trees of the *assA* fragments also revealed that each *assA* gene fragment clustered within a *Smithella* subclade with *assA* sequences retrieved from other alkane and/or crude oil-amended studies (see Figure 10.3), supporting the notion that *Smithella* is a chief alkane fermenter (Cheng et al., 2013; Tan et al., 2013; 2015a; 2015b; Gründger et al., 2015; Wawrik et al., 2016; Oberding and Gieg, 2017). Maximum likelihood trees were also generated for the recovered *bssA* gene fragments but were distributed within a more undefined clade containing several *assA* genes from unclassified *Clostridiales* (see Figure 10.3). The *bssA* clade was phylogenetically distinct from published fumarate addition genes belonging to cultured aromatic hydrocarbon degraders (<77% sequence identity), thus we assessed their taxonomic affiliations to previously characterized enrichment cultures and environmental strains. The *bssA* fragments shared 90–91% sequence identity with a presumed *Desulfotomaculum* sp. previously recovered from a methanogenic toluene-degrading enrichment culture (Bacterium bssA-1; Edwards and Grbić-Galić, 1994; Washer and Edwards, 2007), and 92–99% identity with several uncultured prokaryotic clones retrieved from a northern Alberta crude oil tailings pond (Mildred Lake Settling Basin) and corresponding hydrocarbon-degrading enrichment cultures (Siddique et al., 2012; Tan et al., 2013; 2015a).

Further evidence that *Desulfotomaculum* was a key aromatic hydrocarbon degrader in the oil-amended PW samples was shown using time course–dependent 16S rRNA gene sequencing. Both the light and heavy oil-amended PW cultures saw substantial shifts in microbial community composition during the 17-month incubation period; many of these shifts were similar across both cultures. The most apparent increase in

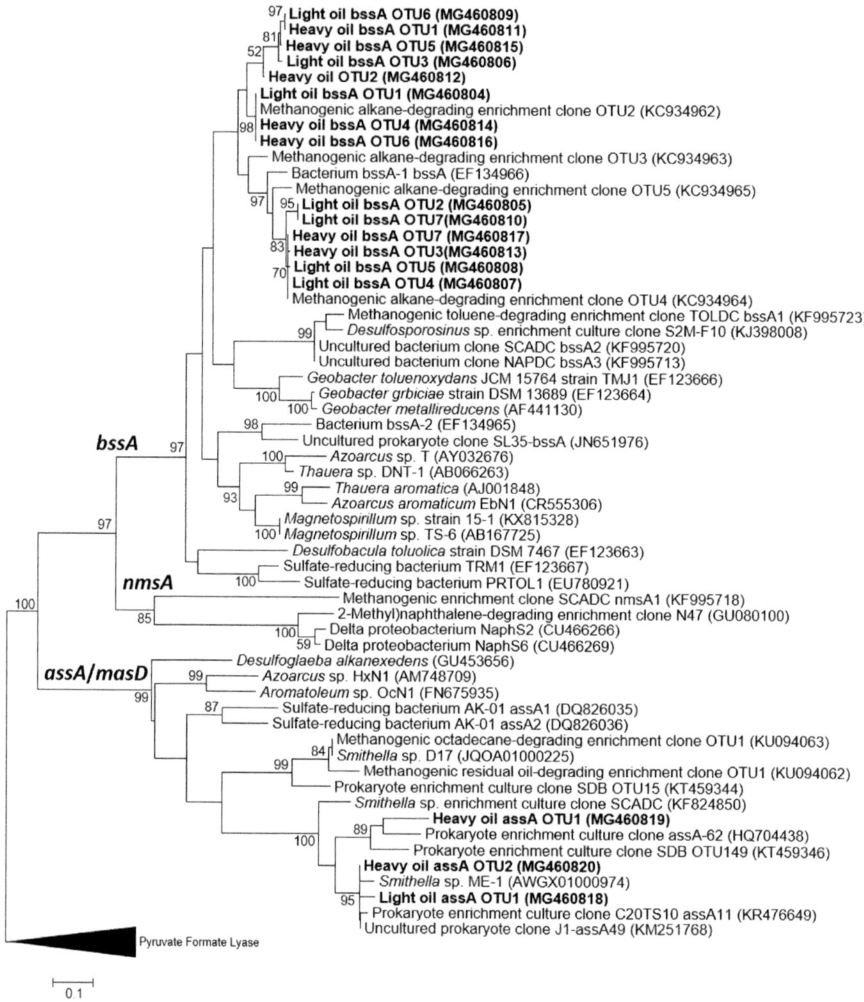

FIGURE 10.3 Maximum likelihood tree (500 bootstraps) showing the affiliation of recovered *assA* and *bssA* gene fragments (bold) with previously published reference strains, enrichment cultures, and environmental samples. Pyruvate formate lyase (pfl) sequences were used to root the consensus tree (collapsed in figure).

abundance over time was seen within the Firmicutes and Euryarchaeota phyla, comprising 60% and 74% of quality sequence reads in both oil-amended cultures by T_{17}, respectively (see Figure 10.4; Table 10.2). This represented a 20- to 25-fold increase in abundance from reported T_0 values, which on average comprised less than 3% of total reads. Archaeal sequences assigned to hydrogenotrophic (*Methanocalculus*, *Methanoculleus*, and *Methanolinea*) and acetotrophic (*Methanosaeta*) representatives also became enriched in the oil-amended PW samples (see Figure 10.4; Table 10.2). The predominant Firmicute members enriched over time affiliated most

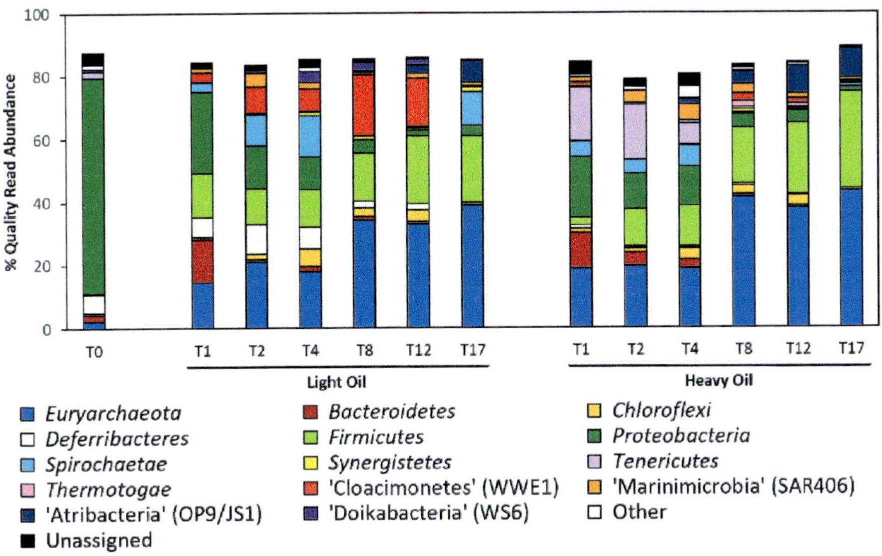

FIGURE 10.4 Phylum level microbial community composition dynamics of the crude oil-degrading produced water cultures based on 16S rRNA gene sequencing.

closely related to *Desulfotomaculum*, comprising up to 30.5% of reads by the end of the 17-month incubation period (see Table 10.2). The most abundant taxon classified as *Desulfotomaculum* shared >97% 16S rRNA sequence identity with an uncultured *Peptococcaceae* bacterium clone recovered from the Mildred Lake Settling Basin (accession no. EU22655; Siddique et al., 2012), supporting the functional evidence that this organism participates in methanogenic aromatic hydrocarbon degradation. Based on its enrichment over time and the identification/proliferation of several *bssA* genes aligning to presumed *Desulfotomaculum*-derived genes, we proposed that this organism was the chief aromatic hydrocarbon degrader in the experimental PW cultures (see Figure 10.5).

TABLE 10.2
Taxon Distribution of Putative Hydrocarbon-Degrading Bacteria and Candidate Phyla (% Abundance) in the Crude Oil-Amended Produced Water Cultures over 17 Months of Incubation

		Light Oil						Heavy Oil					
Taxon	T_0	T_1	T_2	T_4	T_8	T_{12}	T_{17}	T_1	T_2	T_4	T_8	T_{12}	T_{17}
Desulfotomaculum	**0.4**	0.0	0.0	0.0	**14.5**	**21.2**	**21.0**	0.0	0.0	**0.1**	**17.2**	**22.4**	**30.5**
Smithella	0.0	0.0	**2.7**	**2.8**	0.0	0.0	0.2	0.0	**2.7**	**6.1**	0.0	0.0	**0.6**
Syntrophus	**0.4**	**1.3**	**0.1**	**0.2**	**0.4**	**0.4**	0.0	**1.3**	**0.3**	**1.0**	**1.8**	**1.4**	**0.1**
"Cloacimonetes" (WWE1)	0.0	**3.2**	0.0	0.0	**19.4**	**15.4**	0.0	**1.4**	0.0	0.0	**2.5**	**1.6**	0.0
"Atribacteria" (OP9/JS1)	**0.6**	0.0	0.0	0.0	**0.4**	**2.8**	**7.0**	0.0	0.0	**1.4**	**4.1**	**8.8**	**9.1**
"Marinimicrobia" (SAR406)	0.0	**1.4**	**4.5**	**2.2**	**0.9**	**1.5**	**1.1**	**1.5**	**3.9**	**5.1**	**3.2**	**1.6**	**0.9**

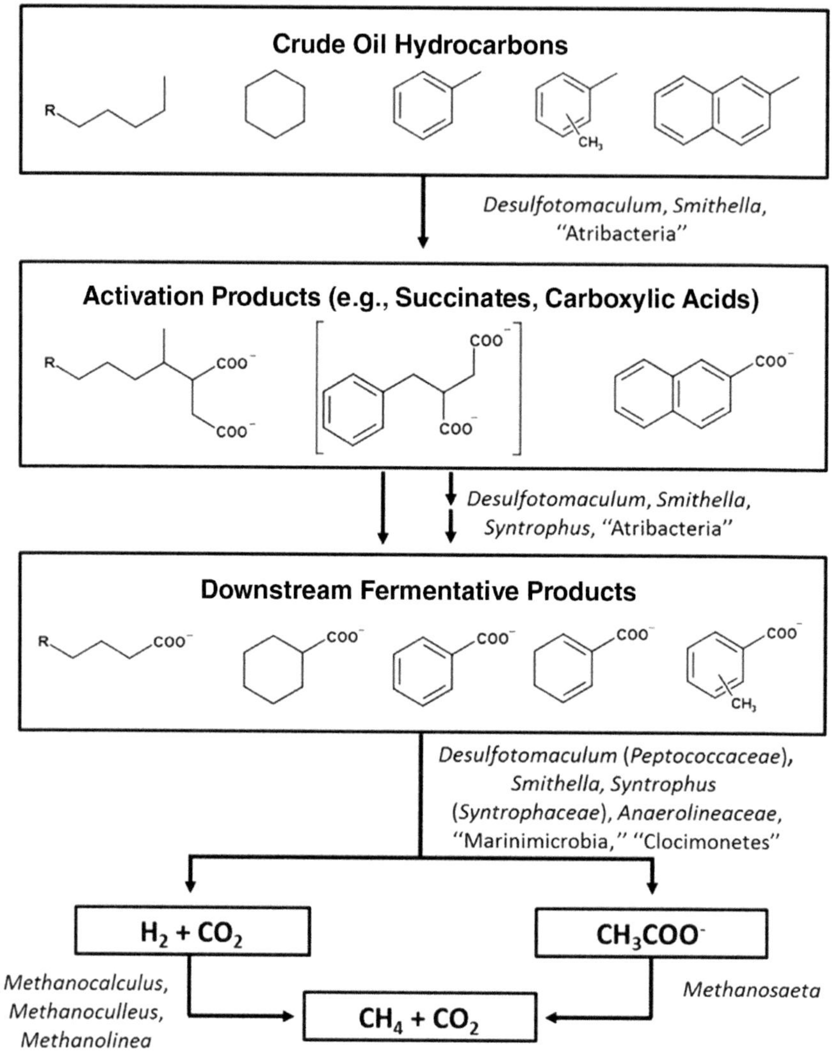

FIGURE 10.5 Working model for the methanogenic biodegradation of saturates and aromatic hydrocarbons by MHGC produced water-associated consortia based on evidence collected in the presented case study. Susceptible hydrocarbons are proposed to be activated by addition to fumarate by *Smithella* (*assA*) or *Desulfotomaculum* (*bssA*), yielding corresponding alkylsuccinates and benzylsuccinates (not detected, though their production is inferred based on results from other tests, e.g., oil analyses, qPCR assays, and detection of downstream metabolite products such as benzoate and toluic acids). Members of the candidate phylum "Atribacteria" may also participate in hydrocarbon activation; carboxylation of two-ringed PAHs may be catalyzed by an unknown organism. Subsequent degradation steps are proposed to be carried out by several putative syntrophic oxidizers (e.g., *Peptococcaceae*, *Syntrophaceae*, and candidate phyla) presumably yielding H_2/CO_2 and short-chain fatty acids (e.g., acetate) that are consumed by hydrogenotrophic and acetotrophic methanogens. Schematic modeled after Jiménez et al. (2016).

Sequences matching to other hydrocarbon degradation-associated bacterial genera included *Smithella* (up to 6.1% of reads) and *Syntrophus* (up to 1.8% of reads), a key syntrophic organism that participates in the transformation of alkane and aromatic hydrocarbon intermediates to products (e.g., H_2/CO_2, acetate) consumed by methanogens (see Table 10.2) (Fowler et al., 2012; Cheng et al., 2013; Wawrik et al., 2016). The identification of the *Smithella* phylotype, coupled with the detection of fumarate addition genes (*assA*) and metabolite products (C_1-C_9 alkylsuccinates), strongly suggests that this organism participated in methanogenic alkane activation in the PW cultures (see Figure 10.5). Taxa typically associated with nitrate reduction (e.g., *Thauera*, *Rhodocyclaceae*) and/or the metabolism of detritus components such as proteins, fatty acids, or phospholipids (e.g., *Pelobacter*, *Bacteroidales*, *Spirochaetes*) became only minor community constituents after T_4, coinciding with the proliferation of fumarate addition activity and the enrichment of methanogenic hydrocarbon-degrading taxa (see Figure 10.4; Toth and Gieg, 2018). Thus, metabolic activities coordinated by these organisms are likely independent of hydrocarbon degradation. Interestingly, several candidate phyla (e.g., "Atribacteria," "Cloacimonetes," and "Marinimicrobia") were detected in the PW cultures, comprising up to 20.8% total read abundance during the incubation period (see Figure 10.4; Table 10.2). Reads assigned to "Atribacteria" became enriched over time in a similar manner to *Desulfotomaculum*, whereas the abundance of the remaining two candidate phyla appeared to be more dynamic. All three of these phyla are of key significance, as recent studies have proposed roles for each of them in methanogenesis from crude oil. Hu et al. (2016) hypothesized that "Atribacteria" (formerly OP9/JS1) could participate in the activation of both alkanes and aromatic hydrocarbons based on the presence of genes putatively encoding several subunits from a fumarate addition enzyme observed in draft genomes. The authors also prepared a draft genome for "Marinimicrobia" (SAR406) and identified an Fe-containing hydrogenase, an enzyme typically associated with syntrophic hydrogen transfer, thus this phylotype may participate in syntrophic interactions with methanogens. Members of the phylum "Cloacimonetes" (WWE1) were recently identified in fluids collected from coalbed methane wells (Kirk et al., 2015) and in anaerobic sludge digesters (Limam et al., 2014), also indicating a putative hydrocarbon-degrading role for these phylotypes. Most convincingly, Cheng et al. (2013) identified members of the "Cloacimonetes" that participate in hexadecane biodegradation within a methanogenic oilfield enrichment consortium (along with *Syntrophaceae*), based on incorporation of ^{13}C from ^{13}C-hexadecane into their DNA in SIP experiments. It is likely, therefore, that each of these candidate phyla is also contributing to the degradation of susceptible hydrocarbons in our PW incubations (see Figure 10.5).

10.5 CONCLUSIONS

This case study of methanogenic crude oil biodegradation in laboratory microcosms confirmed that fumarate addition is a possible mechanism of methanogenic hydrocarbon activation by oilfield-associated microbial communities in the MHGC oilfield and by extrapolation can feasibly function as a metabolic route contributing to heavy oil formation in petroleum reservoirs. Time course–based functional gene analyses and microbial community 16S rRNA gene sequencing identified *Desulfotomaculum* and *Smithella* as key bacteria catalyzing fumarate activation of susceptible hydrocarbons under methanogenic conditions in the examined oilfield, and that the degradation of these compounds was occurring simultaneously in both cultures (see Figure 10.5). Members of several candidate divisions may also play important roles in methanogenic hydrocarbon degradation, expanding on existing knowledge of the diversity of hydrocarbon-degrading consortia. From the collective evidence garnered in this study, we propose a working model for methanogenic crude oil biodegradation by MHGC oil reservoir-associated microorganisms (see Figure 10.5). To assess the metabolic potential of "Atribacteria" and other putative hydrocarbon degraders enriched in this study, shotgun metagenomic profiling of the oil-amended PW cultures is currently underway. We intend to use these metagenomic data to help uncover new functional evidence of biodegradative genes and anaerobic hydrocarbon degradation pathways in these consortia leading towards a greater understanding of the microorganisms and mechanisms coordinating crude oil methanogenesis in petroleum reservoirs.

This case study has also demonstrated that using multiple lines of evidence can help compensate for detection and specificity limitations of any one chemical or functional assay. For example, despite a lack of metabolite evidence indicating benzylsuccinate formation (<10 nM), an increase in *bssA* gene copies as methanogenesis progressed, coinciding with aromatic hydrocarbon loss and an increase in the abundance of *Desultofomaculum*, supported that fumarate addition was a likely mechanism of aromatic hydrocarbon activation in our incubations. Further, our time course experiment demonstrated that metabolites can accumulate transiently before methanogenic activity is established, offering the best window of opportunity to search for hydrocarbon activation products in microcosm experiments. In all, this study underlines the importance of utilizing a multi-pronged approach when collecting evidence for anaerobic hydrocarbon-degrading metabolic processes in environmental samples.

REFERENCES

Abu Laban, Dao A, Foght J. (2015a) DNA stable-isotope proving of oil sands tailings pond enrichment cultures reveals different key players for toluene degradation under methanogenic and sulfidogenic conditions. *FEMS Microbiol Ecol*. 91:fiv039.

Abu Laban N, Dao A, Semple K, Foght J. (2015b) Biodegradation of C_7 and C_8 *iso*-alkanes under methanogenic conditions. *Environ Microbiol*. 17:4894–4915.

Agrawal A, Park HS, Nathoo S, Gieg LM, Jack TR, Miner K, Ertmoed R, Benko A, Voordouw G. (2012) Toluene depletion in produced oil contributes to souring control in a field subjeccted to nitrate injection. *Environ Sci Technol*. 46:1285–1292.

Aitken CM, Jones DM, Maguire MJ, Gray ND, Sherry A, Bowler BFJ, Ditchfield AK, Larter SR, Head IM. (2013) Evidence that crude oil alkane activation proceeds by different mechanisms under sulfate-reducing and methanogenic conditions. *Geochim Cosmochim Acta*. 109:162–174.

Anderson RT, Lovley DR. (2000) Hexadecane decay by methanogenesis. *Nature*. 404:722–723.

Beller HR, Edwards EA. (2000) Anaerobic toluene activation by benzylsuccinate synthase in a highly enriched methanogenic culture. *Appl Environ Microbiol*. 66:5503–5505.

Berdugo-Clavijo C, Dong X, Soh J, Sensen CW, Gieg LM. (2012) Methanogenic biodegradation of two-ringed polycyclic aromatic hydrocarbons. *FEMS Microbiol Ecol.* 81:124–133.

Berdugo-Clavijo C, Gieg LM (2014) Conversion of crude oil to methane by a microbial consortium enriched from oil reservoir production waters. *Front Microbiol.* 5:197.

Bian X-Y, Mbadinga SM, Liu, YF, Yang, SZ, Liu, JF, Ye, R-Q, Gu J-D, Mu B-Z. (2015) Insights into the anaerobic biodegradation pathway of *n*-alkanes in oil reservoirs by detection of signature metabolites. *Sci Rep.* 5:9801.

Cai M, Jiménez N, Krüger M, Guo H, Jun Y, Straaten N, Richnow HH. (2015) Potential for aerobic and methanogenic oil biodegradation in a water flooded oilfield (Dagang oilfield). *Fuel.* 141:143–153.

Callaghan AV. (2013) Enzymes involved in the anaerobic oxidation of *n*-alkanes: from methane to long-chain paraffins. *Front Microbiol.* 4:89.

Cheng L, Ding C, Li Q, He Q, Dai L-R, Zhang H. (2013) DNA-SIP reveals that *Syntrophaceae* play an important role in methanogenic hexadecane degradation. *PLoS One.* 8:e66784.

Duncan KE, Gieg LM, Parisi VA, Tanner RS, Tringe SG, Bristow J, Suflita JM. (2009) Biocorrosive thermophilic microbial communities in Alaskan North Slope oil facilities. *Environ Sci Technol.* 43:7977–7984.

Edwards EA, Grbić-Galić D. (1994). Anaerobic degradation of toluene and *o*-xylene by methanogenic consortium. *Appl Environ Microbiol.* 60:313–322.

Foght J. (2008) Anaerobic biodegradation of aromatic hydrocarbons: pathways and prospects. *J Mol Microbiol Biotechnol.* 15:93–120.

Fowler SJ, Dong X, Sensen CW, Suflita JM, Gieg LM. (2012) Methanogenic toluene metabolism: community structure and intermediates. *Environ Microbiol.* 14:754–764.

Fowler SJ, Gutierrez-Zamora M-L, Manefield M, Gieg LM. (2014) Identification of toluene degraders in a methanogenic enrichment culture. *FEMS Microbiol Ecol.* 89:625–636.

Gieg LM, Alumbaugh RE, Field J, Jones J, Istok JD, Suflita JM. (2009) Assessing *in situ* rates of anaerobic hydrocarbon bioremedation. *Microbiol Biotechnol.* 2:222–233.

Gieg LM, Davidova IA, Duncan KE, Suflita JM. (2010) Methanogenesis, sulfate reduction and crude oil biodegradation in hot Alaskan oilfields. *Environ Microbiol.* 12:3074–3086.

Gieg LM, Duncan KE, Suflita, JM. (2008) Bioenergy production via microbial conversion of residual oil to natural gas. *Appl Environ Microbiol.* 74:3022–3029.

Gieg LM, Kolhatkar RV, Tanner RS, Harris Jr. SH, Sublette KL, Suflita JM. (1999) Intrinsic bioremediation of petroleum hydrocarbons in a gas condensate-contaminated aquifer. *Environ Sci Technol.* 33:2550–2560.

Gieg LM, Suflita JM. (2002) Detection of anaerobic metabolites of saturated and aromatic hydrocarbons in petroleum-contaminated aquifers. *Environ Sci Technol.* 36: 3755–3762.

Gieg LM, Toth CRA. (2017). Signature metabolite analysis to determine *in situ* anaerobic hydrocarbon biodegradation. In Timmis KN (ed.), Boll M (vol ed) *Handbook of Hydrocarbon and Lipid Microbiology*, 2nd ed., New York: Springer, pp. 1–30.

Gray ND, Sherry A, Grant RJ, Roawn AK, Hubert CRJ, Callbeck CM, Aitken CM, Jones DM, Adams JJ, Larter SR, Head IM. (2011) The quantitative significance of *Syntrophaceae* and syntrophic partnerships in methanogenic degradation of crude oil alkanes. *Environ Microbiol.* 13:2957–2975.

Gründger F, Jiménez N, Thielemann T, Straaten N, Lüders T, Richnow H-H, Krüger M. (2015) Microbial methane formation in deep aquifers of a coal-bearing sedimentary basin, Germany. *Front Microbiol.* 6:200.

Head IM, Gray ND, and Larter SR. (2014) Life in the slow lane; biogeochemistry of biodegraded petroleum containing reservoirs and implications for energy recovery and carbon management. *Front Microbiol.* 5:566.

Head IM, Jones DM, Larter SR. (2003) Biological activity in the deep subsurface and the origin of heavy oil. *Nature*. 426:344–352.

Hu P, Tom L, Singh A, Thomas BC, Baker BJ, Piceno VM, Ansersen GL, Banfield JF. (2016) Genome-resolved metagenomic analysis reveals roles for candidate phyla and other microbial community members in biogeochemical transformations in oil reservoirs. M*Bio*. 7:e01669–15.

Imachi H, Sekiguchi Y, Kamagata Y, Loy A, Qiu Y-L, Hugenholtz P, Kimura N, Wagner M, Ohashi A, Harada H. (2006) Non-sulfate-reducing, syntrophic bacteria affiliated with *Desulfotomaculum* cluster I are widely distributed in methanogenic environments. *Appl Environ Microbiol*. 72:2080–2091.

Jiménez N, Richnow HH, Vogt C, Treude T, Krüger M. (2016) Methanogenic hydrocarbon degradation: evidence from field and laboratory studies. *J Mol Microbiol Biotechnol* 26:227–242.

Jones DM, Head IM, Gray ND, Adams JJ, Rowan AK, Aitken CM, Bennett B, Huang H, Brown A, Bowler BFJ, Oldenburg T, Erdmann M, Larter SR. (2008) Crude-oil biodegradation via methanogenesis in subsurface petroleum reservoirs. *Nature*. 451:176–180.

Kirk, MF, Wilson BH, Marquart KA, Zeglin LH, Vinson DS, Flynn TM. (2015) Solute concentrations influence microbial methanogenesis in coal-bearing strata of the Cherokee basin, USA. *Front Microbiol*. 6:1287.

Larter S, Huang H, Adams J, Bennett B, Jokanola O, Oldenburg T, Jones M, Head I, Riediger C, Fowler M. (2006) The controls on the composition of biodegraded oils in the deep subsurface: Part II — Geological controls on subsurface biodegradation fluxes and constraints on reservoir-fluid property prediction. *AAPG Bull*. 90:921–938.

Larter SR, Adams JJ, Gates I, Bennett B, Huang H. (2008) Oil viscosity heterogeneity of tar sand and heavy oil reservoirs: prediction and impact of the production fluid properties. *J Can Pet. Technol*. 47:52–61.

Li W, Wang L-Y, Duan R-Y, Liu J-F, Gu J-D, Mu B-Z. (2012) Microbial community characteristics of petroleum reservoir production water amended with *n*-alkanes and incubated under nitrate-, sulfate-reducing and methanogenic conditions. *Int Biodeterior Biodegrad*. 69:89–96.

Li X-X, Mbadinga SM, Liu J-F, Zhou L, Yang S-Z, Gu J-D, Mu B-Z. (2017) Microbiota and their affilitation with physiochemical characteristics of different subsurface petroleum reservoirs. *Int Biodeterio Biodegrad*. 120:170–185.

Liang B, Wang L-Y, Mbadinga SM, Liu J-F, Yang S-Z, Gu J-D, Mu B-Z. (2015) *Anaerolineaceae* and *Methanosaeta* turned to be the dominant microorganisms in alkanes-dependent methanogenic culture after long-term of incubation. *AMB Expr*. 5:37.

Limam RD, Chouari R, Mazéas L, Wu T-D, Li T, Grossin-Debattista J, Guerquin-Kern J-L, Saidi M, Landoulsi A, Sghir A, Bouchez T. (2014) Members of the uncultured bacterial candidate division WWE1 are implicated in anaerobic digestion of cellulose. *Microbiologyopen*. 3:157–167.

Mbadinga SM, Li K-P, Zhou L, Wang L-Y, Yang S-Z, Liu J-F, Gu J-D, Mu B-Z. (2012) Analysis of alkane-dependent methanogenic community derived from production water of a high-temperature petroleum reservoir. *Appl Microbiol Biotechnol*. 96, 531–542.

Musat F. (2015) The anaerobic degradation of gaseous, nonmethane alkanes – From *in situ* processes to microorganisms. *Comput Struct Biotechnol J*. 13:222–228.

Oberding LK, Gieg LM. (2017) Methanogenic paraffin biodegradation: alkylsuccinate synthase gene quantification and dicarboxylic acid production. *Appl Environ Microbiol*. 84:e01773-17.

Rabus R, Boll M, Heider J, Meckenstock RU, Buckel W, Einsle O, Ermler U, Golding BT, Gunsalus RP, Kroneck PMH, Krüger M, Lueders T, Martins BM, Musat F, Richnow HH, Schink B, Seifert J, Szaleniec M, Treude T, Ullmann GM, Vogt C, von Bergen M, Wiles H. (2016) Anaerobic microbial degradation of hydrocarbons: from enzymatic reactions to the environment. *J Mol Microbiol Biotechnol*. 26:5–28.

Siddique T, Penner T, Klassen J, Nesbø C, Foght JM. (2012) Microbial communities involved in methane production from hydrocarbons in oil sands tailings. *Environ Sci Technol.* 46:9802–9810.

Tan B, Dong X, Sensen CW, Foght J. (2013) Metagenomic analysis of an anaerobic alkane-degrading microbial culture: potential hydrocarbon-activating pathways and inferred roles. *Genome.* 611:599–611.

Tan B, Fowler SJ, Abu Laban N, Dong X, Sensen CW, Foght J, Gieg LM. (2015a) Comparative analysis of metagenomes from three methanogenic hydrocarbon-degrading enrichment cultures with 41 environmental samples. *ISME J.* 9:2028–2045.

Tan B, Semple K, and Foght J. (2015b) Anaerobic alkane biodegradation by cultures enriched from oil sands tailings ponds involves multiple species capable of fumarate addition. *FEMS Microbiol Ecol.* 91:fiv042.

Toth CRA, Gieg LM. (2018) Time course-dependent methanogenic crude oil biodegradation: dynamics of fumarate addition metabolites, biodegradative genes and microbial community composition. *Front Microbiol.* 8:2610.

Voordouw G, Grigoryan AA, Lambo A, Lin S, Park HS, Jack TR, Coombe D, Clay B, Zhang F, Ertmoed R, Miner K, Arensdorf JJ. (2009) Sulfide remediation by pulsed injection of nitrate into a low temperature Canadian heavy oil reservoir. *Environ Sci Technol.* 43:9512–9518.

Wang L-Y, Gao C-X, Mbadinga SM, Zhou L, Liu J-F, Gu J-D, Mu B-Z. (2011) Characterization of an alkane-degrading methanogenic enrichment culture from production water of an oil reservoir after 274 days of incubation. *Int Biodeterio Biodegrad.* 65:444–450.

Washer CE, Edwards EA. (2007) Identification and expression of benzylsuccinate synthase genes in a toluene-degrading methanogenic consortium. *Appl Environ Microbiol.* 73:1367–1369.

Wawrik B, Marks CR, Davidova IA, McInerney MJ, Pruitt S, Duncan KE, Suflita JM, Callaghan AV. (2016). Methanogenic paraffin degradation proceeds via alkane addition to fumarate by "*Smithella*" spp. mediated by a syntrophic coupling with hydrogenotrophic methanogens. *Environ Microbiol.* 18:2604–2619.

Wawrik B, Mendivelso M, Parisi VA, Suflita JM, Davidova IR, Marks CR, Van Nostrand JD, Liang Y, Zhou J, Huizinga BJ, Strąpoć D, Callaghan AV. (2012) Field and laboratory studies on the bioconversion of coal to methane in the San Juan Basin. *FEMS Microbiol Ecol.* 81:26–42.

Zengler K, Richnow HH, Rosselló-Mora R, Michaelis W, Widdel F. (1999) Methane formation from long-chain alkanes by anaerobic microorganisms. *Nature.* 401:266–269.

Zhou L, Li K-P, Mbadinga SM, Yang S-Z, Gu, J-D, Mu B-Z. (2012) Analyses of *n*-alkanes degrading community dynamics of a high-temperature methanogenic consortium enriched from production water of a petroleum reservoir by a combination of molecular techniques. *Ecotoxicology.* 21:1680–1691.

11 Culturable Microbiome and Biodegradation Activity of Diesel B5 and Biodiesel (B100) in a Contaminated Soil Bioremediation Study

Guilherme Matos, Ana Oliveira, Diogo Coutinho, Thais Oliva, Clenilson Junior, Ben Stenuit, Spiros N. Agathos, Márcia Lutterbach, and Eliana Sérvulo

CONTENTS

11.1 INTRODUCTION

11.1.1 Biodiesel in the Energy Landscape and Its Environmental Fate

The global impacts due to environmental pollution have driven the development of technologies towards substitution of petroleum in the energy landscape, with the focus on renewable sources (Vianna 2006). Some of the most promising alternatives are fuels obtained from plant biomass, such as biodiesel (Rathman et al. 2005).

This fuel, basically composed of fatty acid methyl esters (FAME) mainly produced by vegetable oils or animal fats transesterification, has given rise to environmental advantages, such as cleaner burning, higher biodegradability potential and lower environmental toxicity (Suarez et al. 2007). Therefore, it emerges as a potential substitute for mineral diesel, in principle partially through its inclusion in blends (diesel/biodiesel), called diesel B. The biodiesel percentage is expressed, for example, as diesel B5 (95% diesel and 5% biodiesel) up to B100 (100% biodiesel).

Many countries have presented public incentives for the production and use of biodiesel. The main world program is in the European Union (EU), the largest consumer and producer, with rapeseed as the main source (68%). Studies have shown that the final aim of these countries is to achieve blends with 20% biodiesel by 2020 (Sorda et al. 2010; Klessmann et al. 2011).

Thereby, biodiesel and its blends with diesel are subject to the same kind of accidents commonly associated with the total petroleum hydrocarbons (TPH) industry, which can compromise the environment, through soil and water pollution. In contrast to diesel, biodiesel, as a low structural complexity compound, has been identified as more susceptible to degradation by microbial action (Fernández-Alvarez et al. 2007). However, further studies are needed for a satisfactory understanding of the environmental fate of leaks or spills of pure biodiesel, as well as of diesel/biodiesel blends, at different concentrations.

11.1.2 (Bio)diesel Bioremediation

Several technologies are available for remediation of contaminated soils by spills or fuel leaks toward reduction of concentration and/or toxicity contaminants reduction (USEPA 2012). An efficient alternative suggested for the remediation of organic pollutants in soils is the use of bioremediation techniques. The treatment of contaminated areas using biological processes can present lower cost and equivalent efficiency to physicochemical methods, besides being environmentally safe (Khan et al. 2004).

Bioremediation is the degradation of organic pollutants soil, water or sediment by the microorganisms catabolism of complex molecules, either by mineralizing them or by transforming them into less toxic compounds (Alexander 1999; Crawford and Crawford 1996). Depending on the microorganisms involved in the process and the environmental conditions in the contaminated site, bioremediation may occur under aerobic, anaerobic or anoxic conditions. In general, the aerobic treatment is more conducive to the contaminant mineralization and reduces the degradation time (Österreicher-Cunha et al. 2012). However, for the fuels studied in this chapter, additional information is still needed for the effective application of this process, especially in relation to the microbial populations involved, biodegradation time and favorable environmental conditions.

Autochthonous microbial communities can conduct intrinsic bioremediation (IB), which can promote the removal or reduction of contaminants naturally by the action of microbial metabolism associated with other spontaneous physico-chemical routes (Singh et al. 2009). Filamentous fungi and, mainly, bacteria (due to faster growth and enzymatic complex system) are more efficient for contaminants degradation. Not only the microbial diversity, but also the population density and the contaminated site conditions are key points to the success of biodegradation. In addition, there is no doubt about the importance of understanding the microbial community interactions into the contaminated site during the process (Dragun 1998; Watanabe 2001; Singh et al. 2006; Li et al. 2008; Muyzer and Stams 2008).

Diesel B bioremediation features in soil, including the optimum nutritional and environmental conditions for biodegradation, as well as an extensive microbial description and dynamics (bacteria and fungi) are not yet fully established. For example, some studies conducted in soil laboratory micro/mesocosms suggest an increase in the biodegradation of blends due to biodiesel addition. However, they diverge about the influence of biodiesel concentration, microbial diversity, its biomass density throughout the process and soil characterization (Junior et al. 2009; Horel and Schiewer 2011; Silva et al. 2012).

11.2 OBJECTIVES AND STUDY DESIGN

This chapter discusses the characterization of *ex situ* (bio)degradation of FAME components of biodiesel, as well as TPH associated with FAME in the B5 blend, at different soil depths. Parameters related to the reduction of the contaminant concentration and the culturable degrader microbiota dynamics (fungi and bacteria) present throughout the test will be examined. The NA bioremediation efficiency will be evaluated by the remaining ecotoxicity degree in the contaminated soil at the end of the incubation period.

The aim is to contribute to a better understanding of the participation of the native microbiota in terrestrial ecosystems contaminated by diesel/biodiesel blends and to advance bioremediation action strategies.

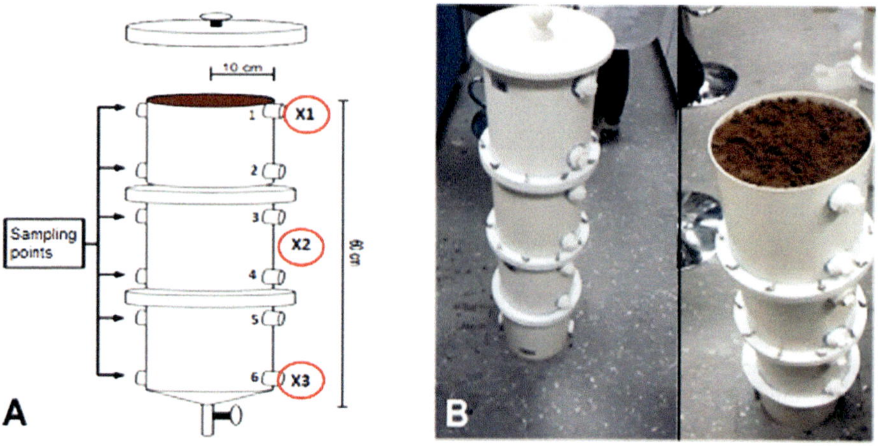

FIGURE 11.1 Microcosm experiments in PVC reactor columns. This system is composed of three cylindrical parts (cells) with 10 cm radius and 20 cm height each, placed on top of each other and allowing contact and fluid flow between adjacent compartments. Each cell is provided with four lateral openings, allowing samplings at six different depths. Samples were taken at three points, as indicated in the diagram: X1 (5 cm), X2 (30 cm, sample composed of points 3 and 4) and X3 (60 cm) (A). Incubation conditions: 21 kg clay loam soil, 20% humidity, 10% (v/m) contaminant concentration, outer temperature: room (25–28°C), incubation time: 60 days. Material previously mixed with microcosm assembly.

The bioremediation assays discussed herein were conducted in polyvinyl chloride (PVC) reactors as detailed in Figure 11.1. The soil used was from an urban area with no history of previous contamination. Granulometric, chemical and physico-chemical analyses classified the soil as clay loam and with conditions compatible with microbial activity. Fuels used as contaminants, biodiesel (B100) and diesel (B0), were donated by a distribution center for gas stations in the metropolitan region of Rio de Janeiro, Brazil. The diesel/biodiesel blend (B5) was produced in the laboratory.

Experimental tests and controls (abiotic degradation determination) are described in Table 11.1, which presents the substrate components of each microcosm. The

TABLE 11.1
Laboratory Microcosms

Assay	Contaminant	Description
EXP B5	B5	Soil contaminated by B5 blend (10% (v/w) contamination) *in natura*
EXP B100	B100	Soil contaminated by B100 blend (10% (v/w) contamination) *in natura*
CS B5	B5	Sterile soil contaminated by B5 blend* (10% (v/w) contamination)
CS B100	B100	Sterile soil contaminated by B5 blend* (10% (v/w) contamination)

**Soil sterilization:* Three times autoclaving and 2% (v/w) formaldehyde. Fuel sterilization: 0.45 µm filtration.

operational conditions were established so that they were compatible with the microbial activity and the contaminant percolation.

Each microcosm was assembled with the total amount of soil (sterile or contaminated), previously humidified. The test was set up to provide the least possible disturbance to the system during the bioremediation period. At the beginning and monthly during the treatment, the soils were analyzed for pH and moisture, always verifying the need to adjust the water content to 20% (v/m). The outer temperature was measured every two days, and the soil temperature at the beginning and after 30 and 60 days. No other intervention was performed throughout the experiments.

11.3 ANALYTICAL METHODOLOGY

During the experimental period (60 days), the soil samples were analyzed as summarized in Table 11.2. It is possible to observe the variation on the periodicity of the tests, whereas the ecotoxicity evaluation was carried out only at the end of the bioremediation assay.

11.3.1 Quantification of Culturable Microbiome

In order to estimate the culturable microbiome biomass (bacteria and fungi), sample suspensions (pristine soil, fuels and contaminated soil) were obtained using saline solution (0.85% m/v) plus 0.1% of Tween™, homogenized at 300 rpm for 30 minutes. After obtaining this suspension, serial dilutions (1:10) were performed for cultivation assays.

TABLE 11.2
Methodological Analysis Applied to Monitored Natural Attenuation (MNA) Bioremediation Assays

Parameter	Methodology	Objective	Period (days)
Culturable Bacteria	Aerobic and Anaerobic MPN	(An)aerobic biomass quantification of culturable heterotrophic bacteria	0/30/60
	Spread plate cultivation	Prevalent bacterial groups isolation–5 cm depth	
	16S rRNA gene sequencing of isolated cultures	Genera identification of putatively bacteria-biodegrader	0/30/60
	Mixed cultures *Denaturing Gradient Gel Eletrophoresis* (DGGE)	Community molecular profiles evaluation	0/60
Culturable Fungi	*Spread plate cultivation*	Biomass quantification and fungi/yeast isolation	0/30/60
	ITS gene sequencing of isolated cultures	Genera/species identification of putative biodegrader fungi	
Fuel Concentration	Gas Chromatography – Flame Ionization Detector	Fuel (bio)degradation assesment	0/60
Soil Toxicity	*Eisenia fetida* mortality tests	Soil (bio)remediation outcome	0/60

Biomass quantifications of culturable heterotrophic bacteria populations were estimated by Most Probable Number (MPN) technique in tubes with nutrient broth.[1] A similar method was employed to measure the bacterial density of culturable heterotrophic anaerobic biomass. For this, serial dilutions (1:10) were obtained in reducing solution[2] and the cultivation was performed in Fluid Thioglycollate Medium (SIGMA™), using sealed flasks. The anaerobic environment was maintained by N_2 purge. Incubation temperature was 30°C and the microbial growth verified after 48 hours. The results were expressed in MPN per gram of soil.

Fungal biomass estimation was carried out by direct counting of spread plate cultivation on Sabouraud medium (OXOID™), with chloramphenicol addition (SIGMA™), using 0.1 mL of the initial suspension and serial dilutions up to 10^{-5}. Incubation was performed at 30°C for 7 to 14 days. After that period, the total and morphologically distinct and prevalent colony counts were performed, and the results expressed in terms of colony forming units (CFU) per gram of dry soil. Phenotypically distinct colonies were isolated in specific media, including Malt Extract Agar (MEA – OXOID™), Dextrose Potato Agar (BDA – OXOID™) and Czapek Yeast Extract Agar (CYA – Santos et al. 1998). Afterwards, microcultures were performed on slides for identification of characteristic structures of fungal genera. In relation to yeasts, only presence/absence and their identification were verified.

11.3.2 Identification of Bacteria and Fungi Populations

Prevalent aerobic bacterial group isolation on microcosm surfaces (5 cm) after 30 and 60 days and in pristine soil was carried out cultivating 1 mL of each serial dilution by spread plate on PCA (Plate Count Agar, Himedia™), under the same incubation conditions as mentioned above.

After that, DNA extraction from isolated colonies was performed by thermal lysis: 10 sec inside liquid nitrogen, followed by 10 min 60°C dry bath at 3 cycles.

Molecular identification of the isolated microorganisms was followed by gene ribosomal 16S gene subunit (16S rRNA) amplification, approximately 1500 bp. Standard PCR (Thermocycler, Applied Biosystems) was employed PCR (Thermocycler PCR System 9700, Applied Biosystems) and Taq PCR Master Mix Kit (Qiagen) plus 0.2 μM of specific primers, Sadir (foward: 5′-AGAGTTTGATCATGGCTCAGA-3′) and S17 (reverse: 5′-GTTACCTTGTTACGACTT-3′), following the cycling conditions: 94°C – 4 min; 94°C – 30 sec; 50°C – 30 sec; 72°C – 30 sec (30 cycles); 72°C – 5 min (Altschul et al. 1990).

DNA extraction from isolated fungi was performed using DNeasy Power Soil Kit™ (Qiagen). The identification was followed by amplification of the gene encoding the 5.8S subunit ribosomal RNA and the ITS I and II regions (Internal Transcribed Spacers), fragment of approximately 600 bp. This procedure followed similar conditions as previously described, differing in the cycling conditions (94°C – 5 min; 94°C – 30 sec; 50°C – 30 sec; 72°C – 30 sec (30 cycles); 72°C – 30 sec), as well as in the primers employed: ITS5 forward 5′-GGAAGTAAAAGTCGTAACAAGG-3′ and ITS4 (reverse) 5′-TCCTCCGCTTATTGATATGC-3′ (White et al. 1990).

The amplicons were submitted for sequencing using BigDye Terminator v3.1 Cycle Sequencing Standard™ and using an automatic sequencer (AB 3500 Genetic Analyzer, Applied Biosystems). The final sequences, obtained after editing using Chromas Lite™

and Bioedit softwares, were submitted to similarity comparison using BLAST and GenBank online (www.ncbi.nlm.nih.gov). The analysis provided the gene fragment identification (genus or species), considering a minimum of 98% of similarity.

11.3.3 Molecular Characterization of Culturable Heterotrophic Bacterial Community

Molecular characterization by DGGE used DNA extracted by thermal lysis (aforementioned) from aerobic and anaerobic liquid cultivation of pristine soil and after 60 days of bioremediation, at the three depths and in both microcosms (EXPB100 and EXPB5). The methodology was adapted from Muyzer and colleagues (1993), involving 16S ribosomal RNA subunit specific amplification. However, with insertion of a high GC "clamp" (5 min CGCCCGCCGCGCGCGGCGGGCGGGGCGGGGCACGGGGGG-3′) added to the 5 min portion of the U968 primer (forward: 5′-GCACGGGGGACGCGAAGAACCTTAC-3′), in addition to the other primer L1401 (reverse: 5′-GCGTGTGTACAAGACCC-3′). The cycling conditions used were 94°C – 3 min; 94°C – 30 sec; 50°C – 30 sec; 72°C – 30 sec (30 cycles); 72°C – 30 sec.

The amplified products were subjected to 6% polyacrylamide gel (acrylamide and bis-acrylamide) electrophoresis in a tris-acetate buffer (pH 8.3), with a denaturing gradient (urea and formamide, 40% to 60%), according to Aboim et al. (2004). Dendrograms were constructed based on the Jaccar similarity coefficient using Bionumerics v 6.0 (Applied Maths) software. Statistical analysis of non-metric multidimensional scaling (NMS) was also performed using online software "PAleontological STatistics - version 2.16".

11.3.4 Fuel Concentration Analysis

The soil contaminants were extracted using an adapted protocol from Schwartz, Ben-Dor and Eshel (2012) and USEPA 3550C (2007). In summary, 1 gram of soil was resuspended in n-hexane:acetone solution (50:50 v/v) followed by an ultrasonic bath for 60 min. After this, the final supernatant was filtered through Nylon™ membrane for injection in the gas chromatograph.

Chromatographic runs for TPH detection followed adapted protocols described in Weisman (1998), Petit (2004) and Institut Scientifique de Service Public (2010). Briefly, an extract aliquot was injected into a gas chromatograph (HP 5890 Series II, Hewlett-Packard™) equipped with a flame ionization detector (FID). The column used was a DB-5 (30 m by 0.25 mm) with hydrogen carrier gas at a flow rate of 20 mL min^{-1} and air at a flow of 310 mL/min^{-1}. The temperature program used was 40°C 2 min^{-1}, 150°C 10 min^{-1}, 270°C 20 min^{-1}.

The FAME analysis was carried out in a gas chromatograph (ThermoQuest, ThermoFinnigan™) also equipped with a FID, based on methodology adapted from McCurry (2012), Huppel and Huybrighs (2012), Grama et al. (2014) and Standard EN14105:2011. It used a column Rt – 2560 (100 m by 0.25 mm) with helium carrier gas at a flow rate of 35 mL min^{-1} and air at a flow of 350 mL min^{-1}. The temperature program applied was 80°C 25 min^{-1}, 175°C 10 min^{-1}, 205°C 10 min^{-1}; 225°C 20 min^{-1}.

The final chromatograms were analyzed according to molecular standards (SIGMA) of aliphatic hydrocarbons contained in the c8-c24 range, for TPH analyses;

and for EMAG, esters were used: c17: 0 (Marginic Acid), c16: 0 (Palmitic Acid), c18: 0 (Stearic Acid), c18: 1 (Oleic Acid), c1 (Linoleic acid) and c1 3 (α-Linolenic acid).

11.3.5 Ecotoxicity Evaluation

Ecotoxicity analyses were conducted in accordance with standard NBR 15.537 (ABNT, 2007), in triplicate and over a period of 14 days. The NA bioremediation efficiency was verified by comparing the following trials: pristine soil, B5 or B100 (10% m/v contamination) contaminated soil and B5 or B100 bioremediated soils, from the superficial and deepest cells of both microcosms.

11.4 RESULTS AND DISCUSSION

11.4.1 (Bio)degradation Profile of TPH

The diesel B5 and B100 biodegradation in artificially contaminated clay loam soil was determined by estimating the total reduction percentages by discounting the abiotic ones obtained in the control microcosms. Figure 11.2 shows the percentage values of TPH contaminant reduction discriminated for abiotic control and total degradation (experimental reactor), with the difference attributed to biodegradation.

It was observed that the bioremediation process was only effective at the intermediate depth (30 cm), with the microorganisms contributing to the hydrocarbon content reduction by only around 10%. This rate could be due to the retention of compounds of higher molecular mass and great recalcitrance on the reactor surface. In this way, intermediate (partially degraded) compounds and other fractions with lower molecular mass would be percolated, increasing the concentration of more easily biodegradable compounds in such depth. It could also be considered that possibly the oxygen availability is still satisfactory for aerobic biodegradation up to a depth of 30 cm.

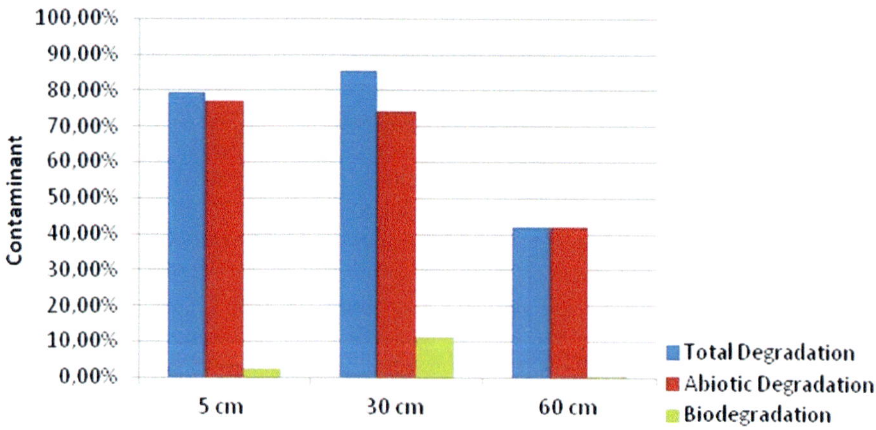

FIGURE 11.2 Biodegradation profile of the TPH fraction determined by the comparison between the decrease in contaminant levels observed in EXP B5 and CS B5 after 60 days of IB, at three different depths.

The rate described here can be considered low in relation to previous studies. Li et al. (2007) and Boopathy (2004) found that the natural biodegradation of pure diesel was around 20% and 4%, in studies with similar characteristics to those presented here, both for aerobic and anaerobic conditions. Additionally, considering the incubation time, it is quite plausible that biodegradation occurred only for less recalcitrant compounds in B5. In fact, in bioremediation assays by biopiles, Chemlal et al. (2013) described that the catabolism sequence of diesel starts with the simpler alkanes and ends with the aromatics. The TPH evaluated in this study were introduced in blends with biodiesel, i.e., with more easily biodegradable molecules. According to Prince et al. (2008), studying the B20 biodegradation, FAME present in the blend are readily degraded, that is, their catabolism precedes that of the simpler alkanes. It is supposed that FAME structures enable attack by various enzymes, which result in the oxidation of the molecules into acetic acid and other fatty acids. The latter, through a series of reactions, can be completely mineralized to CO_2 and H_2O (Jakeria et al. 2014).

11.4.2 (Bio)degradation Profile of FAME

The FAME biodegradation percentages after 60 days of NA can be observed in Figure 11.3. It shows FAME (bio)degradation values for soil contaminated by B5 (see Figure 11.3A) and B100 (see Figure 11.3B), at the three depths.

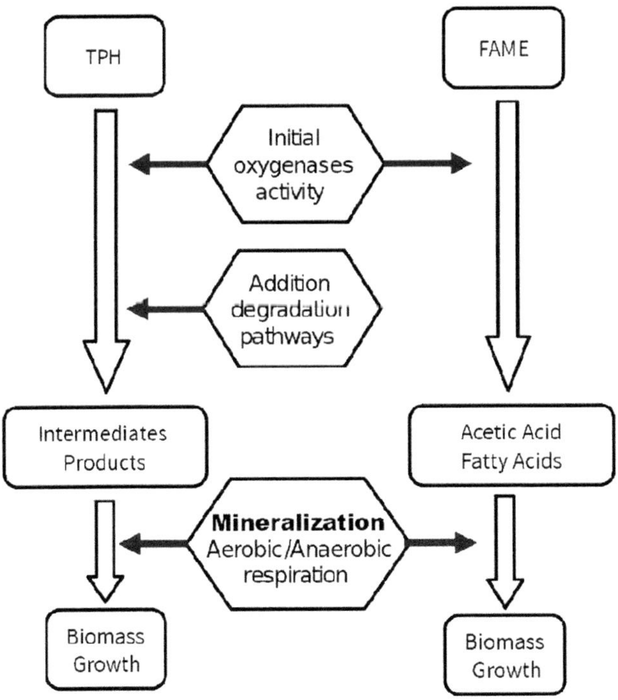

FIGURE 11.3 Relationship between suggested biodegradation pathways for diesel/biodiesel blends in impacted soil.

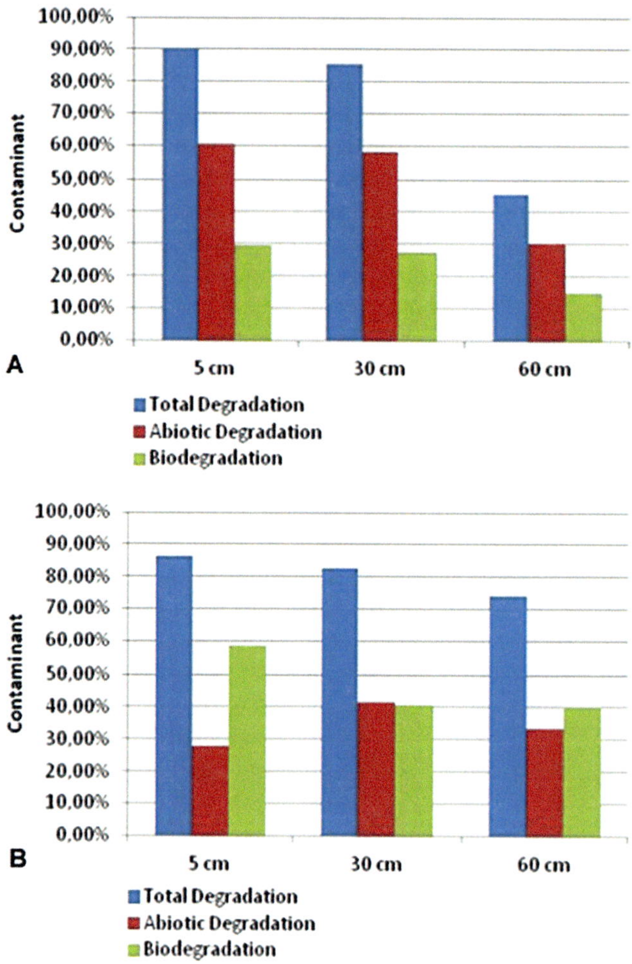

FIGURE 11.4 FAME fraction (bio)degradation profile was determined by comparing the decreases in contaminants observed in EXP B5 and CS B5 (A) and in EXP B100 and CS B100 (B) after 60 days of IB, at three different depths.

In this way, Figure 11.3A suggests the influence of column depth on the efficiency of FAME bioremediation (in B5 blend), especially when comparing rates at 5/30 cm (30%) to rates at 60 cm (15%). However, Figure 11.3B shows that FAME biodegradation was more effective in pure biodiesel (B100) than in B5 blend. Soil contamination by B100 presented biodegradation percentages of approximately 60% (5 cm) and 40% (30 and 60 cm).

FAME biodegradation occurred up to 60 cm in depth in both microcosms. However, higher bioremediation rates were observed in the soil surface horizon and, therefore, in the region where probably there is greater oxygen availability.

The (bio)degradation profiles indicate reduction in the bioremediation efficiency in soil contaminated by FAME and TPH concomitantly, when compared to B100 and with literature data for diesel biodegradation. This decrease could be a consequence of similarities between the metabolic biodegradation pathways of hydrocarbon and FAME in a blend, mainly for the initial stages of aerobic metabolism (Jakeria et al. 2014), as shown in Figure 11.4. Based on this model, both co-stimulation—due to abundance of substrates and common intermediates as well as competition could be expected between the metabolic pathways.

The results presented here agree with some data previously described in literature, which demonstrate that FAME are primarily metabolized by microorganisms in blends with diesel, aerobically and/or anaerobically. However, these studies have described an increase in general fuel biodegradation with no influence on the TPH metabolism (Prince 2008; Owsianiak et al. 2009; Cyplick et al. 2011; Wu et al. 2015).

11.5 CULTURABLE HETEROTROPHIC BACTERIA (CHB)—BIOMASS DENSITY AND MOLECULAR COMMUNITY STRUCTURE BY DENATURING GRADIENT GEL ELECTROPHORESIS (DGGE)

Figure 11.5 shows the culturable heterotrophic bacteria (CHB) levels. It is possible to observe that the cell concentrations were equal to or higher than the values detected in pristine soil, in EXPB5 (see Figure 11.5 A/C) and EXPB100 (see Figure 11.5 B/D), even in aerobic or anaerobic cultivation. The biomass evaluation in EXPB5 (see Figure 11.5 A/C) suggests that the fuel catabolism had intensified after 30 days, probably after selection and acclimatization of native biodegrader populations. The

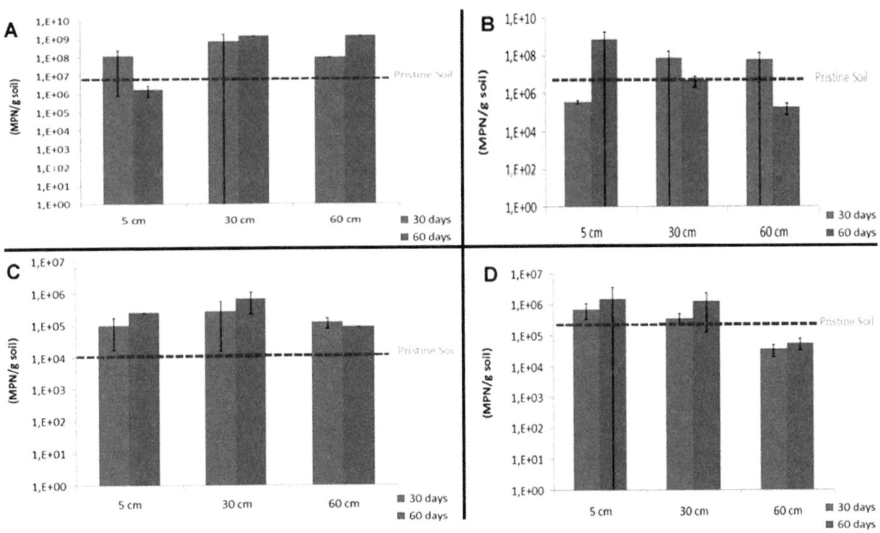

FIGURE 11.5 Culturable aerobic/anaerobic heterotrophic bacteria biomass from samples collected at EXPB5 (11.5 A/C) and EXPB100 (11.5 B/D) microcosms, as a function of depths and times. [Dashed line: cell concentration in the pristine soil.]

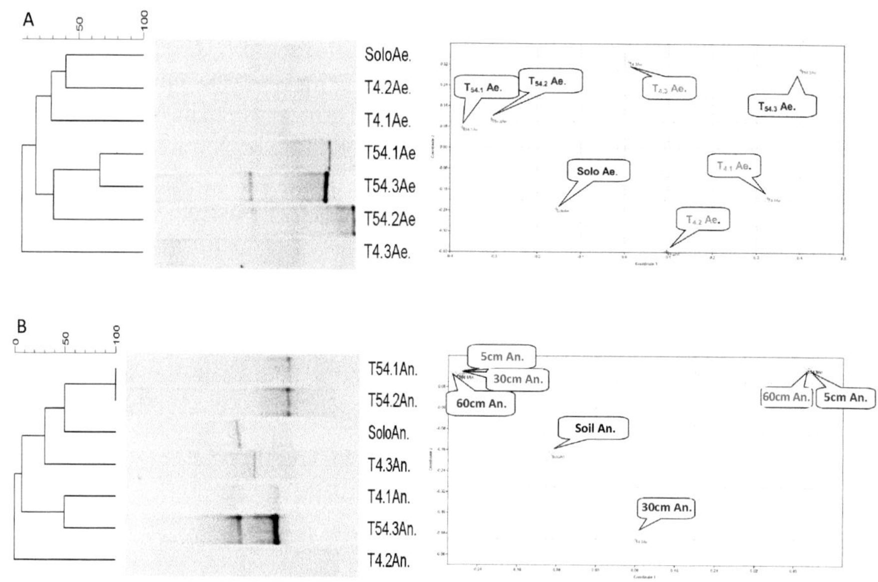

FIGURE 11.6 DNA fingerprinting by Denaturing Gradient Gel Eletrophoresis (DGGE) followed by "Multidimensional Scaling" of culturable bacterial communities under aerobic (A) and anaerobic conditions (B), after 60 days. Blue – EXP B100 samples and Red – EXP B5 samples.

results of the cell growth agree with those verified in the biodegradation (TPH and FAME) analyses: low rates at 60 cm of reactor depth and the highest one in the intermediate zone (30 cm).

The CHB growth during B100 bioremediation is shown in Figures 11.5 B/D, in aerobic and anaerobic cultures, respectively. The present data suggest populations are established quite early in this system and likely are influenced by aeration. It is possible to observe that after 30 days, aerobic growth decreases inversely to the anaerobic cultivation, starting from a depth of 30 cm. In addition, the highest values were observed in the superficial layer (Figure 11.5 B/D), which is consistent with the high FAME biodegradation at this site (around 60%). This dataset suggests that biodiesel biodegradation probably resulted in shorter periods of bioremediation.

11.6 CULTURABLE HETEROTROPHIC BACTERIA (CHB)—MOLECULAR COMMUNITY STRUCTURE BY DENATURING GRADIENT GEL ELECTROPHORESIS (DGGE)

Molecular biology techniques have been used to better characterize communities involved in the bioremediation of biodiesel and blends (Meyer et al. 2014; Soriano et al. 2015). Figure 11.6 analyzes the molecular community structure of the putative culturable biodegrader populations after 60 days of treatment. The molecular profiles by DGGE, dendrograms and Non-Metric Multidimensional Scaling (NMS) plots were obtained from pristine soil, EXPB5 and EXPB100, cultivated under aerobic (11.6 A) and anaerobic (11.6 B) conditions.

The data strongly contribute to the hypothesis of differential community structure, as consequence primarily of the fuel inserted in the microcosm. In relation to B5, that consists mainly of hydrocarbons, a distinct bacterial community is observed compared to that established for B100 contaminated soil. It is important to consider the toxic effect of TPH on the native bacteria, as well as the fact that biodiesel is subject to an easier catabolism, which allows the appearance of greater microbial diversity along the column. Figure 11.6 also suggests that the presence of different contaminants is a selection factor for bacterial populations more relevant than depth, although it is possible to infer that this latter factor can contributes to the community structuring with greater influence for B100 than for B5.

11.6.1 Culturable Heterotrophic Bacteria—Diversity at the Microcosm Surface

As previously observed, the reactor surfaces (5 cm) presented the highest biodegradation rates for FAME. This fact in association with adequate biomass growth curves for CHB could suggest autochthonous bioremediation activity. In this way, the present work also sought to identify some of the prevalent bacterial genera that are putative biodegraders at this depth.

Nine bacterial genera were isolated among all the substrates after 60 days. The prevalence of *Arthrobacter*, *Bacillus* and *Micrococcus* genera in the pristine soil could be noticed. After changes in the microenvironments caused by the fuels biodegradation, diversification of the microbial populations prevalent in microcosms could be observed. After 30 days, in the EXP B5 *Agromyces* sp., *Lysobacter* sp., *Sinomonas* sp. and *Arthrobacter* sp. (lasting up to 60 days) were observed, while in the EXP B100, *Bacillus* sp. and *Burkholderia* sp., appeared by the end of the incubation.

In relation to the populations described in EXPB5, it is worth highlighting those belonging to the genus *Arthrobacter*, which is widely cited in the literature for its participation in the biodegradation of hydrocarbons and derivatives (including diesel B) in different environments and laboratory assays (Prince 2005; Chen et al. 2015; Soriano et al. 2015). According to Paria (2008), the biodegradation efficiency of several recalcitrant compounds by *Arthrobacter* species is due to their capacity to produce different classes of biosurfactants. In this same way, the biodegradation activity of the genera isolated in the EXP B100 have been mostly described for hydrocarbons, as the metabolism of aliphatics has been attributed to *Bacillus* sp. (Van Beilen and Withold 2005; Chen et al. 2015), that of PAHs to *Burkholderia* sp. (Haritash and Kaushis 2009; Chen et al. 2015; Keskin et al. 2015), since data on the biodegradation of biodiesel are still scarce.

11.7 CULTURABLE MICROBIOME OF FUNGI

Fungi are also widely present in soils; therefore they are fundamental for the microbiota characterization during the TPH and FAME biodegradation process. Figure 11.7 shows the fungal biomass variation in the microcosms EXP B5 (see Figure 11.7 A) and EXP B100 (see Figure 11.7 B), respectively. The values related to the concentrations of fungi determined in the pristine soil and in those contaminated with fuels (diesel and biodiesel) are also indicated. Contaminant species from fuels have a common characteristic of robustness, given their ability to survive in an adverse environment, and possibly by the

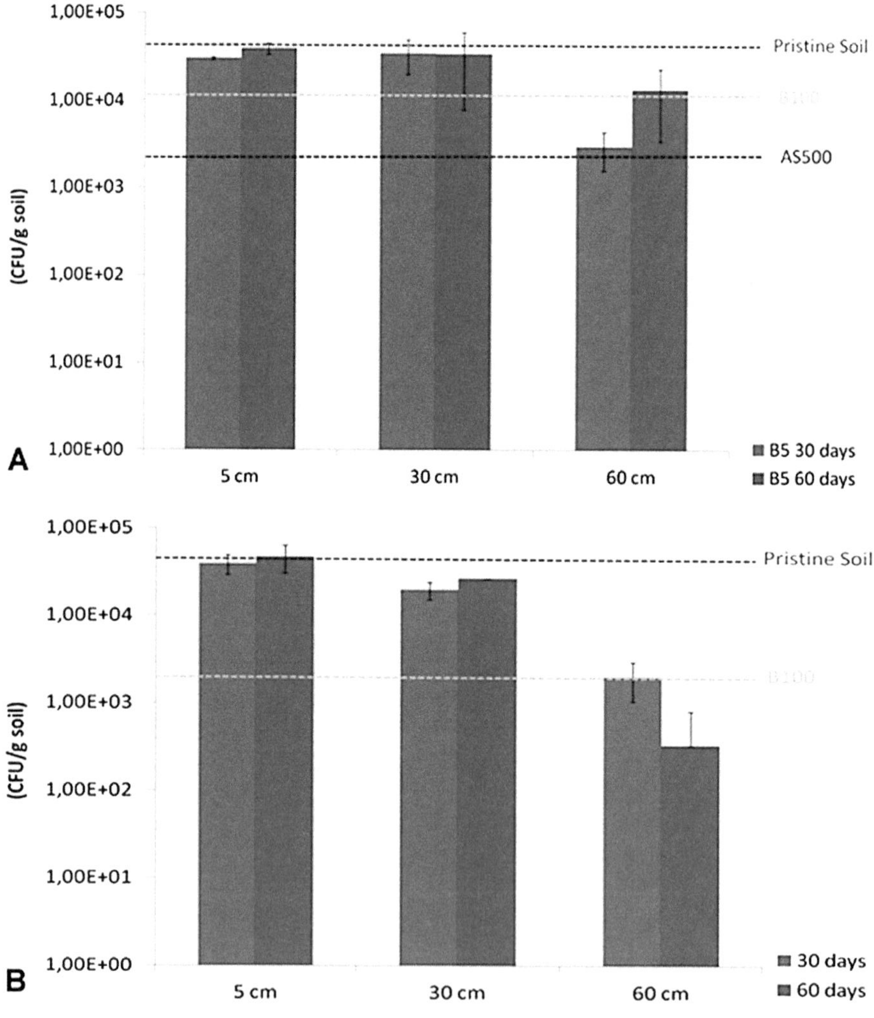

FIGURE 11.7 Quantification of culturable mycelial fungi, (A) EXP B5 and (B) EXP B100.

capacity to use these compounds as carbon sources. Therefore, besides the soil, the fungal microbiota of each microcosm may have had its origin in the fuels.

The fungi density in both microcosms remained, in general, around 10^4 CFU/g of soil during the experimental period, both in the superficial layer (5 cm) and in the intermediate layer (30 cm). Considering that this is the density of fungi in the pristine soil, it could be suggested that they participated in the fuels' degradation. However, in the deepest soil layer (60 cm), there was a decrease in the fungal population, most significantly in EXPB100 (Figure 11.7 B). This could be possibly due to the influence of oxygen availability to the fungal microbiota growth, which could limit its overall metabolic activity. Most mycelial fungi are obligate aerobes so they could indirectly indicate the soil oxygenation profile.

The prevalent fungi in the EXP B5 and EXP B100 microcosms at different depths (5 cm, 30 cm and 60 cm) and treatment times (30 and 60 days) are representative of the *Ascomycota* (mostly) and *Basidiomycota* phyla. The first phylum populations were present mainly in the pristine soil and diesel, remaining in higher proportions in the reactors. However, the main source of the *Basidiomycota* in the experimental microcosms was the B100 itself (data not shown).

Regarding identification, 19 genera of culturable fungi in pristine soil, fuels and microcosms were characterized. Through general data analyses of both microcosms, it was possible to verify that the main genus participants and, possibly, with highest catabolism capacity of contaminants, were *Fusarium* sp. and *Talaromyces* sp. These were the most prevalent, both in 30 and 60 days, in the different depths studied.

Particularly regarding the EXP B5 microcosm, it was observed that an initial effective selection of populations remains stable until the end of the assay. This means that the stable structure of the fungal community occurs early in the assay. From 30 days onwards, it is possible to determine the genera resistant to the toxic action of hydrocarbons and possibly responsible for TPH and FAME biodegradation present in the blend. In this context, the presence of *Fusarium* sp., *Penicillium* sp. and *Talaromyces* sp. populations in the superficial layers can be highlighted.

In the B100 bioremediation it was possible to observe an exclusive dynamics of fungal genera. It can be suggested that there was no stabilization of the local fungal community throughout the period of 60 days. The most prominent genera were *Aspergillus*, *Fusarium* and *Talaromyces*, originating from the soil; and *Phanerochaete* only initially detected in fuels (diesel and B100). It can be suggested that biodiesel, being less toxic and more easily metabolized than diesel, allowed the establishment of a greater fungal diversity culturable during the bioremediation.

According to Harms, Schlosser and Wick (2011), most xenobiotic degrading fungi belong to the *Ascomycota* and *Basidiomycota* phyla. Among the genera known as degraders of aliphatic hydrocarbons, polyaromatic hydrocarbons, chlorophenols and pesticides, are *Acremonium*, *Fusarium*, *Penicillium* and *Phanaerochaete* (Chang 2008; Harms et al. 2011). The information about the biodiesel degrading fungal genera, presented here, is still scarce and more studies in this direction are extremely necessary.

Figure 11.8 shows the most prevalent fungal species observed in the microcosms. In total, 36 species were isolated and identified, the putative biodegrader species originating mainly from the soil. The greatest diversity was verified for the genera *Aspergillus*, *Talaromyces*, *Penicillium*, *Fusarium* and *Cladosporium*.

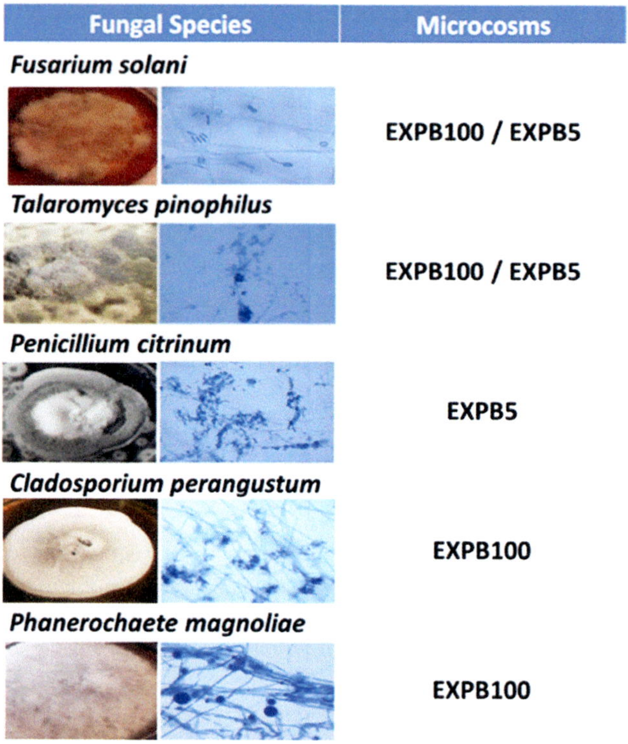

FIGURE 11.8 Most frequently fungal species isolated after 60 days of assay.

Each microcosm revealed distinct profiles. The most frequently isolated species were *Fusarium solani* and *Talaromyces pinophilus*, suggesting they are the main biodegraders in the systems analyzed. These species were observed acting from the superficial to the deepest soil layers, where the fuels were detected in the highest concentration. *F. solani* is highlighted in the literature because it has biodegradative activity for different hydrocarbons, alone or in consortium (Veignie et al. 2004; Rafin et al. 2009; Wu et al. 2010; Thion et al. 2012). However, there is a lack of information about its ability to metabolize particularly pure diesel and biodiesel, as well as their blends.

Particularly in EXP B5, *Penicillium citrinum* shows a certain activity in the B5 biodegradation, having been isolated throughout the soil column. This species has already been reported as degrading hydrocarbons (Oudot et al. 1993; April et al. 2000). In relation to EXP B100, *Cladosporium perangustum* and *Phanerochaete magnoliae* were present also at the different reactor depths. *P. magnoliae* was also observed originally in B100, which suggests its importance for biodiesel biodegradation, since it appeared as a colonizer of such fuel. In fact, the literature describes that some species of this genus can express the ability to degrade xenobiotics (Prenafeta-Boldu et al. 2006; Pinedo-Rilla et al. 2009; Harms et al. 2011; Gadd 2016).

In order to complement the culturable fungal microbiome description, the presence of yeasts was verified. They were detected in all substrates used to probe the microcosms composition (soil and fuels). In this context, it is possible to highlight the populations of the genera *Cryptococcus* and *Aureobasidium* in EXPB5. Both have already been characterized by their capacity for biodegradation of hydrocarbons either by direct metabolism or mediated by biosurfactants (Van Hamme et al. 2003; Van Beilen & Witholt 2005; Prince 2005; Corradi da Silva et al. 2006).

11.8 ECOTOXICITY ASSESSMENT

The bioremediation process reported in this chapter was a consequence of TPH and/or FAME reduction mediated by autochthonous microbial populations. However, this biodegradation of the contaminants is not necessarily relatated to the decrease of toxicity after treatment, mainly considering petroleum hydrocarbons (Dorn and Salanitro 2000).

In order to verify the aforementioned relationship, the degree of soil toxicity was evaluated immediately after the contamination and after 60 days of bioremediation. The model used for the ecotoxicity tests was the mortality of earthworms (*Eisenia fetida*). This organism has been found to be a very sensitive indicator to the presence of soil contamination.

Earthworm mortality rates for soil immediately after contamination and at the end of NA (60 days) were compiled in Figure 11.9 for microcosms B5 (see Figure 11.9A) and B100 (see Figure 11.9B). It can be noticed that the same test was done simultaneously with the uncontaminated pristine soil (negative control), that showed 0% mortality. Comparative analysis revealed the highest toxicity degree for TPH (see Figure 11.9A), although FAME contamination alone caused 63% mortality (see Figure 11.9B). The 14-day mortality readings show that there was a decrease in the toxicity of the contaminated soil. It must be pointed out that in EXP B100, after 60 days of bioremediation, toxicity reached values equal to those of the negative control.

11.9 FINAL REMARKS AND CONCLUSIONS

The ever-increasing participation of biodiesel in the worldwide energy landscape has made available more knowledge about the fate of this fuel, especially for diesel/biodiesel blends, in cases of environmental contamination occurrence. This chapter has demonstrated the dynamics of culturable microbial populations participating in the B5 and B100 biodegradation in artificially contaminated soil. The monitoring of

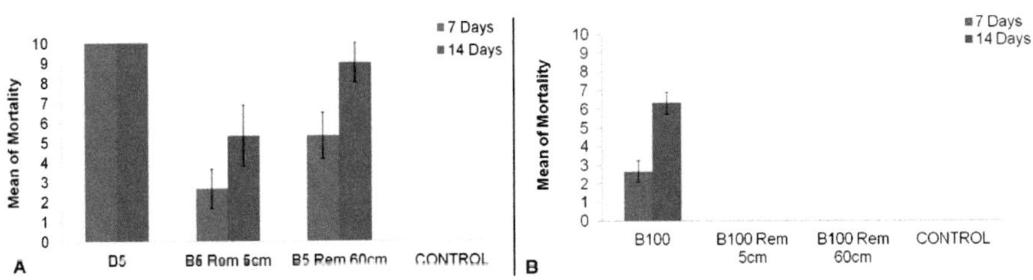

FIGURE 11.9 Ecotoxicological evaluation using *Eisenia fetida*.

the natural attenuation bioremediation was adopted as a treatment strategy in order to simulate the effect of blending petroleum hydrocarbons and fatty acid methyl esters on the autochthonous microbial community profile and in the biodegradation activity under natural conditions.

These results demonstrate the participation of yeasts and, mainly, bacteria and filamentous fungi in the NA bioremediation of contaminated soil up to 60 cm deep. The majority of the microbial species originated from the pristine soil, but the fuels also contained microrganisms that participate in the biodegradation. Effective bioremediation was observed for the contaminated soils, for pure biodiesel (B100) and low concentration blends (B5). However, the efficacy of microbial biodegradation for B100 was higher, promoting a reduction of almost 60% of the initial mass of FAME in the soil in 60 days. The presence of hydrocarbons (B5) negatively influenced FAME biodegradation, in the same way that the presence of FAME, even in low concentrations, seems to impair TPH degradation.

This observation is probably due to the differentiation of the microbial community observed in the soils as a consequence of contaminant action. Possibly these populations, even including TPH biodegraders, preferentially metabolized FAME as a carbon source, as seen in the (bio)degradation profile, which may have influenced the efficiency of the biodegradation capacity in the EXPB5 microcosm. This finding raises questions concerning previous studies that had observed an increase in biodegradation of blend contaminated systems directly related to the biodiesel addition under aerobic (Bücker et al. 2011; Lisiecki et al. 2014) and anaerobic conditions (Sørensen et al. 2011).

The data presented here are intended to assist in the monitoring and evaluation of the impacts on soils caused by contamination from biodiesel and diesel/biodiesel blends. In addition, more suitable procedures to remediate areas contaminated by such fuels can be better defined by predicting the pathways of their biotransformation in association with the microbiota involved. In this way, more improved techniques of removal of these soil contaminants, including bioaugmentation and/or biostimulation, could be applied more accurately and efficiently.

NOTES

1. Nutrient Broth: Meat peptone (10 g/L); Meat extract (3 g/L); K_2HPO_4 (1 g/L); NaCl (8.5 g/L).
2. Reducing solution: Sodium thioglycollate (0.124 g/L); Ascorbic acid (0.10 g/L); Resazurin (4.0 mL); NaCl (8.5 g/L).

REFERENCES

Aboim, M.C.R., et al. 2004. *Avaliação de diversidade microbiana em amostras de solos: técnica do PCR/DGGE (Protocolo Laboratorial)*. Rio de Janeiro: Embrapa Solos, n. 68, p. 31.

Associação Brasileira de Normas Técnicas (ABNT). 2007. *NBR 15537: Ecotoxicologia terrestre–Toxicidade aguda–Método de ensaio com minhocas (Lumbricidae)*. Rio de Janeiro, p. 11.

Alexander, M. 1999. *Biodegradation and Bioremediation*. San Diego, California: Academic Press Inc., p. 302.

Altschul, S.F., et al. 1990. Basic local alignment search tool. *J Mol Biol.*, v. 215, n. 3, pp. 403–410.

April, T.M., Foght, J.M., and Currah, R.S. 2000. Hydrocarbon-degrading filamentous fungi isolated from flare pit soils in Northern and Western Canada. *Can J Microbiol.*, v. 46, n. 1, pp. 38–49.

Boopathy, R. 2004. Anaerobic biodegradation of no. 2 diesel fuel in soil: a soil column study. *Bioresource Technol.*, v. 94, n. 2, pp. 143–151.

Bücker, F., Santestevan, N.A., Roesh, L.F., et al. 2011. Impact of biodiesel on biodeterioration of stored Brazilian diesel oil. *Int Biodeter Biodegr.*, v. 65, n. 1, pp. 172–178.

Chang, Y.S.J. 2008. Recent developments in microbial biotransformation and biodegradation of dioxins. *J Mol Microbiol Biotechnol.*, v. 15, n. 2–3, pp. 152–171.

Chemlal, R., Abdi, H., Lounici, N., et al. 2013. Modeling and qualitative study of diesel biodegradation using biopile process in sandy soil. *Int Biodeter Biodegr.*, v. 78, pp. 43–48.

Chen, M., Xu, P., Zeng, G., et al. 2015. Bioremediation of soils contaminated with polycyclic aromatic hydrocarbons, petroleum, pesticides, chlorophenols and heavy metals by composting: applications, microbes and future research needs. *Biotechnol Adv.*, v. 33, pp. 745–755.

Crawford, R., and Crawford, D. 1996. *Bioremediation: Principles and Applications.* Cambridge, England: Cambridge University Press, p. 406.

Cyplick, P., Schmidt, M., Szulc, A., et al. 2011. Relative quantitative PCR to assess bacterial community dynamics during biodegradation of diesel and biodiesel fuels under various aeration conditions. *Bioresour Technol.*, v. 102, n. 6, pp. 4347–4352.

Dorn, P.B., and Salanitro, J.P. 2000. Temporal ecological assessment of oil contaminated soils before and after bioremediation. *Chemosphere*, v. 40, n. 4, pp. 419–426.

Dragun, J. 1998. *The Soil Chemistry of Hazardous Materials.* Amherst, Massachusetts: Amherst Scientific Publishers Ed., 1998. p. 830.

EN 14105:2011-Fatty Acid Methyl Esters (FAME). 2011. *Determination of free and total glycerol and mono-di- and triglyceride contents.* p. 26.

Fernández-Alvarez, P., Vila, J., Garrido, J.M., et al. 2007. Evaluation of biodiesel as bioremediation agent for the treatment of the shore affected by the heavy oil spill of the Prestige. *J Hazard Mater*, v. 147, pp. 914–922.

Gadd, G.M. 2016. Fungi and industrial pollutants. In: Druzhinina, I. S.; Kubicek, C. P. (Eds.). *Environmental and Microbial Relationships.* Basel, Switzerland: Springer International Publishing.

Grama, B.S., et al. 2014. Characterization of fatty acid and carotenoid production in an *Acutodesmus* microalga isolated from the Algerian Sahara. *Biomass Bioenerg.*, v. 69, pp. 265–275.

Haritash, A.K., and Kaushik, C.P. 2009. Biodegradation aspects of polycyclic aromatic hydrocarbons (PAHs): a review. *J Hazard Mater.*, v. 169, n. 1–3, pp. 1–15.

Harms, H., Schlosser, D., and Wick, L.Y. 2011. Untapped potential: exploiting fungi in bioremediation of hazardous chemicals. *Nature Rev Microbiol.*, v. 9, n. 3, pp. 177–192.

Horel, A., and Schiewer, S. 2011. Influence of constant and fluctuating temperature on biodegradation rates of fish biodiesel blends contaminating Alaskan sand. *Chemosphere*, v. 83, n. 5, pp. 652–660.

Huppel, T., and Huybrighs, T. 2012. Fatty Acid Methyl Esters in B100 Biodiesel by Gas Chromatography (Modified EN 14103). *Application Note–Gas Chromatography.* Downers Grove, IL: PerkinElmer, Inc.

Institut Scientifique de Service Public (ISSeP). 2010. Fractionnement des hydrocarbures aromatiques et aliphatiques semi-volatils et quantification par chromatographie en phase gazeuse. S-III-&7V3.

Jakeria, M.R., Fazal, M.A., Haseeb, A.S.M.A., et al. 2014. Influence of different factors on the stability of biodiesel: a review. *Renew Sust Energ Rev.*, v. 30, pp. 154–163.

Junior, J.S., Mariano, A.P., and Angelis, D.F. 2009. Biodegradation of biodiesel/diesel blends by *Candida viswanathii*. *Afr J Biotechnol.*, v. 8, n. 12, pp. 2774–2778.

Keskin, N.O.S., Han, D., Ozkan, A.D., et al. 2015. Production and structural characterization of biosurfactant produced by newly isolated *Staphylococcus xylosus* STF1 from petroleum contaminated soil. *J Petrol Sci Eng.*, v. 133, pp. 689–694.

Khan, F.I., Husain, T., and Ramzi, H. 2004. An overview and analysis of site remediation technologies. *J Environ Manage.*, v. 71, pp. 95–122.

Klessmann, C., Held, A., Rathmann, M., Ragwitz, M., et al. 2011. Status and perspectives of renewable energy policy and deployment in the European Union - What is needed to reach the 2020 targets? *Energ Policy*, v. 39, pp. 7637–7657.

Li, H., Zhang, Y., Kravchenko, I., Xu, H., and Zhang, C. 2007. Dynamic changes in microbial activity and community structure during biodegradation of petroleum compounds: a laboratory experiment. *Journal of Environmental Sciences*, v. 19, n. 8, pp. 1003–1013.

Li, X., Li, P., Lin, X., et al. 2008. Biodegradation of aged polycyclic aromatic hydrocarbons (PAHs) by microbial consortia in soil and slurry phases. *J Hazard Mater*, v. 150, n. 1, pp. 21–26.

Lisiecki, P., Chrzanowski, L., Szulc, A., et al. 2014. Biodegradation of diesel/biodiesel blends in saturated sand microcosms. *Fuel*, v. 116, n. 15, pp. 321–327.

Meyer, D.D., Beker, S.A., Bücker, F., et al. 2014. Bioremediation strategies for diesel and biodiesel in oxisol from southern Brazil. *Int Biodeter Biodegr.*, v. 95, part B, pp. 356–363.

McCurry, J.D. 2012. GC Analysis of Total Fatty Acid Methyl Esters (FAME) and Methyl Linolenate in Biodiesel Using the Revised EN14103:2011 Method. *Application Note*. Agilent Technologies, Inc. USA.

Muyzer, G., and Stams, A.J.M. 2008. The ecology and biotechnology of sulphate-reducing bacteria. *Nature Rev Microbiol.*, v. 6, n. 6, pp. 441–454.

Muyzer, G., De Waal, E.C., and Uitterlindern, A.G. 1993. Profiling of complex microbial populations by denaturing gradient gel electrophoresis analysis of polymerase chain reaction-amplified genes coding for 16S rRNA. *Appl Environ Microbiol.*, v. 59, n. 3, pp. 695–700.

Österreicher-Cunha, P., Vargas, Jr, E.A., Antunes, F.S., et al. 2012. Influence of soil and climate on carbon cycling and microbial activity of a heterogeneous tropical soil. *Geomicrobiol J.*, v. 29, pp. 399–412.

Oudor, J., Duport, J., Haloui, S., and Roquebert, M.F. 1993. Biodegradation potential of hydrocarbon assimilating tropical fungi. *Soil Biology and Biochemistry*, v. 25, pp. 1167–1173.

Owsianiak, M., Chrzanowski, Ł., Szulc, A., et al. 2009. Biodegradation of diesel/biodiesel blends by a consortium of hydrocarbon degraders: effect of the type of blend and the addition of biosurfactants. *Bioresour Technol.*, v. 100, n. 3, pp. 1497–1500.

Paria, S. 2008. Surfactant-enhanced remediation of organic contaminated soil and water. *Adv Colloid Interface Sci.*, v. 138, n. 1, pp. 24–58.

Petit, J.F. 2004. *Analyse des Hydrocarbures Aromatiques Polycycliques dans les sols– Pratiques de laboratoire d'analyses.* In: Comparaison de méthodes d'analyse des éléments traces et métalliques (ETM) et des hydrocarbures aromatiques polycycliques (HAP) sur les sols et les végétaux. Laboratoire Régional des Ponts et Chaussées de Lille.

Pinedo-Rilla, C., Aleu, J., and Collado, L.G. 2009. Pollutants biodegradation by fungi. *Curr Org Chem.*, v. 13, pp. 1194–1214.

Prenafeta-Boldú, F.X., Summerbell, R., and Sybren de Hoog, G. 2006. Fungi growing on aromatic hydrocarbons: biotechnology's unexpected encounter with biohazard? *FEMS Microbiol Rev.*, v. 30, n. 1. pp. 109–130.

Prince, R.C. 2005. The microbiology of marine oil spill bioremediation. *Petroleum Microbiology*, American Society of Microbiology, Washington, pp. 317–335.

Prince, R.C., Haitmanek, C., Lee, C.C., et al. 2008. The primary aerobic biodegradation of biodiesel B20. *Chemosphere*, v. 71, n. 8, pp. 1446–1451.

Rafin, C., Veignie, E., Faveulle, A., Surpateanu, G., et al. 2009. Benzo [a] pyrene degradation using simultaneously combined chemical oxidation, biotreatment with *Fusarium solani* and cyclodextrins. *Bioresource Technol.*, v. 100, n. 12, pp. 3157–3160.
Rathmann, R., Benedetti, O., Plá, J.A., Padula, A.D., et al. 2005. *Biodiesel*: Uma Alternativa Estratégica Na Matriz Energética Brasileira? Página Oficial do Biodiesel do Governo Federal do Brasil. Brasília.
Santos, I.M., Venâncio, A., and Lima, N. 1998. *Fungos Contaminantes na Indústria Alimentar.* Ed. Micoteca da Universidade do Minho. Centro de Engenharia Biológica. Edit. Sociedade Gráfica, S.A. Braga. p. 130.
Schwartz, G., Ben-dor, E., and Eshel, G. 2012. Quantitative analysis of total petroleum hydrocarbons in soils: comparison between reflectance spectroscopy and solvent extraction by 3 certified laboratories. *Appl Environ Soil Sci.*, 11 pages (Article ID 751956, http://dx.doi.org/10.1155/2012/751956).
Silva, G.S., Marques, E.L.S., Dias, J.C.T., et al. 2012. Biodegradability of soy biodiesel in microcosm experiments using soil from the Atlantic Rain Forest. *Appl Soil Ecol.*, v. 55, pp. 27–35.
Silva, M.L.C. da, Martinez, P.F., Izeli, N.L., et al. 2006. Caracterização química de glucanas fúngicas e suas aplicações biotecnológicas. *Quim Nova.*, v. 29, n. 1, pp. 85–92.
Singh, A., Kuhad, R.C., and Ward, O.P. 2009. *Advances in Applied Bioremediation.* New York: Springer Dordrecht, v. 17, pp. 173–187.
Singh, R., Paul, D., and Jain, R.K. 2006. Biofilms: implications in bioremediation. *Trends Microbiol.*, v. 14, n. 9, pp. 389–397.
Sorda, G., Banse, M., and Kemfert, C. 2010. An overview of biofuel policies across the world. *Energy Policy*, v. 38, n. 11, pp. 6977–6988.
Sørensen, G., Pedersen, D.V., Nørgaard, A.K., Sørensen, K.B., Nygaard, S.D., et al. 2011. Microbial growth studies in biodiesel blends. *Bioresour Technol.*, v. 102, n. 8, pp. 5259–5264.
Soriano, A.U., Martins, L.F., Ventura, E.S., et al. 2015. Microbiological aspects of biodiesel and biodiesel/diesel blends biodeterioration. *Int Biodeter Biodegr.*, v. 99, pp. 102–114.
Suarez, P.A.Z., Meneghetti, S.M.P., Meneghetti, M.R., Wolf, C.R., et al. 2007. Transformação de triglicerídeos em combustíveis, materiais poliméricos e insumos químicos: algumas aplicações da catálise na oleoquímica. *Quím Nova*, v. 30, n. 3, pp. 667–676.
Thion, C., Cébron, A., Beguiristain, T., Leyval, C., et al. 2012. PAH biotransformation and sorption by *Fusarium solani* and *Arthrobacter oxydans* isolated from a polluted soil in axenic cultures and mixed co-cultures. *Int Biodeter Biodegr.*, v. 68, pp. 28–35.
United States Environmental Protection Agency (USEPA). 2007. *Method 3550C: Ultrasonic Extraction.*
United States Environmental Protection Agency (USEPA). 2012. *A Citizen's Guide to Soil Vapor Extraction and Air Sparging.* EPA/542/F-12/018, U.S. Environmental Protection Agency, Office of Solid Waste and Emergency Response, Washington, D.C.
Van Beilen, J.B., and Witholt, B. 2005. Diversity, function, and biocatalytic applications of alkane oxygenases. In: Ollivier, B., and Magot, M. (Eds.) *Petroleum Microbiology.* Washington: ASM press, pp. 277–299.
Van Hamme, J.D., Singh, A., and Ward, O.P. 2003. Recent advances in petroleum microbiology. *Microbiol Mol Biol Rev.*, v. 67, n. 4, pp. 503–549.
Veignie, E., Rafin, C., Woisel, P., Cazier, F., et al. 2004. Preliminary evidence of the role of hydrogen peroxide in the degradation of benzo [a] pyrene by a non-white rot fungus *Fusarium solani. Environ Pollut.*, v. 129, n. 1, pp. 1–4.
Vianna, F. C. 2006. *Análise de ecoeficiência:* avaliação do desempenho econômico-ambiental do biodiesel e petrodiesel. 2006. Dissertação (Mestrado em Engenharia) – Escola Politécnica, Universidade de São Paulo, São Paulo, 182 pages.

Watanabe, K. 2001. Microorganisms relevant to bioremediation. *Curr Opin Biotechnol.*, v. 12, n. 3, pp. 237–241.

Weisman, W.H. 1998. *Analysis of Petroleum Hydrocarbons in Environmental Media - Total Petroleum Hydrocarbon Criteria Working Group Series.* Amherst, MA: Amherst Scientific Publishers, v. 1.

White, T. J., et al. 1990. Amplification and direct sequencing of fungal ribosomal RNA genes for phylogenetics. In: Innis M.A., Gelfand, D.H., Sninsky, J.J., White, T.J. (Eds.). *PCR Protocols: A Guide to Methods and Applications.* New York, N.Y: Academic Press, Inc., pp. 315–322.

Wu, S., Yassine, M.H., Suidan, M.T., Venosa, A.D., et al. 2015. Anaerobic biodegradation of soybean biodiesel and diesel blends under methanogenic conditions. *Water Res.*, v. 87, pp. 395–402.

Wu, Y.-R., Luo, Z.-H., and Vrijmoed, L.L.P. 2010. Biodegradation of anthracene and benz [a] anthracene by two *Fusarium solani* strains isolated from mangrove sediments. *Bioresource Technol.*, v. 101, n. 24, pp. 9666–9672.

12 Bioremediation of Crude Oil by Indigenous Bacteria in Bohai Bay

*Weiwei Chen, Jun Min, and Xiaoke Hu**

CONTENTS

12.1 INTRODUCTION

As strategic energy sources and chemical industry materials, crude oil plays an important role in modern society. Crude oil is a natural, heterogeneous mixture and contains potentially 20,000 chemical components (Marshall and Rodgers, 2004). However, oil spills, due to natural or anthropogenic factors (Mapelli et al., 2017; McKew et al., 2007), including crude oil production and refining, natural seepage from reservoirs, freshwater and terrestrial run-off, and routine ship operations, have been causing noticeable environmental sustainability and human health problems (Gu and Wang, 2015; Li et al., 2012). It is becoming increasingly urgent and crucial to search for effective methods to clean up petroleum pollution (Mbadinga et al., 2011; Ouyang et al., 2005).

So far, many technologies have been developed for the treatment of crude oil spills, including physical, chemical and biological methods, such as sorption (Hussein et al., 2008), thermal treatment (Li et al., 2009), surfactants solutions (Agarwal and Liu, 2015) and so on. Compared to physical and chemical methods, bioremediation,

*Corresponding author: Xiaoke Hu; Email: xkhu@yic.ac.cn

using microorganisms to remove and degrade crude oil, is recognized as more economical, efficient and environmentally friendly (McGenity, 2014; Wang et al., 2011), which generally includes natural attenuation, biostimulation and bioaugmentation (Mapelli et al., 2017; Bento et al., 2005). Seawater and sediments always harbor a large number of microorganisms, which form the basis of the food web. These microorganisms contribute to global biogeochemical cycles and bioremediation of pollutants in marine ecosystems (Gomez et al., 2007; McGenity et al., 2012).

To date, numerous crude oil-degrading microorganisms have been isolated from different habitats, such as *Alcanivorax*, *Acinetobacter*, *Pseudomonas*, *Dietzia*, etc. (Lin et al., 2014a; Zhang et al., 2011; Liu et al., 2010), and they have been well studied in laboratory. However, little is known about the functions of individual bacterial taxa and co-acclimations between different bacteria in the degrading process (Wang et al., 2016). With the development of the genomics and metabolomics, studies on the genetic background of petroleum hydrocarbons degradation and metabolic mechanisms are supposed to understand the function of interactions between microorganisms and hydrocarbons, which need to be verified in more bioremediation application cases (Jones et al., 2011; Head et al., 2006). As one of the largest marine oil pollution accidents, the Deepwater Horizon oil spill released approximately 779 million liters of crude oil in 2010. Around 2 million gallons of aerial and subsurface dispersants were applied to improve the bioavailability of crude oil and speed up the bioremediation process (Atlas and Hazen, 2011).

Although a considerable amount of researches on microbial petroleum hydrocarbon degradation have been reported (Chen et al., 2017; Li et al., 2017; Yang et al., 2015; Teira et al., 2007), including the isolation, physiology and metabolism, until recently we had limited knowledge about which bacteria are the most important hydrocarbon degraders in the environment (Head et al., 2006). Furthermore, we have lacked a solid theoretical basis to support and predict the behaviour and interactions of hydrocarbon-degrading bacterium or consortium *in situ* environment. Thus, this chapter focuses on our growing understanding of the bacteria that are responsible for petroleum hydrocarbons removal and how indigenous bacteria can be exploited and applied in bioremediation of marine crude oil pollution.

12.2 THE STRATEGIC IMPORTANCE OF BOHAI BAY AND CRUDE OIL POLLUTION SITUATION

The Bohai Sea, with an average water depth of 18 metres, is a semi-enclosed marginal sea on the northern coast of China (Liu et al., 2015). Generally, the Bohai Sea is often divided into three major bays, including the Bohai Bay in the west, the Liaodong Bay in the north and the Laizhou Bay in the south, and the remaining area is called the Central Area (Gao et al., 2014). Due to the good ability of shelter condition and wave attenuation, Bohai Bay becomes an important marine ecosystem in northern China (Sun et al., 2010), which is also the first selection for port construction and marine aquaculture. Bohai Bay is rich in oil and gas resources and surrounded by a series of industrial zones known as the Bohai Sea Economic Rim

(BER) (Gao et al., 2014). The BER has gone through major changes and transformations in manufacturing, heavy industry and chemical industry, which complied with the development of reform and opening-up policy in China at the end of 1970s. The BER is about 5.2×10^5 km^2 in area and 230 million in population, accounting for about 5.4% and 17.5% of total land area and population in China, respectively (Gao et al., 2014; Gao and Chen, 2012). The BER is rising as a northern economic "powerhouse," competing with the Pearl River Delta Economic Rim in the south and the Yangtze River Delta Economic Rim in the east (Gao et al., 2014; Gao and Chen, 2012).

The ecosystem health of semi-enclosed bays is related to economic development at both regional and larger scales (Chen and Yu, 2011). Meanwhile, the development of economy is always accompanied by environmental pollution and destruction (Pan and Wang, 2012). Along with the exploitation of crude oil and accelerated economic development of the BER, especially the fast-growing industrialization and urbanization progress, the Bohai Bay has been facing serious threats and challenges from petroleum and other pollution (Gao et al., 2014; Gao and Chen, 2012). In June 2011, a serious crude oil spill occurred in Penglai 19-3 oilfield of Bohai Bay, which was operated by ConocoPhillips China Inc. (COPC). According to COPC, a sum of approximate 723 barrels (115 cubic metres) of crude oil flowing into the Bohai Sea and 2620 barrels (416 cubic metres) of mineral oil-based drilling mud stranding under the sea. By June 17, it was reported that 840 square kilometres of clean coastal water were polluted and it was also the first large-scale oil spill from a submarine oil well in China (Liu et al., 2015; Liu and Zhu, 2014). This accident was considered not only an ecosystem disaster, which diminished species diversity, reduced sediments quality and deteriorated aquacultures, but also a complex event resulting in much concern from the public and official governments including both the North Korea and the South Korea. Korea media expressed that the spill would lead to a cross-country environmental damage to Korea, if the spilled oil flow into the Yellow Sea. More detailed information can be found at http://en.wikipedia.org/wiki/2011_Bohai_Bay_oil_spill.

The BER's high economic development rate put the Bohai under more environmental pollution pressure compared with other sea areas in China. Bohai Bay is the only recipient of the wastewater from Beijing, Tianjin and Hebei province. The semi-enclosed structure and relatively slow water exchange with the ocean further aggravate the situation. It is expected that the ecosystem health of Bohai Bay has gradually deteriorated. China Marine Environmental Quality Bulletin for the Year 2012 reported that the BER brought huge pressure of pollution and habitat destruction to the Bohai Sea; the ecosystem of estuaries and bays in the Bohai Sea were all in sub-healthy or unhealthy state (SOA, 2012). The Chinese government pay close attention to the ecological environment protection of Bohai sea. For example, the Chinese government launched a 15-year program—"Bohai Blue Sea Action Plan," with a budget of about 55.5 billion RMB ($9.1 billion US) in 2001, and a further 40 billion RMB ($6.6 billion US) in 2008, to protect Bohai Sea ecosystem (Gao et al., 2014). However, nothing definite has been accomplished and improving the ecological environment of Bohai sea still has a long way to go.

12.3 CURRENT OIL SPILLS REMEDIATION METHODS IN MARINE ENVIRONMENTS

An oil spill in the ocean is subject to several processes, including spreading, drifting, emulsification, evaporation, photo-oxidation, dissolution, adsorption, biodegradation and sinking into the sea (McGenity et al., 2012). Among the processes, lightweight saturated hydrocarbons evaporate easily and lightweight aromatic hydrocarbons dissolve into the water, whereas the majority of heavy oil remains attached to the sediments. The superficial crude oil could slowly break down into simpler compounds via air oxidation (USEPA, 1999). Compared with evaporation and oxidation, natural biodegradation of oil is a very slow process requiring months or even years (Atlas and Hazen, 2011). Moreover, as a result of the toxicity of crude oil, biodegradation by microorganisms occurs slowly at the initial stages (USEPA, 1999). Developing novel and effective technologies for separating and removing oil from oil-contaminated areas is of considerable environmental and economic benefits. In the past few decades, various physical, chemical and biological technologies have been developed for oil removal from marine waters and sediments (Couto et al., 2009; Shin and Shen, 2011). Some of these approaches have also been studied and demonstrated in site (Chien, 2012; Lee et al., 2014).

12.3.1 Physical Remediation

The physical remediation methods are always used as first-response mitigation technologies when crude oil spills occur in the ocean, such as booms, skimmers and absorbent materials, which collect the crude oil rapidly. However, these methods are expensive and cannot accomplish complete cleanup of thin and lightweight oil film, which could be the emergency managements in general (Agarwal and Liu, 2015; Mapelli et al., 2017). The flotation technique has been commonly used to separate asphaltene from oil-contaminated sediments in the petroleum industry (Zhang et al., 2001), which has a high oil removal efficiency and inexpensive. In addition, several studies on the utilization of cationic, anionic as well as nonionic surfactants for removing oil from oil-contaminated areas with both the flotation separation and washing methods have been reported (Ceschia et al., 2014; Duffield et al., 2003). Thermal desorption is a simple and rapid method to remove crude oil in a short time and it saves the cost of excavation and conveyance while avoiding secondary pollution, which is properly suitable *in situ* remediation (Falciglia et al., 2011). There are some other physical methods to solve crude oil pollution, including ultrasonic desorption (Agarwal and Liu, 2015), porous medium filtration and coalescence including porous beds (Huang and Lim, 2006) and membranes (Chakrabarty et al., 2010).

12.3.2 Chemical Remediation

By contrast, chemical remediation is extremely fast, effective and cheap, with no need for on-site excavation and off-site treatments (USEPA, 1998). The chemical dispersants and oil coagulating agents play a particularly significant role in treating oil spill at sea, especially to the thin oil film. To date, several chemical oxidants such as dispersants, hydrogen peroxide including ozone, persulphate and permanganate have been utilized in the remediation of crude oil pollution (Do et al., 2010; Ferguson et al., 2004; Tsai and Kao, 2009). Utilizing hydrogen peroxide or ozone to remediate oil pollution could not only convert the contaminants to more biodegradable components but also provide oxygen to the microbial community, thus improving remediation (Goi et al., 2006). As a strong oxidizing agent, persulphate could generate more powerful free sulphate radicals with a redox potential of 2.40 V on being activated by heat or ultraviolet (UV) light, which has been increasingly preferred for *in situ* chemical oxidation of oil-contaminated sediments (Yen et al., 2011). Nevertheless, chemical methods, particularly dispersants, are only allowed at depths greater than 15 metres as their overall effectiveness is doubtful and potential toxicity effects on marine organisms are higher (Kirby and Law, 2008). Chemical remediation methods combined with bioremediation are considered to be cost-effective and environmentally friendly for the remediation of oil-contaminated areas.

12.3.3 Bioremediation

Bioremediation technique uses hydrocarbon-degrading microorganisms such as fungi and bacteria to degrade oil contaminants into non-toxic and simple inorganic compounds, mostly carbon dioxide and water (Kostka et al., 2011). Hydrocarbon-degrading microorganisms feed on oil contaminations and take them as carbon and energy sources to maintain growth and reproduction. Compared with physical and chemical remediation, bioremediation is deemed to more effective, versatile, economical and environmentally friendly to eliminate oil pollution (Margesin et al., 2013; McGenity, 2014).

As the remediation study develops, there are some typical and valuable oil-degrading bacteria. Most notable is the genus *Alcanivorax*, which degrade straight-chain and branched alkanes (Head et al., 2006), followed by genus *Cycloclasticus*, utilizing polycyclic aromatic hydrocarbons (PAHs) with high efficiency (Teira et al., 2007). These species can efficiently degrade a wide range of saturated, aromatic hydrocarbons and PAHs into simpler compounds. Furthermore, as the indigenous bacteria, bacterial communities make a difference after a crude oil spill occurs. Yang et al., (2015) found that α- and β-diversity of the bacterial communities were significantly affected by crude oil pollution, especially the BTEX (Benzene, Toluene, Ethylbenzene, Xylene) and the bacterial communities have hydrocarbon biodegradation potential in Bohai sea.

For faster recovery of oil-contaminated fields, biostimulation and bioaugmentation are two common strategies used to accelerate the natural biodegradation of oil. Biostimulation is the addition of nutrients (mainly N&P), oxygen or biosurfactants to promote indigenous bacteria growth (Coulon et al., 2007; Whang et al., 2008). Thanks to the diversity of microorganisms and their metabolites, microorganisms produce many kinds of biosurfactants with different structure and physico-chemical properties, such as glycolipid, phospholipids, lipopeptides, fatty acids, neutral lipids and so on (Bharali and Konwar, 2011). As a result of the excellent surface activities, lower toxicity, easy availability and high biodegradability (Qian and Shao, 2010; Wang et al., 2013), biosurfactants are considered as potential candidates for the bioremediation of oil spills. The biosurfactants could reduce the interfacial tension, thus facilitating the availability of oil for enhanced biodegradation (Szulc et al., 2014). For instance, Wang et al. (2014) found that *Dietzia maris* As-13-3, isolated from deep sea hydrothermal field, could produce di-rhamnolipid as biosurfactant in *n*-hexadecane degradation assay. The di-rhamnolipid lowered the surface tension of water from 74 ± 0.2 to 38 ± 0.2 mN m^{-1}. Furthermore, bioaugmentation involves inoculation of an enriched functional microbial culture into the polluted environment (Thompson et al., 2005). Both laboratory and field studies have shown that bioaugmentation could speed up oil degradation (Mishra et al., 2001; Szulc et al., 2014). For instance, in a field study, the addition of specific microorganisms enhanced the total petroleum hydrocarbons removal by 76% compared with the 3.6% removed in the control groups (Mukherjee and Bordoloi, 2011). Although the effectiveness of bioaugmentation is still debatable, the supplement of oil-degrading microorganisms has been proposed to assist indigenous bacteria for degradation of more crude oil components (Bouchez et al., 2000).

12.4 INDIGENOUS BACTERIA PLAY IMPORTANT ROLE IN BOHAI BAY BIOREMEDIATION

So far, some indigenous bacteria have been isolated from Bohai Sea and studied in the laboratory, such as *Acinetobacter*, *Pseudomonas*, *Dietzia*, *Bacillus* and so on (Chen et al., 2017; Li et al 2016; Chen et al., 2014; Lin et al., 2014a). Li et al., (2016) isolated five crude oil-degrading strains in the Bohai Sea, and two strains (*Bacillus* and *Castellaniella*), with different degradation abilities to middle-chain *n*-alkanes and aromatics, were chosen to construct a consortium, which was capable of degrading about 51.87% of crude oil (2%, w/v) after one week of incubation in saline MSM medium. In another study, *Acinetobacter* sp. XM-02 and *Pseudomonas* sp. XM-01 were isolated from soil samples polluted by crude oil at Bohai offshore. Compared with pure culture, mixed culture of the two strains showed higher capability in degrading crude oil (87%) (Chen et al., 2014).

In our group, two salt-tolerant bacteria, *Acinetobacter* sp. HC8-3S and *Dietzia* sp. CN-3, were isolated from petroleum-contaminated sediments in the Penglai 19-3 oil spill field of Bohai Bay, China. *Acinetobacter* sp. HC8-3S showed relatively high biodegradation (>70%) of saturated hydrocarbons in the pH range of 5.6–8.6, especially in the alkaline condition. The degradation activities were not affected significantly when the NaCl concentration reached 90 g/L, which showed the excellent salinity tolerance (Lin et al., 2014a). *Dietzia* sp. CN-3 has the ability to degrade a wide range of alkanes (C_{14}–C_{31}), aromatic compounds and crude oil, with the efficiency degradation rate (91.6%) to the alkanes, which is higher than previously reported rates (66.6–82.5%), even for mixed bacterial consortia (Gao et al., 2015; Von der Weid et al., 2007).

Although different petroleum hydrocarbons degrading effects have been well studied in the laboratory by pure cultures, little is known about functions of individual bacterial taxa and co-acclimations between different bacteria in the degrading process (Wang et al., 2016). These microorganisms form part of an ecological network, which involves many direct and indirect interactions with other community members and the environments. McKew et al., (2007) investigated the key bacterial taxa involved in aerobic degradation of specific hydrocarbons in a previous work. The results indicated that bacteria in the genera *Talassolituns* and *Roseobacter* were responsible for *n*-alkane degradation; bacteria in the genus *Alcanivorax* dominated the process of the degradation of branched alkane, and *Cycloclasticus* was the dominant genus in PAHs degradation. Our group successfully revealed successions of bacterial communities and co-acclimation between different functional bacterial taxa under stresses of hydrocarbons with different structures (tetradecane, pristane, pyrene, hexacosane and phenanthrene) (Wang et al., 2016). Our study indicated that bacteria in an unclassified group in the family of *Erythrobacteraceae* was proposed as a new generalist for degrading different hydrocarbons, while bacteria in the genera *Aequorivita* and *Hyphomonas* were mainly responsible for stresses of pristane and PAHs, respectively. According to the network construction and functional prediction by KEGG database and PICRUSt, we could also demonstrate that functional bacteria co-acclimated to the changing environments and conducted biodegradation via interactions with each other. However, further investigations should be conducted on revealing individual functions of different bacterial taxa and interactions between different microorganisms in the process of biodegradation. More questions need to be explained in the future research. For example, how would bacterial community respond to different hydrocarbons or crude oil and what are their individual roles in the degradation process? That will provide useful information to guide the bioremediation of crude oil pollution.

12.5 PETROLEUM HYDROCARBONS DEGRADATION IN THE GENOMIC AGE

Because the saturated hydrocarbons constitute the largest fraction of the 20,000 chemical components of crude oil by mass, and aromatic hydrocarbons are toxic and persistent pollutants, the degradation of hydrocarbons are recognized to be the most important process in bioremediation of crude oil contamination (Head et al., 2006). So far, traditional molecular genetics has already offered many significant and valuable insights into the genetic basis theory of hydrocarbons degradation (Rojo, 2009; Head et al., 2006).

For the aerobic degradation of alkanes, bacteria use O_2 as a reactant for the activation of the alkane molecule, which is the first and key step. Aerobic pathways for the *n*-alkanes with two or more carbon atoms, usually including the terminal and subterminal oxidation (Rojo, 2009). The terminal oxidation happened from the teminal methyl group to a primary alcohol, which is further oxidized to the corresponding aldehyde and finally converted into a fatty acid. For another, the product generated a secondary alcohol, converted into the corresponding ketone and finally transformed into a fatty acid in the subterminal oxidation process. The process needs several classes of enzymes, among which alkane hydroxylases play a crucial role (Van Beilen and Funhoff, 2007). Three classes of alkane hydroxylases have been distinguished in microbial degraders. Methane monooxygenase (MMO) enzymes were responsible for short-chain-length alkanes (C_1–C_4). Medium-chain-length alkanes (C_5–C_{16}) degradation usually rely on soluble cytochrome P450 enzymes belonging to the CYP153 family (Munro et al., 2013; Rojo, 2009), and integral-membrane non-heme di-iron monooxygenase (*alk*B) (Van Beilen and Funhoff, 2005). The oxidation of long-chain-length alkanes (>C_{16}) depends on *Alm*A and *Lad*A, which were reported in *Acinetobacter* sp. DSM 17874 and *Geobacillus thermodenitrificans* NG80-2, respectively (Feng et al., 2007; Throne-Holst et al., 2007). However, these genes responsible for different chain-length alkanes may do not co-exist in the same bacterium and it is not so clear whether they cooperate with each other in alkane degradation. For example, the *Lad*A gene, found in *Geobacillus thermodenitrificans* NG80-2, oxidizes C_{15}–C_{36} alkanes, but does not utilize chain lengths smaller than C_{14} (Feng et al., 2007). In our study, CYP153 and *alk*B genes were cloned in *Dietzia* sp. CN-3, but no *Lad*A and *Alm*A genes were detected (Chen et al., 2017). The transcript levels of the *alk*B and CYP153 genes were detected by RT-qPCR when strain CN-3 induced by *n*-alkanes (C_{14}, C_{16} and C_{26}) and a branched alkane (pristane). The results indicated that CYP153 gene was dramatically induced by C_{14} (31.6 folds), C_{16} (61.3 folds), C_{26} (85.8 folds) and pristane (7.8 folds), but *alk*B gene was only weakly detected (Figure 12.1), which may be explained that the CYP153 gene likely played the predominant role in alkane degradation and two genes, especially the CYP153 gene, were responsible for the long-chain-length alkane degradation pathway in *Dietzia* sp. CN-3. Besides, several bacteria able to use alkanes as carbon source in the absence of O_2 have been described in the last few years (Widdel and Rabus, 2001). These strains use nitrate or sulfate as electron acceptor. Although growth is significantly slower than that of aerobic alkane degraders, anaerobic degradation of alkanes plays an important role in the degradation of hydrocarbons, especially in the deep-sea, thermal spring, oil reservoirs or other extreme environmental habitats (Rojo, 2009; Widdel and Rabus, 2001).

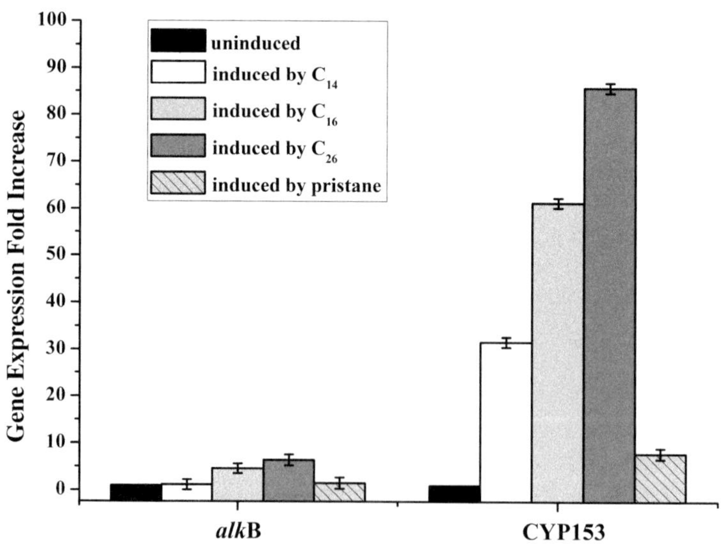

FIGURE 12.1 Transcription of the alkane degradation genes *alk*B and CYP153 in *Dietzia* sp. CN-3 grown on C_{14}, C_{16}, C_{26} and pristane.

For the aromatic hydrocarbons degradation, considerable efforts based on traditional cultivation-dependent methods have focused on isolating and identifying aromatic hydrocarbons degraders to explore the degradation efficiency (Li et al., 2017). Low-molecular-weight polycyclic aromatic hydrocarbons (PAHs), such as naphthalene, phenanthrene and anthracene, are usually degraded by bacteria in soil and under laboratory conditions (Lin et al., 2014b; Peng et al., 2008). However, less is known about the bacteria capable of utilizing PAHs containing five or more rings as a carbon and energy source, such as benz[α]anthracene and benzo[α]pyrene (Peng et al., 2008). Furthermore, functional genes associated with PAHs degradation, such as PAH-ring hydroxylating dioxygenases (PAH-RHDs) and PAH-ring cleaving dioxygenases (PAH-RCDs) (Cebron et al., 2011; Moser et al., 2001), including catechol dioxygenase and protocatechuate dioxygenase (Thomas et al., 2016; Peng et al., 2008), have been identified. Factually, it is difficult to obtain all the PAHs-degrading isolates by the cultivation-based approach and it fails to explain the complicated interactions among individuals within microbial communities in the environments (Jones et al., 2011).

With the developments of the sequencing technology, the whole genomic sequences of numerous organisms have become publicly available, such as the aerobic saturated hydrocarbon-degrading bacterium *Alcanivorax borkumensis* SK2 (Golyshin et al., 2003), aromatic hydrocarbons-degrading bacterium *Pseudomonas putida* KT2440 (Nelson et al., 2002) and *Azoarcus* sp. EbN1 (Rabus et al., 2005). Relying on genomic-data resource of whole genome sequencing and molecular genetics studies, the metagenomic studies of hydrocarbon-degrading bacteria could be investigated and carried out. In our study, the whole genomic data of *Acinetobacter* sp. HC8-3S were sequenced by pyrosequencing technology, which revealed the organization of the genes involved in the alkane oxidation (the whole genome sequence has not been submitted to GenBank). Genes involved in alkane metabolism in strain HC8-3S include alkane monooxygenases, rubredoxin, rubredoxin reductases, alcohol dehydrogenases, aldehyde dehydrogenases, acyl-CoA synthetases and regulator of *alk* gene expression. However, the gene coding for the alkane hydroxylase maps separately from those of the rubredoxin and rubredoxin reductase, and the genes coding for alcohol and aldehyde dehydrogenases map separately as well. The regulator of *alk* gene expression belongs to the AraC superfamily, which was not map adjacent to the *alk* gene. The wise application of genomic technologies, connected with more traditional biochemical and microbial community analyses, provides excellent opportunities for increasing our understanding and awareness about hydrocarbon degradation in the environments.

12.6 UTILIZATION OF IMMOBILIZATION TECHNIQUE TO ENHANCE DEGRADATION EFFICIENCY

There is the fact that the utilization of lab-selected bacterial strains, with high degradation rates and broad substrate ranges, may fail in natural environments. The bacteria must have the ability to overcome biotic and abiotic stresses in the environment. Macnaughton et al. (1999) demonstrated the early disappearance of a microbial consortium introduced in a natural polluted site. The microbial cells could provide valuable degradation in lab experiments but were not suited to industrial applications. Especially, in an open ocean system, how to maintain the quantities and functional stability of bacteria, as well as to increase the bioavailability of crude oil are the key factors to enhance the effectiveness of bioremediation (McGenity et al., 2012; Thompson et al., 2005).

The use of inoculants immobilized on carrier material, particularly in packed or fluidized bed reactors could minimize the disadvantages of the variable environment conditions. Carrier materials could generally provide protective niche to immobilized bacteria, either physically, via the provision of a protective surface, or nutritionally, via the supplement of a specific substrate (Gentili et al., 2006). Importantly, the carriers should be innoxious and non-polluting, have defined structure and maintain constant quality. There are two categories of hydrogel material for bacteria cells immobilization, natural and synthetic. The natural hydrogel materials have low mechanical strength and durability, such as agar, agarose, cotton, polyacrylamides, chitosan, alginate and so on. In contrast, strength is one of the main superiorities of synthetic materials (Cunningham et al., 2004), such as poly(carbamoyl)sulphonate (PCS), poly (hydroxyethylmethacrylate) (polyHEMA), polyacrylamide (PAM) and polyvinyl alcohol (PVA) etc. (Jen et al., 1996). Compared to planktonic bacteria, immobilized bacteria have more advantages to suit the natural environments. The carriers could prevent bacteria from being washed away by current and reduce perturbance of environmental conditions, such as toxic substances (Wang et al., 2012). As a result, immobilization has been proved to be a valid strategy to apply indigenous bacteria for bioremediation. Wang et al. (2012) compared the degradation efficiency of free and immobilized cultures of petroleum hydrocarbon-degrading bacteria. The microbial communities with high tolerance and degrading capability of crude oil were obtained and then immobilized in sodium alginate-diatomite beads and sodium alginate beads. The immobilized cells in both beads resulted in higher removal rate of crude oil than free cells at the same condition.

Due to the advantages of low price, high crude oil adsorption, high surface roughness and high binding strength with bacterial cells, cotton fibers were selected as biocarrier for *Acinetobacter* sp. HC8-3S in our research. Figure 12.2 shows SEM

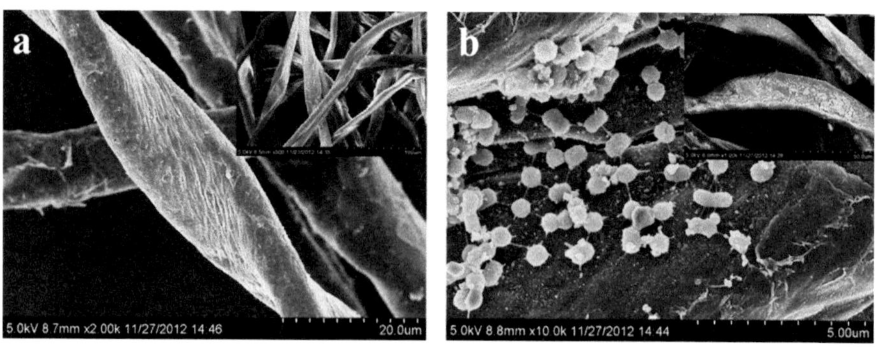

FIGURE 12.2 SEM showing the typical adhesion of the bacterial isolates *Vibrio* sp. HC8-3B (a) and *Acinetobacter* sp. HC8-3S (b) to the surface of cotton fibers after 24 h incubation.

of *Vibrio* sp. HC8-3B and *Acinetobacter* sp. HC8-3S immobilized onto the surface of cotton fibers after washing with phosphate buffer solution. Figure 12.2a revealed that HC8-3B was attached on the biocarrier, which was thought to be washed away by phosphate buffer solution. In comparison, a strong interaction was discovered between the strain HC8-3S and the cotton fibers (Figure 12.2b). Interestingly, the coherent mass of HC8-3S grown on the surface of cotton fibers, and the interconnection between individual bacterial cells, suggested that HC8-3S was likely to produce extracellular polysaccharides, which needs to be further investigated. In addition, the crude oil biodegradation efficiency of the immobilized bacteria on the cotton fibers was 30% higher than that of the planktonic bacterial cells after five days of incubation. The biodegradation rates of both planktonic and immobilized bacteria were above 60% and the pH ranged from 5.6 to 8.6; the degradation activity was not influenced significantly when the NaCl concentration reached 70 g/L (Lin et al., 2014a). In a word, cotton fibers proved to be an excellent crude oil adsorbent and bacterial biocarriers, which could be used as immobilization material in the petroleum bioremediation.

12.7 FIELD-SCALE APPLICATION TO REMEDIATE OIL-POLLUTED MARINE SEDIMENTS IN BOHAI BAY

Marine sediments represent an important sink for petroleum hydrocarbons after oil spills. Different chemical, physical and biological processes contribute to the transport of petroleum hydrocarbons from a buoyant state in the water column down to the seabed, such as weathering, adsorption and chemical dispersants (Mapelli et al., 2017). In the sediments, petroleum hydrocarbons may persist because of the popular anoxic conditions that drastically limit the occurrence of oxidative biodegradation processes, and the low bioavailability resulting from the strong sorption onto hydrophobic sedimentary materials. *In situ* bioremediation is typically regarded as one of the most efficient and sustainable strategies to clean up contaminated sediments (Mapelli et al., 2017; Ron and Rosenberg, 2014).

Up to now, the field-scale application of the bioremediation technique for the marine sediments oil spills was still very limited. This is because effective bioremediation is dependent on various factors, such as natural environmental conditions, petroleum components and concentrations, bioavailability of the substrates and metabolic capability of microorganisms (Boopathy, 2000; Van Hamme et al., 2003). Of these, natural environmental factors (temperature, salinity, pH, oxygen, nutrients) play a key role in deciding the fate of bioremediation (Nikolopoulou et al., 2013). Suitable environmental conditions accelerate the growth of microorganisms, thus leading to higher biodegradation rate. Some other factors influence the bioremediation effectiveness may include: use of a single organism, focus on biodegrading strains only, inadequate dispersion or access to the pollutants and lack of protection (e.g., from grazers) (McGenity et al., 2012).

In order to overcome the above obstacles of the bioremediation technique, autochthonous bioaugmentation (ABA) technology has been put forward and applied. ABA was defined that utilize entirely indigenous microorganisms in the sites (soil, sand and water) to be decontaminated, and enrichments of contaminant-degrading bacteria are conducted under the same or comparable conditions of the sites that ABA will be carried out (Nikolopoulou et al., 2013). To date, it has been overwhelming that using a consortium of functional microorganisms rather than a single strain greatly enhances the chances of successful bioremediation (Head et al., 2006). A well designed microbial consortium will have complementary degradation and metabolic pathways, as well as the potential to improve the bioavailability of petroleum (Gallego et al., 2007; Thompson et al., 2005).

In our study, industrial-scale fermentation of oil-degrading bacterium was conducted primarily in order to meet the production needs. Fermentation parameters were optimized by lowering salinity and inorganic salts in the medium and the optimum conditions for one-ton scale fermentation were 30°C, pH 7.4–7.8, rotation speed 90 rpm (Liu et al., 2016). Two crude oil-degrading strains, isolated from the sediments of Penglai 19-3 platform, were developed as engineered composite microbial product (patent protection). 368 tons of engineered composite microbial products were obtained by industrial-scale fermentation. After immobilized onto the zeolite and embedded by polyglutamic acid, the engineered composite microbial product was put into Penglai 19-3 drilling platform area in Bohai Bay. The seawater and sediments samples were collected regularly to evaluate the bioremediation performance. Our assessment results revealed that the additive bacteria were able to survive well in marine during the remediation process and the amount of additive bacteria were higher in the contaminated area than those in the nonpolluting area; organic carbons, total carbons and petroleum hydrocarbons contents in the sediments declined significantly after bioremediation, which displayed the positive effect of the composite microbial product. After four months of bioremediation, the toxicity of sediments against photobacteria decreased significantly, apparently due to the removal of crude oil. Compared to the group without bioremediation, the experimental group with bioremediation had a reduced abundance of phytoplankton and increased Shannon diversity and evenness after the bioremediation. In a word, the remediated area (67 hectares) meets the criteria for class I water quality and class I sediment standard according to the third-party evaluation, which revealed the ecological environment was significantly improved and recovered. Noticeably, it is the first application of *in situ* bioremediation of oil-contaminated marine sediments in China and achieved great success. This bioremediation case will create considerable social and economic benefits in the future, which was supported by the project of on-site sediment microbial remediation of public area of central Bohai bay, North China Sea Branch of State Oceanic Administration.

12.8 CONCLUDING REMARKS AND FUTURE PERSPECTIVES

Due to increasing demand for energy worldwide, the exploration of ocean oilfields has increased, accompanied by the disastrous oil spills occurred in the marine ecosystems. The Penglai 19-3 oil spill in Bohai Bay was such a disaster, which not only seriously destroyed the marine ecosystem, reducing the quality of seawater and sediments, species diversity, aquacultures and other aspects, but a concern from the public and official governments. In our research, by industrial-scale fermentation and immobilization carrier (zeolite), wrapping agents (polyglutamic acid), the engineered composite microbial product was produced and applied in the Penglai 19-3 drilling platform of Bohai Bay, which was a first successful case for *in situ* bioremediation of marine oil pollution sediments in China. Marine indigenous bacteria are the ultimate petroleum hydrocarbons degraders and play a significant role in cleanup oil pollution. The integrated availability of advanced molecular technologies, chemical analysis technology and genomic sequencing information of hydrocarbon degrading indigenous bacteria, provides a new horizon and insight to understand the function of these interactions between microorganisms and hydrocarbons, as well as help to identify the key species or processes that control contaminant degradation. Nevertheless, the indigenous bacteria can be exploited and applied more effectively *in situ* bioremediation if the fundamental principle on their metabolism, physiology and ecology are clear and environmental parameters are open. Nowadays, a series of issues on both basic and application aspects still remain. For example, how do combinations of environmental stressors (temperature, hydrostatic pressure, nutrition, etc.) impact the response of autochthonous bacteria in polluted marine environments? What is the structure and dynamics of the functional and metabolic cooperation between microbial components of degrading microbial consortia? What is the functional importance of uncultured hydrocarbon degraders that are emerging from genomic sequencing? All these questions are the severe challenges for the *in situ* bioremediation. For certain, novel discoveries and attempts in crude oil pollution in marine ecosystems will inspire and lead to more efficient and sustainable microbial biotechnology approaches to remediate the marine crude oil pollution.

REFERENCES

Agarwal, A., Liu, Y. (2015). Remediation technologies for oil-contaminated sediments. *Mar Pollut Bull* 101, 483–490.

Atlas, R.M., Hazen, T.C. (2011). Oil biodegradation and bioremediation: a tale of the two worst spills in U.S. history. *Environ Sci Technol* 45, 6709–6715.

Bento, F.M., Camargo, F.A., Okeke, B.C., Frankenberger, W.T. (2005). Comparative bioremediation of soils contaminated with diesel oil by natural attenuation, biostimulation and bioaugmentation. *Bioresour Technol* 96, 1049–1055.

Bharali, P., Konwar, B.K. (2011). Production and physico-chemical characterization of a biosurfactant produced by *Pseudomonas aeruginosa* OBP1 isolated from petroleum sludge. *Appl Biochem Biotechnol* 164, 1444–1460.

Boopathy, R. (2000). Factors limiting bioremediation technologies. *Bioresour Technol* 74, 63–67.

Bouchez, T., Patureau, D., Dabert, P., Juretschko, S., Doré, J., Delgenès, P., et al. (2000). Ecological study of a bioaugmentation failure. *Environ Microbiol* 2, 179–190.

Cebron, A., Louvel, B., Faure, P., France-Lanord, C., Chen, Y., Murrell, J.C., et al. (2011). Root exudates modify bacterial diversity of phenanthrene degraders in PAH-polluted soil but not phenanthrene degradation rates. *Environ Microbiol* 13, 722–736.

Ceschia, E., Harjani, J.R., Liang, C., Ghoshouni, Z., Andrea, T., Brown, R.S., Jessop, P.G. (2014). Switchable anionic surfactants for the remediation of oil contaminated sand by soil washing. *RSC Adv* 4, 4638–4645.

Chakrabarty, B., Ghoshal, A.K., Purkait, M.K. (2010). Cross-flow ultra filtration of stable oil-in-water emulsion using polysulfone membranes. *Chem Eng J* 165, 447–456.

Chen, T., Yu, K. (2011). P/Ca in coral skeleton as a geochemical proxy for seawater phosphorus variation in Daya Bay, northern South China Sea. *Mar Pollut Bull* 62, 2114–2121.

Chen, W.W., Li, J.D., Sun, X.N., Min, J., Hu, X.K. (2017). High efficiency degradation of alkanes and crude oil by a salt-tolerant bacterium *Dietzia* species CN-3. *Int Biodeterior Biodegrad* 118, 110–118.

Chen, Y., Chen, L., Zhou, Z.X., Wen, J.P., You, X.Y., Mao, Y.Z., et al. (2014). Enhanced biodegradation of alkane hydrocarbons and crude oil by mixed strains and bacterial community analysis. *Appl Biochem Biotechnol* 172, 3433–3447.

Chien, Y.C. (2012). Field study of *in situ* remediation of petroleum hydrocarbon contaminated soil on site using microwave energy. *J Hazard Mater* 199–200, 457–461.

Coulon, F., McKew, B.A., Osborn, A.M., McGenity, T.J., Timmis, K.N. (2007). Effects of temperature and biostimulation on oil-degrading microbial communities in temperate estuarine waters. *Environ Microbiol* 9, 177–186.

Couto, H.J.B., Massarani, G., Biscaia Jr., E.C., Sant'Anna Jr., G.L. (2009). Remediation of sandy soils using surfactant solutions and foams. *J Hazard Mater* 164, 1325–1334.

Cunningham, C.J., Ivshina, I.B., Lozinsky, V.I., Kuyukina, M.S., Philp, J.C. (2004). Bioremediation of dieselcontaminated soil by microorganisms immobilized in polyvinyl alcohol. *Int Biodeterior Biodegrad* 54, 167–174.

Do, S.H., Kwon, Y.J., Kong, S.H. (2010). Effect of metal oxides on the reactivity of persulfate/Fe(II) in the remediation of diesel-contaminated soil and sand. *J Hazard Mater* 182, 933–936.

Duffield, A.R., Ramamurthy, R.S., Campanelli, J.R. (2003). Surfactant enhanced mobilization of mineral oil within porous media. *Water Air Soil Pollut* 143, 111–122.

Falciglia, P.P., Giustra, M.G., Vagliasindi, F.G.A. (2011). Low-temperature thermal desorption of diesel polluted soil: influence of temperature and soil texture on contaminant removal kinetics. *J Hazard Mater* 185, 392–400.

Feng, L., Wang, W., Cheng, J.S., Ren, Y., Zhao, G., Gao, C.X., et al. (2007). Genome and proteome of long-chain alkane degrading *Geobacillus thermodenitrificans* NG80-2 isolated from a deep-subsurface oil reservoir. Proc. *Nat Acad Sci* 104, 5602–5607.

Ferguson, S.H., Woinarski, A.Z., Snape, I., Morris, C.E., Revill, A.T. (2004). A field trial of *in situ* chemical oxidation to remediate long-term diesel contaminated Antarctic soil. *Cold Reg Sci Technol* 40, 47–60.

Gallego, J.L.R., Garcia-Martinez, M.J., Llamas, J.F., Belloch, C., Pelaez, A.I., Sanchez, J. (2007). Biodegradation of oil tank bottom sludge using microbial consortia. *Biodegradation* 18, 269–281.

Gao, X., Chen, C.T.A. (2012). Heavy metal pollution status in surface sediments of the coastal Bohai Bay. *Water Res* 46, 1901–1911.

Gao, X.X., Gao, W., Cui, Z.S., Han, B., Yang, P.H., Sun, C.J., et al. (2015). Biodiversity and degradation potential of oil-degrading bacteria isolated from deep-sea sediments of South Mid-Atlantic Ridge. *Mar Pollut Bull* 97, 373–380.

Gao, X., Zhou, F., Chen, C.T.A. (2014). Pollution status of the Bohai Sea: an overview of the environmental quality assessment related trace metals. *Environ Int* 62, 12–30.

Gentili, A.R., Cubitto, M.A., Ferrero, M., Rodriguéz. M.S. (2006). Bioremediation of crude oil polluted seawater by a hydrocarbon-degrading bacterial strain immobilized on chitin and chitosan flakes. *Int Biodeterior Biodegrad* 57, 222–228.

Goi, A., Trapido, M., Kulik, N., Palmroth, M.R.T., Tuhkanen, T. (2006). Ozonation and Fenton treatment for remediation of diesel fuel contaminated soil. *Ozone Sci Eng* 28, 37–46.

Golyshin, P.N., Martins Dos Santos, V.A.P., Kaiser, O., Ferrer, M., Sabirova, Y.S., et al. (2003). Genome sequence completed of *Alcanivorax borkumensis*, a hydrocarbon-degrading bacterium that plays a global role in oil removal from marine systems. *J Biotechnol* 106, 215–220.

Gomez, M.J., Pazos, F., Guijarro, F.J., de Lorenzo, V., Valencia, A. (2007). The environmental fate of organic pollutants through the global microbial metabolism. *Mol Syst Biol* 3, 114.

Gu, J.D., Wang, Y.S. (2015). Coastal and marine pollution and ecotoxicology. *Ecotoxicology* 24, 1407–1410.

Head, I.M., Jones, D.M., Röling, W.F.M. (2006). Marine microorganisms make a meal of oil. *Nat Rev Microbiol* 4, 173–182.

Huang, X.F., Lim, T.T. (2006). Performance and mechanism of a hydrophobic-oleophilic kapok filter for oil/water separation. *Desalination* 190, 295–307.

Hussein, M., Amer, A.A., Sawsan, I.I. (2008). Oil spill sorption using carbonized pith bagasse: trial for practical application. *Int J Environ Sci Tech* 5, 233–242.

Jen, A.C., Wake, M.C., Mikos, A.G. (1996). Review: hydrogels for cell immobilization. *Biotechnol Bioeng* 50, 254–272.

Jones, M.D., Crandell, D.W., Singleton, D.R., Aitken, M.D. (2011). Stable-isotope probing of the polycyclic aromatic hydrocarbon-degrading bacterial guild in a contaminated soil. *Environ Microbiol* 13, 2623–2632.

Kirby, M.F., Law, R.J. (2008). Oil spill treatment products approval: the UK approach and potential application to the Gulf region. *Mar Pollut Bull* 56, 1243–1247.

Kostka, J.E., Prakash, O., Overholt, W.A., Green, S.J., Freyer, G., Canion, A., et al. (2011). Hydrocarbon-degrading bacteria and the bacterial community response in Gulf of Mexico beach sands impacted by the deepwater horizon oil spill. *Appl Environ Microbiol* 77, 7962–7974.

Lee, H., Lee, Y., Kim, J., Kim, C. (2014). Field application of modified *in situ* soil flushing in combination with air sparging at a military site polluted by diesel and gasoline in Korea. *Int J Environ Res Public Health* 11, 8806–8824.

Li, D., Zhang, Y., Quan, X., Zhao, Y. (2009). Microwave thermal remediation of crude oil contaminated soil enhanced by carbon fiber. *J Environ Sci* 21, 1290–1295.

Li, J.B., Luo, C.L., Song, M.K., Dai, Q., Jiang, L.F., Zhang, D.Y., et al. (2017). Biodegradation of phenanthrene in polycyclic aromatic hydrocarbon-contaminated wastewater revealed by coupling cultivation-dependent and -independent approaches. *Environ Sci Technol* 51, 3391–3401.

Li, W., Wang, L.Y., Duan, R.Y., Liu, J.F., Gu, J.D., Mu, B.Z. (2012). Microbial community characteristics of petroleum reservoir production water amended with *n*-alkanes and incubated under nitrate-, sulfate-reducing and methanogenic conditions. *Int Biodeter Biodegr* 69, 87–96.

Li, X.F., Zhao, L., Adam, M. (2016). Biodegradation of marine crude oil pollution using a salt-tolerant bacterial consortium isolated from Bohai Bay, China. *Mar Pollut Bull* 105, 43 50.

Lin, M., Liu, Y.H., Chen, W.W., Wang, H., Hu, X.K. (2014a). Use of bacteria-immobilized cotton fibers to absorb and degrade crude oil. *Int Biodeter Biodegr* 88, 8–12.

Lin, M., Hu, X.K., Chen, W.W., Wang, H., Wang, C.Y. (2014b). Biodegradation of phenanthrene by *Pseudomonas* sp. BZ-3, isolated from crude oil contaminated soil. *Int Biodeter Biodegr* 94, 176–181.

Liu, D., Zhu, L. (2014). Assessing China's legislation on compensation for marine ecological damage: a case study of the Bohai oil spill. *Mar Policy* 50 (Part A(0)), 18–26.

Liu, X., Guo J., Guo M.X., Hu. X.K., Tang C., Wang C.Y., et al. (2015). Modeling of oil spill trajectory for 2011 Penglai 19-3 coastal drilling field, China. *Appl Math Model* 39, 5331–5340.

Liu, Y.C., Li, L.Z., Wu, Y., Tian, W., Zhang, L.P., Xu, L., et al. (2010). Isolation of an alkane degrading *Alcanivorax* sp. strain 2B5 and cloning of the *alkB* gene. *Bioresour Technol* 101, 310–316.

Liu, Y.H., Hu, X.K., Liu H. (2016). Industrial-scale culturing of the crude oil-degrading marine *Acinetobacter* sp. strain HC8-3S. *Int Biodeter Biodegr* 107, 56–61.

Macnaughton S.J., Stephen, J.R., Vesosa, A.D., Davis, G.A., Chang, Y.J., White, D.C. (1999). Microbial population changes during bioremediation of an experimental oil spill. *Appl Environ Microbiol* 65, 3566–3574.

Mapelli, F., Scoma, A., Michoud, G., Aulenta, F., Boon, N., Borin, Sara., et al. (2017). Biotechnologies for marine oil spill cleanup:indissoluable ties with microorganisms. *Trends Biotechnol* 35, 860–870.

Margesin, R., Moertelmaier, C., Mair, J. (2013). Low-temperature biodegradation of petroleum hydrocarbons (*n*-alkanes, phenol, anthracene, pyrene) by four actinobacterial strains. *Int Biodeter Biodegr* 84, 185–191.

Marshall, A.G., Rodgers, R.P. (2004). Petroleomics: the next grand challenge for chemical analysis. *Accounts Chem Res* 37, 53–59.

Mbadinga, S.M. Wang, L.Y. Zhou, L., Liu, J.F. Gu, J.D., Mu, B.Z. (2011). Microbial communities involved in anaerobic degradation of alkanes. *Int Biodeter Biodegr* 65, 1–13.

McGenity, T.J. (2014). Hydrocarbon biodegradation in intertidal wetland sediments. *Curr Opin Biotechnol* 27, 46–54.

McGenity, T.J., Folwell, B.D., McKew, B.A., Sanni, G.O. (2012). Marine crude-oil biodegradation: a central role for interspecies interactions. *Aquat Biosyst* 8, 10–29.

McKew, B.A., Coulon F., Osborn, A.M., Timmis K.N., McGentity T.J. (2007). Determining the identity and roles of oil-metabolizing maring bacteria from the Thames estuary, UK. *Environ Microbiol* 9, 165–176.

Mishra, S., Jyot, J., Kuhad, R.C., Lal, B. (2001). Evaluation of inoculum addition to stimulate *in situ* bioremediation of oily-sludge-contaminated soil. *Appl Environ Microbiol* 67, 1675–1681.

Moser, R., Stahl, U. (2001). Insights into the genetic diversity of initial dioxygenases from PAH-degrading bacteria *Appl Microbiol Biotechnol* 55, 609–618.

Mukherjee, A.K., Bordoloi, N.K. (2011). Bioremediation and reclamation of soil contaminated with petroleum oil hydrocarbons by exogenously seeded bacterial consortium: a pilot-scale study. *Environ Sci Pollut Res* 18, 471–478.

Munro, A.W., Girvan, H.M., Mason, A.E., Dunford, A.J., McLean, K.J. (2013). What makes a P450 tick? *Trends Biochem Sci* 38, 140–150.

Nelson, K.E., et al. (2002). Complete genome sequence and comparative analysis of the metabolically versatile *Pseudomonas putida* KT2440. *Environ Microbiol* 4, 799–808.

Nikolopoulou, M., Eickenbusch, P., Pasadakis, N.,Venieri, D., Kalogerakis, N. (2013). Microcosm evaluation of autochthonous bioaugmentation to combat marine oil spills. *New Biotechnol* 30, 734–742.

Ouyang, W., Liu, H., Murygina, V., Yu, Y., Xiu, Z., Kalyuzhnyi, S. (2005). Comparison of bio-augmentation and composting for remediation of oily sludge: A field-scale study in China. *Process Biochem* 40, 3763–3768.

Pan, K., Wang, W.X., (2012). Trace metal contamination in estuarine and coastal environments in China. *Sci Total Environ* 42, 3–16.

Peng, R.H., Xiong, A.S., Xue, Y., Fu, X.Y., Gao, F., Zhao, W. (2008). Microbial biodegradation of polyaromatic hydrocarbons. *FEMS Microbiol Rev* 32, 927–955.

Rabus, R., et al. (2005). The genome sequence of an anaerobic aromatic-degrading denitrifying bacterium, strain EbN1. *Arch Microbiol* 183, 27–36.

Rojo, F. (2009). Degradation of alkanes by bacteria. *Environ Microbiol* 11, 2477–2490.

Ron, E.Z., Rosenberg, E. (2014). Enhanced bioremediation of oil spills in the sea. *Curr Opin Biotech* 27, 191–194.

Shin, Y.J., Shen, Y.H. (2011). Remediation of oil-contaminated sand by coal agglomeration using ball milling. *Environ Technol* 32, 1551–1558.

SOA (State Oceanic Administration of China). (2012). Marine Environmental Quality Bulletin of China. http://www.coi.gov.cn/gongbao/huanjing/201304/t20130401_26428.html. [In Chinese].

Sun, J., Guo, F., Geng, X., Wei, J., Li, X., Li, J. (2010). Seasonal changes and diversity of bacteria in Bohai Bay by RFLP analysis of PCR-amplified 16S rDNA gene fragments. *World J Microbiol Biotechnol* 27, 275–284.

Szulc, A., Ambrożewicz, D., Sydow, M., Ławniczak, Ł., Piotrowska-Cyplik, A., Marecik, R., et al. (2014). The influence of bioaugmentation and biosurfactant addition on bioremediation efficiency of diesel-oil contaminated soil: feasibility during field studies. *J Environ Manag* 132, 121–128.

Teira, E., Lekunberri, I., Gasol, J.M., Nieto-Cid, M., Alvarez-Salgado, X.A., Figueiras, F.G. (2007). Dynamics of the hydrocarbon-degrading *Cycloclasticus* bacteria during mesocosm-simulated oil spills. *Environ Microbiol* 9, 2551–2562.

Thomas, F., Lorgeoux, C., Faure, P., Billet, D., Cébron, A. (2016). Isolation and substrate screening of polycyclic aromatic hydrocarbon degrading bacteria from soil with long history of contamination. *Int Biodeterior Biodegrad* 107, 1–9.

Thompson, I.P., Van Der Gast, C.J., Ciric, L., Singer, A.C. (2005). Bioaugmentation for bioremediation: the challenge of strain selection. *Environ Microbiol* 7, 909–915.

Throne-Holst, M., Wentzel, A., Ellingsen, T.E., Kotlar, H.K., Zotchev, S.B. (2007). Identification of novel genes involved in long-chain *n*-alkane degradation by *Acinetobacter* sp. strain DSM 17874. *Appl Environ Microbiol* 73, 3327–3332.

Tsai, T.T., Kao, C.M. (2009). Treatment of petroleum-hydrocarbon contaminated soils using hydrogen peroxide oxidation catalyzed by waste basic oxygen furnace slag. *J Hazard Mater* 170, 466–472.

USEPA. (1998). Chemical Oxidation, field application of *in situ* remediation of technologies. Office of Solid Waste and Emergency Response, Technology Innovation Office, Washington U.S., p. 31.

USEPA. (1999). In: United States Environmental Protection Agency, O.o.E.a.R.R., Understanding oil spills and oil spill response. United States Environmental Protection Agency, Office of Emergency and Remedial Response, p. 48.

Van Beilen, J.B., Funhoff, E.G. (2005). Expanding the alkane oxygenase toolbox: new enzymes and applications. *Curr Opin Biotechnol* 16, 308–314.

Van Beilen, J.B., Funhoff, E.G. (2007). Alkane hydroxylases involved in microbial alkane degradation. *Appl Microbiol Biotechnol* 74, 13–21.

Van Hamme, J.D., Singh, A., Ward, O.P. (2003). Recent advances in petroleum microbiology. *Microbiol Mol Biol Rev* 67, 503–549.

Von der Weid, I., Marques, J.M., Cunha, C.D., Lippi, R.K., Dos Santos, S.C., Rosado, A.S., et al. (2007). Identification and biodegradation potential of a novel strain of *Dietzia cinnamea* isolated from a petroleum-contaminated tropical soil. *Syst Appl Microbiol* 30, 331–339.

Wang, H., Wang, B., Dong, W.W., Hu, X.K. (2016). Co-acclimation of bacterial communities under stresses of hydrocarbons with different structures. *Sci Rep-uk* 6, 34588.

Wang, L.Y., Gao, C.X., Mbadinga, S.M., Zhou, L., Liu, J.F., Gu, J.D., et al. (2011). Characterization of an alkane-degrading methanogenic enrichment culture from production water of an oil reservoir after 274 days of incubation. *Int Biodeter Biodegr* 65, 444–450.

Wang, W.P., Cai, B.B., Shao, Z.Z. (2014). Oil degradation and biosurfactant production by the deep sea bacterium *Dietzia maris* As-13-3. *Front Microbiol* 5, 711.

Wang, X.B., Nie, Y., Tang, Y.Q., Wu, G., Wu, X.L. (2013). *n*-Alkane chain length alters *Dietzia* sp. strain DQ12-45-1b biosurfactant production and cell surface activity. *Appl Environ Microbiol* 79, 400–402.

Wang, Z.Y., Xu, Y., Wang, H.Y., Zhao, J., Gao, D.M., Li, F.M., et al. (2012). Biodegradation of crude oil in contaminated soils by free and immobilized microorganisms. *Pedosphere* 22, 717–725.

Whang, L.M., Liu, P.W.G., Ma, C.C., Cheng, S.S. (2008). Application of biosurfactants, rhamnolipid, and surfactin, for enhanced biodegradation of diesel-contaminated water and soil. *J Hazard Mater* 151, 155–163.

Widdel, F., Rabus, R. (2001). Anaerobic biodegradation of saturated and aromatic hydrocarbons. *Curr Opin Biotechnol* 12, 259–276.

Yang, F.L., Yang, J.S., Deng, C.P., Chen, N., Wang, S.Q., Wang, E.T. (2015). Bacterial communities and their hydrocarbon bioremediation potential in the Bohai Sea, China. *Mar Ecol Prog Ser* 538, 117–130.

Yen, C.H., Chen, K.F., Kao, C.M., Liang, S.H., Chen, T.Y. (2011). Application of persulfate to remediate petroleum hydrocarbon-contaminated soil: feasibility and comparison with common oxidants. *J Hazard Mater* 186, 2097–2102.

Zhang, L., Somasundaran, P., Ososkov, V., Chou, C.C. (2001). Flotation of hydrophobic contaminants from soil. *Colloids Surf A Physicochem Eng Asp* 177, 235–246.

Zhang, Z.Z., Hou, Z.W., Yang, C.Y., Ma, C.Q., Tao, F., Xu, P. (2011). Degradation of *n*-alkanes and polycyclic aromatic hydrocarbons in petroleum by a newly isolated *Pseudomonas aeruginosa* DQ8. *Bioresour Technol* 102, 4111–4116.

13 Modeling the Impact of Dilution on the Microbial Degradation of Dispersed Oil in Marine Environments

Vicente I. Fernandez, Roman Stocker, and Gabriel Juarez

CONTENTS

13.1 INTRODUCTION

Crude oil undergoes natural weathering processes that change its chemical composition and physical properties upon entering the water column (Boehm et al., 1982; National Research Council, 2003; Payne and McNabb, 1984; Tarr et al., 2016). Examples of environmental weathering processes include evaporation, dissolution, dispersion, photo-oxidation, microbial degradation, emulsification, sedimentation, and burial. The timescale associated with each process can range from hours to years and the extent to which any weathering process dominates depends on the location of the oil in the water column. Near the sea surface, oil is susceptible to evaporation and photo-oxidation, which occur over hours to days, whereas deeper in the water column, oil is more susceptible to microbial degradation and sedimentation, which occur over months to years (National Research Council, 2003). Compared to traditional mechanical measures, such as skimming or burning, which typically recover less than 20% of spilled oil (Lessard and DeMarco, 2000; Wadsworth, 1995), microbial degradation is a crucial process in the weathering of unrecovered immiscible hydrocarbons that persist the longest in the marine environment, such as dispersed oil droplets in the water column (Bagby et al., 2016; Head and Swannell 1999; Head et al., 2006).

Spilled oil is mechanically dispersed into oil droplets by shear stresses from wave action, turbulence, and the high-pressure jet flow from a blowout (Delvigne and Sweeney 1988; Johansen et al., 2015). The addition of chemical dispersants increases the efficiency of mechanical dispersion by reducing the oil–water interfacial tension and further reducing the overall droplet size (Li et al., 2008; Li et al., 2009; McAuliffe et al., 1980). Droplets are rendered effectively neutrally buoyant when they are below a diameter of approximately 70 μm (Atlas and Hazen, 2011), as the resuspending effect of turbulence dominates over the rising force of buoyancy. These small, neutrally buoyant oil droplets act as passive fluid particles and are transported by advection and diffusion through the water column. In particular, these small droplets do not rise to the ocean surface and so are typically not exposed to evaporation and photo-oxidation. The ultimate fate of these oil droplets and the extent and mode of their degradation by microorganisms are currently unknown (Passow and Hetland, 2016). Here we apply a new mechanistic computational model to quantify the biodegradation processes of individual droplets of crude oil in naturally diluting environments.

Existing oil spill models vary in their level of complexity, ranging from two-dimensional models that consider the trajectories of oil surface slicks to fully three-dimensional models that take into account a large number of physical, chemical, and biological processes affecting the fate of oil (Huang, 1983; Reed et al., 1999; Spaulding, 1988; Spaulding, 2017). The more processes a model includes, the more realistic yet more computationally intensive and dependent on parameterization it becomes. Therefore, practical oil spill models typically employ a Eulerian-Lagrangian framework to describe the underlying hydrodynamics of the water column at a given location (Eulerian) as well as the transport and trajectories of small parcels of water or representative oil droplets (Lagrangian), with a time step ranging from 10 to 100 minutes (Spaulding, 2017) and a grid spacing ranging from 10 to 1000 meters (French-McCay, 2004; Reed et al., 1999). At these modeling length scales, microscale structure and heterogeneity, such as droplet size or bacterial trajectories, are replaced with empirical relations obtained from field measurements or laboratory experiments. Microbial degradation, for example, is typically modelled using independent first-order degradation equations for each degradable hydrocarbon component, based on the experimentally reported half-lives of hydrocarbon compounds (French-McCay, et al. 2018; French-McCay, 2004). Without an understanding of the fundamental microscale mechanisms that influence oil biodegradation, empirical relations and fitting parameters will introduce unwanted and unnecessary error in model predictions.

In this work, we examine the impact of natural dilution of a dispersed underwater plume by ambient turbulence on the timescale of microbial oil degradation, using an extension of the Microscale Oil Degradation Model, or MODEM. Previously, we have introduced and used MODEM to show that microscale mechanisms, such as the initial encounters between bacteria and oil droplets as well as the droplet size distribution, can be a significant factor in determining the microbial degradation time of dispersed oil droplets in quiescent environments (Fernandez et al., 2019). Unlike laboratory conditions where oil, bacteria, and seawater form a closed system, turbulence in the ocean results in the continuous dilution of any locally high concentration of oil droplets. The average oil concentration of dispersed oil plumes is expected to fall below 10 PPM (parts-per-million) within a few days of a spill, whereas over the ensuing weeks to months—when microorganisms play an important role in degradation—the typical concentration is expected to be much less than 1 PPM (Lee et al., 2013; Macnaughton et al., 2003). Since the impact of encounter times is most significant when oil concentrations are low (Fernandez et al., 2019), these results indicate that estimates of microbial degradation rates based on experimental half-lives that start from high oil concentrations are likely to be strongly overestimated, even when high concentrations are present near the source of oil, such as in the case of an underwater blowout.

13.2 THE MICROSCALE OIL DEGRADATION MODEL—MODEM

To examine the impact of the continuous dilution that oil dispersions are subject to in aquatic environments, we use the same underlying model (MODEM) used in a recent investigation of the optimal droplet size for biological oil degradation (Fernandez et al., 2019). For additional details on the model implementation, see the appendix of this chapter and the supplemental methods in Fernandez et al. (Fernandez et al., 2019) This model focuses on the physical encounters between oil-degrading bacteria and oil droplets as a function of droplet size distribution and oil concentration and predicts the decrease in oil mass as a result of the biological degradation. There are two distinct stages considered in the degradation model: encounter and growth. For a given oil droplet, the first stage lasts until an oil-degrading bacterium first encounters and attaches to the oil droplet. The duration of this stage is inherently stochastic, but depends on the behavior of the bacteria (such as swimming and chemotaxis), the concentration of bacteria in the environment, and the size of the oil droplet. The second stage is assumed to follow deterministic dynamics as bacteria grow on the surface and consume the available carbon in the oil droplet. During this second stage, random encounters with planktonic bacteria continue to occur, but are considered negligible because in nearly all conditions the time between arrivals is much larger than the doubling time of bacteria growing on the oil droplet surface.

The stochastic encounter process of the first oil-degrading bacterium finding an oil droplet is represented using a Poisson arrival process with an arrival rate given by a diffusive encounter kernel. The arrival rate is

$$\lambda = 4\pi\left(r_{bacteria} + r_{droplet}\right)\left(D_{bacteria} + D_{droplet}\right)C_{bacteria}, \tag{13.1}$$

where r, D, and C represent the radii, diffusivity, and concentration of the respective subscripts. In this framework, inter-arrival times, and therefore the time for the first arrival, are exponentially distributed with mean $1/\lambda$. Given that the bulk concentration of bacteria, $C_{bacteria}$, changes as a function of time, we model a large number (10^6) of droplets with a Monte Carlo method, revising the arrival rate at each time step based on the corresponding bulk bacterial concentration.

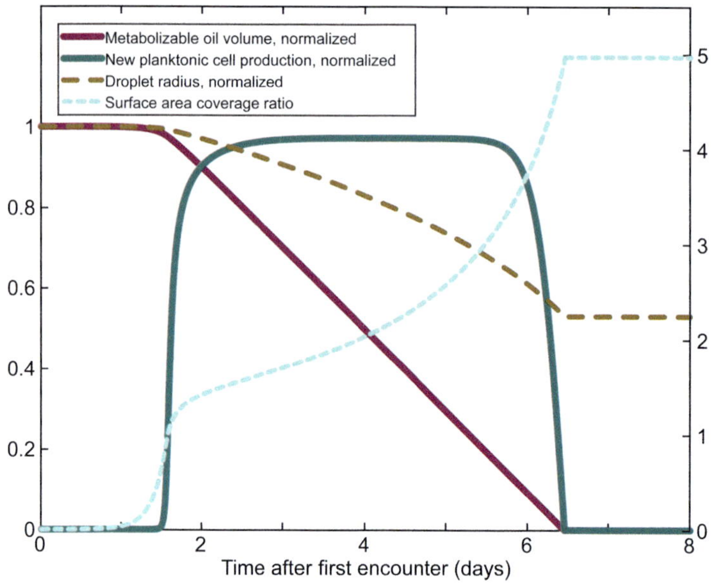

FIGURE 13.1 Droplet degradation dynamics. Elements of the prescribed degradation dynamics for a 62 μm diameter oil droplet are depicted after the initial encounter with an oil degrading bacterium ($t = 0$). The metabolizable oil volume and the oil droplet radius are normalized by their starting values (left axis). The rate of production of new planktonic cells from the droplet is normalized by the maximum production rate (left axis). The surface area coverage ratio (see text) is not normalized (right axis).

Once the first oil-degrading bacterium has attached to an oil droplet, the degradation of the oil droplet and the growth of bacteria follow a prescribed trajectory, such as that shown in Figure 13.1 for the case of a 62 μm diameter droplet and biological parameters according to Table 13.1. Initially, bacteria attached to the droplet surface are able to grow unconstrained by space, carbon, or other nutrients. This results in exponential growth on the surface of the oil droplet, with all new bacteria remaining attached to the oil droplet.

As the surface of the oil droplet is progressively covered in bacteria, it becomes harder for all the bacteria to remain on the surface when individuals divide. The model assumes that there is a maximum number of bacteria that a droplet can host on its surface. Once this capacity is reached, all subsequent growth creates new planktonic bacteria (which are able to colonize other droplets) and the number of bacteria on the original droplet remains constant and equal to the maximum. Based on experimental observations of oil-degrading bacteria at the oil–water interface (unpublished), the maximum capacity is somewhat larger than the number of bacteria it would take to cover the original spherical surface of the droplet (Juarez and Stocker, 2019). This may be due to biofilm formation or other bacterial interactions that allow bacteria to remain attached. This biofilm is also the foundation for why we model the number of bacteria to be constant after it reaches capacity, even though the radius (and surface area) of the droplet decreases (and thus the number of bacteria

TABLE 13.1
Model Parameters: Default Values of the Model Parameters unless Otherwise Stated

Symbol	Description	Value	Units	Reference
$r_{droplet}$	Droplet radius	31	μm	Fernandez et al., 2019
C_o	Initial oil concentration	$10^{-3}-10^{4}$	PPM	
D_{iso}	Isopycnal diffusivity	0.07	m^2/s	Ledwell et al., 1998
D_{dia}	Diapycnal diffusivity	10^{-5}	m^2/s	Ledwell et al., 1998
T	Temperature	283	K	
η	Dynamic viscosity of seawater	1.03×10^{-3}	Pa s	
ρ	Oil density	850	kg/m^3	Daling et al., 1990; Reddy et al., 2012
	Percent carbon in oil	85	-	Reddy et al., 2012; Stewart et al., 1993
$C_{bacteria}$	Concentration of oil-degrading bacteria	10^3 (initial)	cells/mL	Hazen et al., 2010
$r_{bacteria}$	Bacterial radius	0.5	μm	
m_b	Carbon per bacterial cell	100×10^{-15}	gC	Loferer-Krößbacher et al., 1998; Vrede et al., 2002
Γ	Bacterial growth efficiency	0.5	-	
τ_d	Bacterial doubling time	10^4	s	Juarez and Stocker, 2019

fitting on the surface would in principle decrease as well). The saturation at the maximum allowed number of attached cells is implemented as a smooth transition based on the surface area coverage ratio (see Figure 13.1, see Fernandez et al., 2019).

A consequence of the fixed number of attached bacteria once capacity is reached is that the mass of metabolizable oil decreases linearly with time during this final portion of the degradation process: a constant number of bacteria grow at a constant rate, resulting in a steady utilization of the available carbon. This proceeds until all the metabolizable carbon (assumed to be 85% of the initial oil volume (Stewart et al., 1993)) is used. Note that a loss of 85% of the volume would result in a reduction of only approximately 50% in the diameter of the droplet (see Figure 13.1). The limit to the number of bacteria that can attach to the droplet surface is responsible for the relationship between the time it takes for a droplet to degrade (post colonization) and the surface area to volume ratio, which is determined by the size of the droplet. Smaller droplets have a larger surface area to volume ratio and consequently the maximum degradation rate for a given oil volume is significantly higher. This relationship has been recognized and partially motivates the use of dispersants (Lessard and DeMarco, 2000)..

We previously used MODEM to demonstrate that for sufficiently small droplets, encounters with oil-degrading bacteria are the limiting step in the degradation, in contrast to the surface-area constrained degradation after colonization (Fernandez et al., 2019). This results in an optimal oil droplet size for biodegradation (approximately 100 μm in diameter). Although biological and environmental parameters alter the precise value of the optimal droplet size, the existence of a non-zero minimum value for the optimal droplet size was found to be robust.

During the saturated stage of degradation, the bacteria on the droplet continue to grow but new bacteria disperse into the environment. This provides a mechanism for droplets that are degraded early to affect the degradation time of remaining nearby droplets by reducing the time it takes for them to be encountered by oil-degrading bacteria (because bacteria that disperse into the environment cause an increase in $C_{bacteria}$). The increase in the concentration of planktonic oil-degrading bacteria reduces the contribution of the encounter stage to the overall degradation time, leading to a smaller optimal droplet size.

However, the practical impact of the bacteria that are released into the environment from degrading droplets depends on the concentration of oil droplets. At low oil concentrations, the fixed number of new planktonic bacteria that result from the biodegradation of an oil droplet will not meaningfully alter the background concentration $C_{bacteria}$. At the other extreme, at high oil concentrations, a single droplet may increase the local concentration of bacteria to the point of significantly decreasing the time to encounter of nearby droplets, developing into a cascade of increasing bacterial concentration and leading to a large fraction (if not all) of the nearby droplets being colonized in quick succession.

The general results for biodegradation when droplet–bacteria encounters are included hold for monodispersed as well as polydispersed oil droplets. For polydispersions, there are additional interactions between large and small droplets (Fernandez et al., 2019). However, it was found that a monodispersion that maintains the same overall surface area to volume ratio as a polydispersion is a good approximation to the polydispersion in terms of degradation time. For the following discussion of biodegradation under continuous dilution, a monodispersion of oil droplets with a 62 μm diameter will be used, representing an approximation of a theoretical log-normal droplet size distribution with a 34 μm mean diameter (Aman et al., 2015).

13.3 DILUTION IN AQUATIC ENVIRONMENTS

In aquatic environments, volumes of water with higher oil droplet concentrations will naturally mix over time with the surrounding oil-free water, resulting in dilution. At the scales of mixing in natural aquatic environments, a dispersion of small oil droplets can be treated as a dissolved solute for the purposes of characterizing its dilution. Thus, to predict the dilution of an oil droplet dispersion one can harness results from field studies that have tracked artificial tracers spreading over months in the oceans (Ledwell and Watson, 1991; Ledwell et al., 1998) or characterized mixing from hydrodynamic measurements (Hibiya et al., 2006; Polzin et al., 1997), as well as numerical modeling results (Simmons et al., 2004).

While droplets will spread even in completely still water with a diffusivity governed by Brownian motion, most suspended oil droplets are sufficiently large (>10 μm) that this contribution to dilution is minimal. For oil droplets 10 μm in diameter and larger, the Brownian diffusivity given by the Stokes-Einstein equation (Kiorboe, 2008) is less than approximately 2×10^{-14} m²/s, and an individual oil droplet would diffuse less than a millimeter in a day. For objects the size of oil droplets (typically, $10 - 500$ μm diameter), mixing in aquatic environments is driven primarily by the ambient flow and in particular by turbulence. The initial outcome of turbulence is the stirring of a cloud into a complex geometry of thin filaments. As the scale of the filaments decreases, regular diffusion (which acts over short distances) increases in importance and smooths out heterogeneity, resulting in mixing. The overall process, "turbulent diffusion," typically behaves similarly to molecular diffusion, but with significantly higher diffusion coefficients. Away from the surface of a natural water body, any density stratification of the water suppresses mixing in the vertical direction. As a result, mixing occurs at different rates along the water density gradient (diapycnal) and perpendicular to it (isopycnal). Factors such as the dimensions of the water body and roughness of the bottom surface (Polzin et al., 1997) can also have large impacts on the rates of mixing. While dilution is a ubiquitous process affecting locally elevated concentrations of any compound in aquatic environments, its intensity thus depends strongly on the specifics of the environment.

13.4 DILUTION IN THE MICROSCALE OIL DEGRADATION MODEL

Although the dilution of a cloud of oil droplets by turbulence largely behaves diffusively, with regions of high concentration gradients mixing more strongly than regions with lower gradients (such as the center of an oil droplet plume), we will use some simplifications to implement dilution in conjunction with the encounter–growth oil degradation model. The primary simplification is that the model accounting for dilution will consider a single representative oil concentration instead of a range of oil concentrations that smoothly decrease from the center of the plume to the background level. For an instantaneous point source release, the spreading of a compound is given by:

$$C(x,y,z,t)=\frac{M}{(4\pi t)^{3/2}\left(D_xD_yD_z\right)^{1/2}}e^{\left(-\frac{x^2}{4D_xt}-\frac{y^2}{4D_yt}-\frac{z^2}{4D_zt}\right)}, \tag{13.2}$$

where C is the concentration as a function of three-dimensional space and of time, D_j is the turbulent diffusivity in the j direction, M is the total mass of oil injected at time zero at the origin, and t is time. The solution to the 3D diffusion equation is shown in Figure 13.2. The scaling term of the equation $C_{max}(t)=M/(4\pi t)^{3/2}\left(D_xD_yD_z\right)^{1/2}$ gives the maximum concentration in space as the point source spreads. Our approximation for the dilution–degradation model will assume that, at any time t, the entire plume has the concentration $C_{max}(t)$ and a corresponding ellipsoidal volume that, in the absence of degradation, would conserve the initial mass of oil (see Figure 13.2, gray lines). The volume is ellipsoidal as opposed to spherical because the diffusivities are not necessarily equal in all directions. This approximation captures the critical elements of dilution, specifically the temporal dynamics of the reduction of the oil concentration, but significantly simplifies the implementation in MODEM by having only two environmental states at any given time, namely (i) water without oil and (ii) water with a single concentration of oil, $C_{max}(t)$.

In addition to tracking a single representative oil concentration, we will neglect the dependence of the diffusivity on the length scales of the oil plume (Okubo, 1971). Due to the multiple scales of motion that characterize turbulence in aquatic environments,

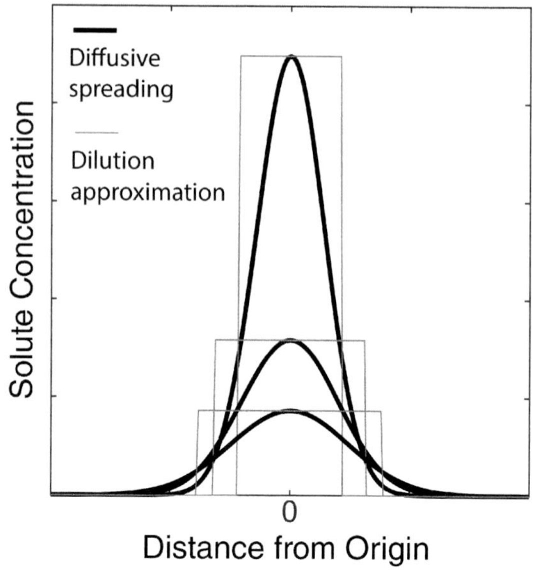

FIGURE 13.2 Simplified dilution model, approximating diffusive spreading while retaining simplicity in the implementation of the encounter–growth model. The black lines illustrate the solution to an instantaneous point source (Eq. 13.2) starting at the origin and spreading diffusively. The gray lines illustrate the simplified dilution model, in which the concentration inside the spreading dispersion of oil droplets is uniform and equal to the maximum of the exact solution. In the simplified model, the width of the spreading distribution (gray lines) is set such that (in the absence of degradation) the total mass of oil is conserved as dilution proceeds.

turbulent diffusivity increases as a plume grows in size, as larger and larger eddies contribute to stirring and mixing. Here, we will neglect this size-dependence and instead use a uniform isopycnal diffusivity ($D_x = D_y = D_{iso} = 0.07$ m²/s) and a uniform diapycnal diffusivity ($D_z = D_{dia} = 10^{-5}$ m²/s) that are both independent of the size of the oil droplet cloud. The specific values are taken from a tracer release experiment performed at 300 m depth in the North Atlantic in 1992 (Ledwell et al., 1998).

The solution for the diffusion of an instantaneous point source (Eq. 13.2) results in a very rapid initial dilution, but this rapid dilution occurring in the first minutes would not be an adequate model for the oil dispersion application being considered. Although a seep in the ocean floor or a leak in a pipe may be very small and reminiscent of a point source, the process that creates the oil droplet dispersion will likely also spread the oil over a volume larger than what can be considered a point source. This initial condition is implemented here by offsetting the time by an amount t_0, $C_{max}(t) = C_{max}(t' + t_0)$, so that the size of the spreading point source at $t' = 0$ matches the desired initial oil plume size. As a result of this temporal offset, the volumetric dilution of oil concentration is significantly slower than $t^{-3/2}$ for $t < t_0$ (see Figure 13.3, black line).

In this numerical study, we focus on the dilution of a cloud of oil droplets that is generated due to a constant influx of oil at depth. We will assume a consistent

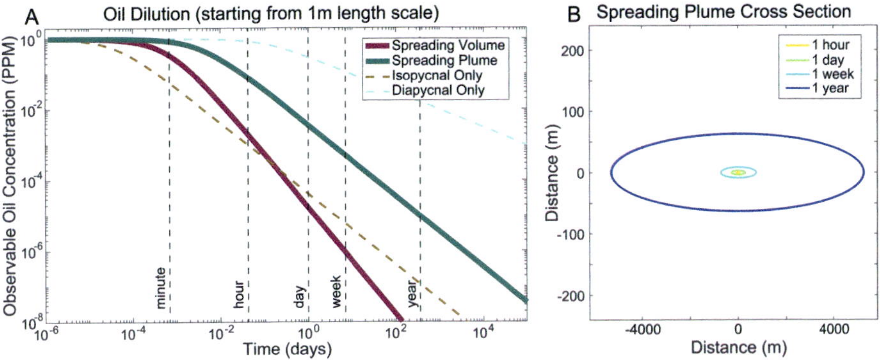

FIGURE 13.3 Dilution of concentrated oil dispersions in the ocean. (A) The concentration of oil (in PPM = parts per million) is plotted over time for different mixing conditions. In all cases, the isopycnal and diapycnal diffusivities are 0.07 m^2/s and 10^{-5} m^2/s, respectively. The volume dilution (solid black line) initially has the volume of a 1 meter diameter sphere. The two-dimensional dilution (solid gray line) of a vertical plume cross section begins with the area of a 1 meter diameter disk, as does the horizontal two-dimensional isopycnal case (black dashed line). The one-dimensional diapycnal-only diffusion (gray dashed line) begins from a 1 meter length (B). The spreading of a plume cross section, corresponding to the solid gray line in (A). Note the different axis scaling.

ambient water current that advects the oil droplet dispersion, generating a plume that increases in width as the distance from the source increases. In such a plume, dilution primarily occurs in the two dimensions perpendicular to the direction of the ambient current. Along the axis of the ambient current there is a much smaller gradient in the oil concentration and therefore mixing does not have a large dilutive impact. Therefore, we can model dilution in a plume from the spreading of a cross section of the plume perpendicular to the ambient current (see Figure. 13.3, gray line). Since mixing occurs in two dimensions only, the dilution scales as t^{-1} rather than as $t^{-3/2}$.

The modeled oil plume cross section will have a constant width during the cross-sectional spreading. This width does not affect the simulation since it is a constant multiplicative factor when calculating the volume of the spreading plume cross section. Since we are modeling the stochastic encounter of individual droplets by directly simulating a number of droplets, the cross-sectional width is chosen for each initial oil concentration to keep the number of simulated droplets fixed at 10^6.

As the oil droplet cloud spreads, new water from the surroundings is entrained into the oil droplet plume. Given that the encounter–growth numerical model is tracking the degradation state of individual oil droplets using a Monte Carlo approach, the increased fluid volume does not directly affect the simulated degradation of the droplets. By summing the total amount of remaining oil volume at any moment and evaluating the total volume that corresponds to the plume cross section at that time, the oil concentration can be straightforwardly calculated.

However, dilution can have a strong indirect impact on the oil degradation by altering the rate at which droplets are colonized. The new entrained water has not recently had an available oil substrate for bacteria, and therefore is assumed to have oil-degrading bacteria at the background concentration. As a result, any increases in the local concentration of planktonic oil-degrading bacteria inside the oil plume due to shedding of cells from droplets colonized to capacity is also being diluted alongside the effect of dilution on the oil droplet concentration. The concentration of oil-degrading bacteria in the diluting oil plume changes over time due to both contributions of new cells from colonized oil droplets and dilution. The impact of dilution on the concentration of bacteria was modeled by directly adding the appropriate number of new bacteria from the new volume mixed into the dispersion (having background bacterial concentration) and renormalizing by the total plume section volume:

$$N_{t+1} = N_t - \frac{\Delta t}{\kappa}\left(N_t - C_{bacteria,0}\ V_t\right) + G_t \Delta t + C_{bacteria,0}\left(V_{t+1} - V_t\right), \qquad (13.3)$$

where

$$V(t) = h\ 4\pi(t + t_0)\sqrt{D_{iso} D_{dia}}, \qquad (13.4)$$

and N_t is the number of bacteria in the plume slice volume at time t, $C_{bacteria,0}$ is the undisturbed background concentration of bacteria, Δt is the time step, κ is the time for the free-swimming bacterial population to turn over (based on growth and loss rates), G_t is the total rate of production of new planktonic bacteria due to oil degradation, h is the width of the cross-sectional slice, and t_0 is the temporal offset for the appropriate starting size. The second term in Equation 13.3 represents the net predation that occurs on the bacterial population and effectively returns the population to the base concentration level in the absence of other elements stimulating growth. The third term quantifies the number of new bacteria produced by all the colonized droplets in the plume section. Since the degradation of a droplet after colonization follows a prescribed path, G_t depends only on how long each simulated droplet has been colonized. Equation 13.4 describes the volume of the simplified dilution model (see Figure 13.2, gray lines) for a two-dimensional vertical cross section of the plume. This volume function is evaluated at the appropriate time steps for Equation 13.3.

With the updated concentration of bacteria within the oil droplet plume $(C_t = N_t / V_t)$, the encounter rate at time t is calculated as in the original microscale oil degradation model (Eq. 13.1) as

$$\lambda_t = 4\pi\left(r_{bacteria} + r_{droplet}\right)\left(D_{bacteria} + D_{droplet}\right)\frac{N_t}{V_t}. \qquad (13.5)$$

Note that, for encounters, we remain interested in the relative diffusivity between the bacteria and the oil droplets as the process that leads to their encounters. This relative motion occurs at scales much smaller than the turbulent mixing that drives the oil dispersion dilution, which originates from velocity gradients (shear) on length scales as small as the Kolmogorov scale. While shear from turbulence or other fluid motion can in principle enhance encounter rates by producing relative motion between the bacteria and the droplet, the effect is small for most realistic conditions(Karp-Boss et al., 1996; Kiorboe, 2008). For the encounter rate, therefore, the diffusivities in Equation 13.5 are the Brownian diffusivities of the bacteria and droplets, respectively.

With these additions, significantly affecting the concentration of bacteria, the microscale oil degradation model is here extended to include continuous dilution, such as that affecting deep-water oil droplet dispersions. The same approach could be used to model dilution near the ocean surface after dispersants and wave action disperse an oil slick; however, that case would be more appropriately modeled as a one-dimensional dilution process with a diffusivity that depends on different processes to those considered here. In the following, we shall examine how dilution in an oil droplet plume impacts the biodegradation of the oil, particularly in comparison to cases without dilution.

13.5 IMPACT OF DILUTION ON OIL DEGRADATION AND APPARENT OIL DEGRADATION

When we compare the temporal dynamics of oil degradation with and without dilution (see Figure 13.4), there is a large shift in the oil concentration needed to sustain a heightened bacterial presence in the ambient fluid and thus shorten the average encounter (and degradation) time. The necessary starting oil concentration is higher in the presence of dilution by three orders of magnitude, approximately $10^{-1.2}$ PPM with dilution and $10^{2.4}$ PPM without dilution. Despite the large shift in the dependence on the starting oil concentration caused by dilution, the two limiting cases of very low and very high starting oil concentration remain unchanged in the presence

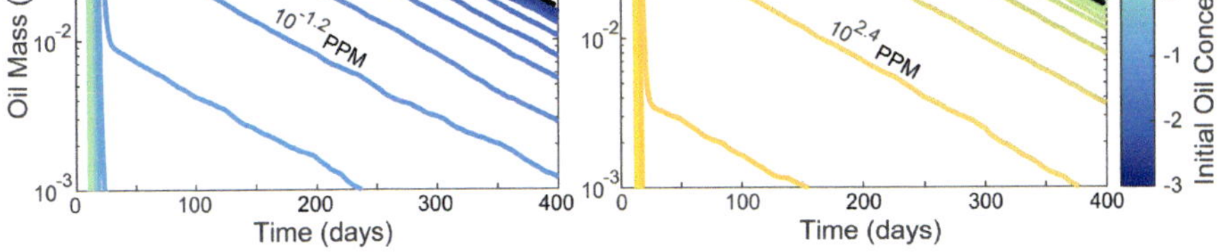

FIGURE 13.4 Comparison of oil degradation with and without dilution. The total oil mass, normalized by its initial value, is shown as oil is degraded over time with (B) and without (A) dilution. Both plots use the same color map to indicate the starting oil droplet concentration (in PPM). In both cases, degradation occurs exponentially for low starting oil concentrations, and the limit for low starting oil concentration is the same (dashed black line). The two cases also show similar dynamics for very high starting oil concentrations. However, for intermediate concentrations, the transition to enhanced degradation rates occurs at much higher oil concentration in the presence of dilution.

and absence of dilution. The dynamics at intermediate oil concentrations are qualitatively similar as well, with an initially elevated degradation rate (see the sharp decline at early times in Figure 13.4) that transitions to the exponential decay. The key effect of dilution is that larger oil concentrations are needed to sustain an increase in the background concentration of bacteria such that degradation rates are increased. Otherwise, the dilution of the oil droplet plume also dilutes any increase in bacterial concentration and returns it to the background levels.

In the absence of dilution and any changes to the background concentration of oil degrading bacteria, the degradation of a population of oil droplets progresses exponentially, as illustrated here for the case of 62 μm diameter droplets (see Figure 13.4, black dashed lines). The rate of the exponential decay is determined by the rate of arrivals in the encounter process, which in turn is dependent on the instantaneous concentration of bacteria in the ambient fluid. This rate is also the low oil concentration limit, since at very low oil concentrations the additional bacteria that are released from degrading oil droplets will not impact the overall bacteria concentration in the plume. Dilution further lowers the concentration of oil over time, but it does not affect degradation in this limiting case. As a result, for starting concentrations of oil below 10^{-2} PPM with dilution and below $10^{1.6}$ PPM without dilution, oil degradation closely follows the same exponential limit.

At very high oil concentrations, the background concentration of bacteria will quickly increase after the first oil droplet is colonized and begins to add bacteria to the environment. Shortly afterwards, all the droplets are colonized and have begun to be degraded, leading to very rapid biodegradation that is constrained by the rate of bacterial degradation and the droplet surface area. The mechanics of this limiting case are also unaffected by dilution, which constantly reduces the bacterial population in proportion to its concentration similar to an additional mortality term, but which is overcome by sufficiently high net growth rates.

While dilution does not qualitatively change the degradation dynamics in the biophysical MODEM model, the increase in the starting oil concentration needed (for the mixing rates considered here) to trigger accelerated degradation has significant implications for bioremediation strategies. As visible in Figure 13.4, there is a very substantial difference in the degradation time when it is exponential and driven by encounters (black dashed lines) in comparison to the rapid (near-linear) degradation that occurs when colonization encounters are enhanced by a large local bacterial concentration. In the example of 62 μm diameter droplets, degradation of 90% of the oil occurs in <20 days when the starting oil concentration is high enough, compared to >400 days when starting at a low concentration. In oil spills, any interventions that shift the degradation dynamics away from the encounter-limited case will accelerate degradation, but this is harder to achieve under conditions with strong dilution.

Dilution affects oil droplet degradation by altering the conditions early in the degradation process. Without dilution, there is an initial lag as the first droplets are encountered and go through the exponential growth phase. At that point, the cascade of increasing bacterial concentration in the environment may or may not begin, depending on the concentration of oil. During the initial lag time, very little if any oil is being degraded. However, with dilution, the initial lag is long enough for the local oil concentration in the plume to drop by two orders of magnitude (see Figure 13.5).

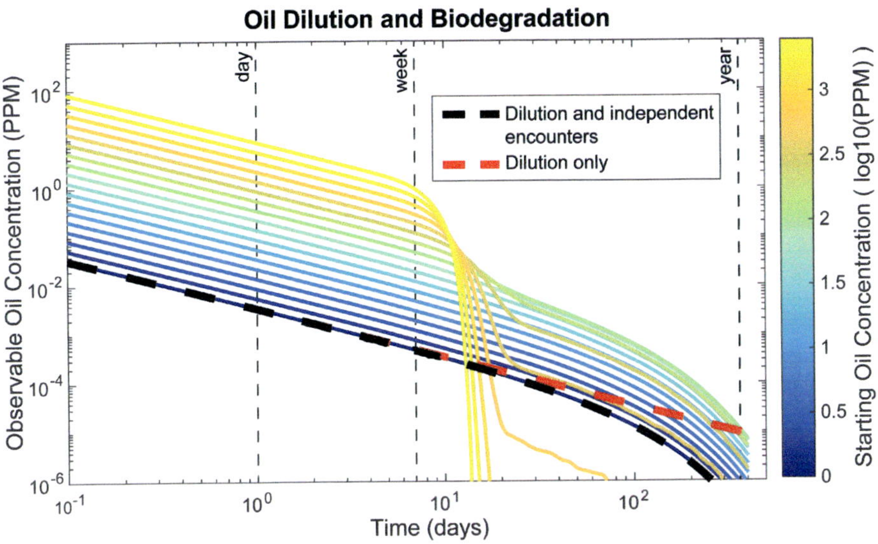

FIGURE 13.5 Evolution of oil concentration with degradation and dilution. The oil concentration in a spreading cross section of a plume is shown over time as it is reduced by both biodegradation and dilution. The color of each curve corresponds to the starting oil concentration, indicated by the color bar. Dilution alone would produce a linear degradation when plotted logarithmically (red dashed line, compare with Figure 13.3). In the absence of any change of the background bacterial concentration, referred to here as the 'independent' limit, the reduction in oil concentration begins to significantly deviate from the dilution-only case after approximately one week (black dashed line). At higher starting oil concentrations, degradation of the oil results in an increase in the background concentration of bacteria, which leads to a faster reduction in oil concentration due to the more rapid colonization of droplets.

Under the simulated conditions, significant degradation of the oil in the plume begins to be noticeable a week into the dilution. In fact, the oil concentration a week after the initial release compares favorably to the starting concentrations without dilution (see Figure 13.4A) for predicting whether encounters will dominate the degradation or not. This suggests that the concentration a week after oil is introduced into the environment determines whether bacteria grown on oil droplets can meaningfully enhance their local background concentration.

Figure 13.5 also makes it apparent that directly observing oil droplet biodegradation in the field may be difficult, and could potentially result in the misinterpretation of dilution. When the conditions allow for a large increase in ambient bacterial concentration such that degradation is enhanced (i.e., starting oil concentrations $>10^{2.4}$ PPM), the oil concentration begins to decrease quite sharply after one week and the oil is largely degraded by bacteria over the following two weeks. In contrast, when the oil concentration is not sufficiently high to directly drive an increase in ambient bacterial concentration, the difference between dilution alone and dilution coupled with microbial degradation may not be noticeable for more than one month (see Figure 13.5, dashed lines). In addition, this evidence of biodegradation would occur at oil concentrations below 10^{-2} PPM, which are difficult to accurately measure in the field (Joint analysis group for the Deepwater Horizon Oil, 2011). As a result, based on our model taking encounters between bacteria and droplets into account, microbial degradation of an oil droplet dispersion should be challenging to directly measure in the field even when contributing significantly to the overall oil degradation.

13.6 INCREASING BIODEGRADATION RATES THROUGH ENCOUNTERS

The Deepwater Horizon spill event in 2010 generated a significant oil plume at a depth of approximately 1100 meters extending approximately 35 km over several months (Camilli et al., 2010), and provides a point of reference and comparison for the conclusions of our dilution-enhanced MODEM study. In analyses and characterization following the spill, some authors reported no significant biodegradation of the Deepwater Horizon oil plume on a timescale of months (Camilli et al., 2010), whereas others found enrichment in oil-degrading bacteria in the plume and signs of reduction in oil concentrations that were tentatively attributed to biodegradation (but acknowledged dilution as a possibility) (Hazen et al., 2010).

Despite the reports of increased bacterial concentrations at the Deepwater Horizon plume, it is difficult to evaluate whether the encounter rates with oil droplets would have been meaningfully impacted. In the study reporting enrichment of bacteria in the plume, the concentration of bacteria (of all types) was found to double from 2.7×10^4 cells/mL to 5.5×10^4 cells/mL (Hazen et al., 2010). At the level of taxa, the highest enrichment was a factor of 3 (Hazen et al., 2010). Given that the encounter rate (see Equation 13.1) is directly proportional to the bacterial concentration, this indicates that the observed enrichment at most tripled the encounter rate of droplet-degrading bacteria with oil droplets in the plume. In addition, it appears that the majority of the microbial enhancement was in response to hydrocarbon gasses in the water (Valentine et al., 2010). This raises the possibility that the bacterial populations that were initially enriched were not able to degrade oil droplets. As a result, studies of the Deepwater Horizon spill leave open the possibility that oil droplets were diluted to sufficiently low concentrations before encountering oil-degrading bacteria, resulting in a prolonged environmental persistence.

Given the likelihood that natural marine conditions lead to encounter-limited degradation due to the effect of dilution, artificially enhancing the droplet encounter rates shortly after oil enters the environment would have a large impact on biodegradation rates. One possibility for increasing encounter rates is to increase oil droplet sizes, as discussed in our previous work on the optimal droplet size (Fernandez et al., 2019). This approach is robust to dilution, since it does not rely on a subsequent enhancement of the bacterial population in the ambient fluid to increase encounter rates. A further approach for reducing the encounter time would consist of increasing the concentration of oil-degrading bacteria that are able to attach to droplets. This could potentially be done by artificially seeding oil-degrading, droplet-attaching bacteria near the source of a spill or by introducing substrates favoring the growth of such bacteria. Based on the MODEM simulations, methods for maintaining higher concentrations of oil for a sufficiently long period of time

(>1 week) would also aid the encounter problem, but that would involve reducing the dilution or increasing the oil concentration.

13.7 CONCLUSIONS

This chapter discussed a biophysical model for hydrocarbon consumption of suspended oil droplets by bacteria under conditions of rapid dilution that occur in natural environments. Regardless of whether it is through the manipulation of droplet size or by increasing the local bacteria concentration, the results of the model indicate that stimulating the initial colonization of droplets is critical for rapid biodegradation of oil droplets in the ocean. We have found in this study that dilution heavily exacerbates the lag until oil-degrading bacteria find oil droplets, because the ambient microbial enhancement is delayed by approximately a week by which point the concentration in an oil plume has fallen. Given the complexity of physical and biological processes involved in biodegradation, the importance of physical encounters between droplets and bacteria in rapidly diluting environments has not been appreciated.

While this work has examined how dilution impacts the encounter process that is a prerequisite for oil droplet degradation, we have abstracted dilution into a single process for a spreading plume well below the ocean surface. It is important to understand microscale microbial encounters with oil droplets in the context of their environments. In natural environments, the strength of the dilution varies greatly with the location, depending on characteristics such as water depth, currents, weather, and distance from the surface. The dilution process is also dependent on how the oil dispersion is created. If an oil slick is dispersed by wave action, much of the dilution will be one dimensional, whereas a brief leak that releases a localized cloud of droplets will spread in three dimensions with stronger dilution than a plume in the same environment. Further work examining this range of conditions is necessary to understand the magnitude of the impact of encounters between droplets and bacteria under realistic oil spill environments.

13.8 APPENDIX

The model (MODEM) used in this study represents the biodegradation of droplets through interactions with a single bacterial species representing the first colonizers of the droplets. The model simplifies the biological complexity of a hydrocarbon-degrading bacterial community to maintain the focus on the impact of the microscale physical environment. The characteristics of the modeled bacteria are described in Table 13.1. In addition, 85% of the starting oil volume is assumed to be biodegradable by these bacteria.

MODEM also assumes that the encounter process between oil droplets and bacteria is driven primarily by relative diffusion and that the impact of fluid flow is comparatively small. This is appropriate for neutrally buoyant small entities that are jointly advected by the local flow. Once colonized by the first viable bacterium, MODEM switches from the encounter stage to the growth stage that neglects subsequent encounters. This assumes that the doubling time of bacteria on a droplet is significantly shorter than the encounter time.

MODEM directly simulates individual oil droplets over time. The primary variables tracked in MODEM are the colonization times of each droplet and the abundance of bacteria in the surrounding fluid that changes over time. For this study, the volume of the simulation also changes over time due to the expanding plume cross section.

In the simulations discussed in this chapter, each of the one million droplets has the same 62 μm diameter, matching the surface area to volume ratio of a more complex distribution of droplet sizes (see Section 13.2). The large number of droplets considered was sufficient that the degradation simulations were highly repeatable, with only small fluctuations developing in the exponential tail of oil volume below 1% of the starting volume (e.g., see Figure 13.4).

At each one-hour time step, uniformly distributed random numbers are generated for each uncolonized droplet and compared to a threshold corresponding to the probability of colonization over 1 hr given the updated encounter rate for that time step. If above the threshold, a droplet is marked as colonized at that time step. This process is repeated, updating the simulation variables at each time step, until the remaining metabolizable oil fraction drops below a critical threshold (0.1% of starting oil volume).

In order to reduce computation time and repetition, the deterministic degradation trajectories for metabolizable oil volume, surface area coverage, and newly generated planktonic bacteria (see Figure 13.1) are precomputed for all droplet sizes using a significantly finer time step than for droplet encounters (30 sec) in order to accurately capture the transitions between stages. For a given time in the simulation, the contribution of each droplet to the change in ambient bacterial abundance is determined from this precomputed information and the difference in time between the current simulation time and the colonization time of each droplet. Coupled with population turnover and new cells from the changing volume (see Equation 13.3), the bacterial abundance is updated each time step. With the further addition of the deterministic update in oil plume volume based on the total time elapsed (see Equation 13.4), the encounter rates can be calculated for each time step (see Equation 13.5) and used to simulate which of the remaining bacteria-free droplets become colonized at that time.

REFERENCES

Aman, Z.M., Paris, C.B., May, E.F., Johns, M.L., Lindo-Atichati, D. (2015). High-pressure visual experimental studies of oil-in-water dispersion droplet size. *Chem Eng Sci* 127, 392–400.

Atlas, R.M., Hazen, T.C. (2011). Oil biodegradation and bioremediation: a tale of the two worst spills in U.S. history. *Environ Sci Technol* 45, 6709–6715.

Bagby, S.C., Reddy, C.M., Aeppli, C., Fisher, G.B., Valentine, D.L. (2016). Persistence and biodegradation of oil at the ocean floor following *Deepwater Horizon*. *Proc Natl Acad Sci* 201610110.

Boehm, P.D., Fiest, D.L., Mackay, D., Paterson, S. (1982). Physical-chemical weathering of petroleum hydrocarbons from the Ixtoc I Blowout: chemical measurements and a weathering model. *Environ Sci Technol* 16, 498–505.

Camilli, R., et al. (2010). Tracking hydrocarbon plume transport and biodegradation at Deepwater Horizon. *Science* 330, 201–204.

Daling, P.S., Brandvik, P.J., Mackay, D., Johansen, O. (1990). Characterization of crude oils for environmental purposes. *Oil Chem Pollut* 7, 199–224.

Delvigne, G.A.L., Sweeney, C.E. (1988). Natural dispersion of oil. *Oil Chem Pollut* 032, 281–310.

Fernandez, V.I., Stocker, R., Juarez, G. (2019). Optimal droplet sizes for microbial degradation of dispersed oil: tradeoffs between physical encounters and consumption by marine bacteria. *Submitted.*

French-McCay, D. (2004). Oil spill impact modeling: development and validation. *Environ Toxicol Chem* 23, 2441–2456.

French-McCay, D.P., et al. 2018. Modeling distribution, fate, and concentrations of Deepwater Horizon Oil in subsurface waters of the Gulf of Mexico. In *Oil Spill Environmental Forensics Case Studies* New York: Elsevier Inc., 683–735.

Hazen, T.C., et al. (2010). Deep-sea oil plume enriches indigenous oil-degrading bacteria. *Science* 330, 204–208.

Head, I.M., Jones, D.M., Röling, W.F.M. (2006). Marine microorganisms make a meal of oil. *Nat Rev Microbiol* 4, 173–182.

Head, I.M., Swannell, R.P.J. (1999). Bioremediation of petroleum hydrocarbon contaminants in marine habitats. *Curr Opin Biotechnol* 10, 234–239.

Hibiya, T., Nagasawa, M., Niwa, Y. (2006). Global mapping of diapycnal diffusivity in the deep ocean based on the results of expendable current profiler (XCP) surveys. *Geophys Res Lett* 33, 2–5.

Huang, J. (1983). A review of the state-of-the-art of oil spill fate/behavior models. *Int Oil Spill Conf* 313–322.

Johansen, Ø., Reed, M., Bodsberg, N.R. (2015). Natural dispersion revisited. *Mar Pollut Bull* 93, 20–26.

Joint analysis group for the Deepwater Horizon Oil, "Review of preliminary data to examine subsurface oil in the vicinity of MC252#1, May 19 to June 19, 2010," *NOAA Tech. Rep.*, no. August, pp. 1–169, 2011.

Juarez, G., Stocker, R. (2019). Buckling and deformation of droplets due to confined growth at oil-water interfaces. *Submitted.*

Karp-Boss, L., Boss, E., Jumars, P. (1996). Nutrient fluxes to planktonic osmotrophs in the presence of fluid motion. *Oceanogr Mar Biol* 34, 71–107.

Kiorboe, T. (2008). *A Mechanistic Approach to Plankton Ecology.* Princeton, New Jersey: Princeton University Press.

Ledwell, J.R., Watson, A.J. (1991). The Santa Monica Basin tracer experiment: a study of diapycnal and isopycnal mixing. *J Geophys Res* 96, 8695.

Ledwell, J.R., Watson, A.J., Law, C.S. (1998). Mixing of a tracer in the pycnocline. *J Geophys Res Ocean* 103, 21499–21529.

Lee, K., Nedwed, T., Prince, R.C., Palandro D. (2013). Lab tests on the biodegradation of chemically dispersed oil should consider the rapid dilution that occurs at sea. *Mar Pollut Bull* 73, 314–318.

Lessard, R.R., DeMarco, G. (2000). The significance of oil spill dispersants. *Spill Sci Technol Bull* 6, 59–68.

Li, Z., Lee, K., King, T., Boufadel, M.C., Venosa, A.D. (2008). Assessment of chemical dispersant effectiveness in a wave tank under regular non-breaking and breaking wave conditions. *Mar Pollut Bull* 56, 903–912.

Li, Z., Lee, K., King, T., Boufadel, M.C., Venosa, A.D. (2009). Evaluating crude oil chemical dispersion efficacy in a flow-through wave tank under regular non-breaking wave and breaking wave conditions. *Mar Pollut Bull* 58, 735–744.

Loferer-Krößbacher, M., Klima, J., Psenner, R. (1998). Determination of bacterial cell dry mass by transmission electron microscopy and densitometric image analysis. *Appl Environ Microbiol* 64, 688–694.

Macnaughton, S.J., Swannell, R., Daniel, F., Bristow, L. (2003). Biodegradation of dispersed forties crude and Alaskan North Slope oils in microcosms under simulated marine conditions. *Spill Sci Technol Bull* 8, 179–186.

McAuliffe, C.D., Johnson, J.C., Greene, S.H., Canevari, G.P., Searl, T.D. (1980). Dispersion and weathering of chemically treated crude oils on the ocean. *Environ Sci Technol* 14, 1509–1518.

National Research Council, *Oil in the Sea III: Inputs, Fates, and Effects.* The National Academies Press, Washington DC, 2003.

Okubo, A. (1971). Oceanic diffusion diagrams. *Deep Res Oceanogr Abstr* 18, 789–802.

Passow, U., Hetland, R.D. (2016). What happened to all of the oil? *Oceanography* 29, 118–125.

Payne, J.R., McNabb Jr., G.D. (1984). Weathering of petroleum in the marine environment. *Marine Technology Society Journal* 18, 24–42.

Polzin, K.L., Toole, J.M., Ledwell, J.R., Schmitt, R.W. (1997). Spatial variability of turbulent mixing in the spatial variability abyssal ocean. *Science* 276, 93–96.

Reddy, C.M., et al. (2012). Composition and fate of gas and oil released to the water column during the Deepwater Horizon Oil spill. *Proc Natl Acad Sci* 109, 20229–20234.

Reed, M., et al. (1999). Oil spill modeling towards the close of the 20th century: overview of the state of the art. *Spill Sci Technol Bull* 5, 3–16.

Simmons, H.L., Jayne, S.R., St. Laurent, L.C., Weaver, A.J. (2004). Tidally driven mixing in a numerical model of the ocean general circulation. *Ocean Model* 6, 245–263.

Spaulding, M.L. (1988). A state-of-the-art review of oil spill trajectory and fate modeling. *Oil Chem Pollut* 4, 39–55.

Spaulding, M.L. (2017). State of the art review and future directions in oil spill modeling. *Mar Pollut Bull* 115, 7–19.

Stewart, P.S., Tedaldi, D.J., Lewis, A.R., Goldman, E. (1993). Biodegradation rates in seawater of crude oil. *Water Environ Res* 65, 845–848.

Tarr, M., Zito, P., Overton, E., Olson, G., Adkikari, P., Reddy, C. (2016). Weathering of oil spilled in the marine environment. *Oceanography* 29, 126–135.

Valentine, D.L., et al. (2010). Propane respiration jump-starts microbial response to a deep oil spill. *Science* 330, 208–211.

Vrede, K., Heldal, M., Norland, S., Bratbak, G. (2002). Elemental composition (C, N, P) and cell volume of exponentially growing and nutrient-limited bacterioplankton. *Appl* 68, 2965–2971.

Wadsworth, T. (1995). Containment and recovery of oil spills at sea. Methods and limitations. *J Waste Manag Environ* 87–90.

Part V

"Omics" and Novel Technologies for the Oil and Gas Industry

14 Environmental Surveillance of Marine Systems Using Genosensors: Application to Offshore Activities

Andrea Bagi, Kamila Knapik, Mari Mæland Nilsen, Adriana Krolicka, Catherine Boccadoro, and Thierry Baussant

CONTENTS

14.1 INTRODUCTION: SURVEILLANCE NEEDS AND GENOSENSORS

Ocean research and monitoring seeks to deepen our fundamental understanding of ecosystems and changes that take place on different spatial and temporal scales. The anthropogenic impact on the function and equilibrium of ocean ecosystems, and biodiversity have come into sharper focus in the last decades. The scientific challenge lies in distinguishing between human impacts and natural variation. Traditionally, this is achieved by collecting discrete samples from shipboard surveys and transporting them to the laboratory for analysis. This cumbersome process is neither cost-effective nor adapted to monitoring for compliance testing. Nor does it support quick decision-making required for operation and management to mitigate the environmental footprint from offshore oil production. Discrete sample collection and aftermath analysis does not lend itself to the new paradigm of ocean observatories, in which scientists seek to collect data, not samples, using *in situ* instruments (Puillat et al. 2016). This paradigm shift is particularly needed for oceanic frontier regions such as deeper waters and the Arctic, the new frontier of future oil and gas exploration and production activities (Niu et al. 2009). It is also relevant for coastal zone management, which demands rapid measurements and feedback for decision and risk assessment pertaining to resources and recreational areas.

In situ ocean observing systems have long been recognized as promising tools to address the problems associated with traditional monitoring approaches (Kroger and Law 2005, Kroger et al. 2009). Novel *in situ* platforms carrying both generic sensor packages (mostly physio-chemical, such as conductivity, temperature, pressure, pH, turbidity, oxygen, flow rate, pigments, chlorophyll, nitrogen and phosphorous sensors) and targeted sensor packages (chemical sensors and gas sniffers, and other detectors including biosensors) are able to rapidly detect changes in the above mentioned environmental indicators and some target compounds (McQuillan and Robidart 2017, Ottesen 2016, Scholin 2010). These instruments can provide early warning systems, for example for hydrocarbon (HC) presence in water, allowing for improved event response opportunities desired by both regulators and industries.

Monitoring of the marine environment presents many inherent peculiarities to be overcome such as harsh conditions, rapid dilution of compounds to low concentration, and vast areas to cover. Despite progress being made in marine monitoring, most methods being developed have focused on fresh water applications, particularly in relation to regulations such as the EU Water Framework Directive (Farre et al. 2009, Rodriguez-Mozaz 2006). There is an increasing demand to adapt these technologies for marine use and thereby address regulatory and industry expectations (Justino et al. 2015a, b). While generic sensors and targeted chemical detection technologies mentioned above have already reached a mature stage in development and are able to provide timely information about the environmental conditions (Mills and Fones 2012), targeted biological sensors able to measure responses on key taxa in marine ecosystems (i.e. microorganisms) are still missing.

Among the wide range of biosensor approaches, here we focus on detection of microorganisms using molecular tools, as promising candidates for monitoring in natural environments (Medlin 2016). Recent advances in genomics and metagenomics have unraveled the importance of microbial processes in the oceans, including microbial responses to contaminants such as oil. These powerful tools can be used to identify targets for monitoring ecosystems and pollution agents, which can then be detected by novel molecular-based techniques based on nucleic acid specific molecular probes (Palchetti and Mascini 2008, Paul et al. 2007, Scholin et al. 2009, Scholin, 2010, Tecon and van der Meer 2008, Wang 2000). These techniques will be further referred to as genosensors. Throughout this chapter, we will maintain a broad definition of *genosensor* and refer to all techniques which employ nucleic acid interactions for recognition of target microorganisms as genosensors.

While many challenges linked to specificity, sensitivity, automatisation, and robustness remain, the integration of genosensors into autonomous sampling and sample processing platforms can enable the identification and/or quantification of specific microbial genes of interest, providing timely and valuable insights into marine ecosystem processes (Ottesen 2016, Scholin 2010). *In situ* genosensing platforms also open the way to meeting some of the legislative requirements for water quality requirements and can contribute to the vision of developing marine observation networks across larger regions for understanding oceans and their changes from natural or human activities (Bohan et al. 2017, Siontorou et al. 2017).

14.2 MICROBIAL RESPONSE TO OIL IN THE SEA

Compounds from crude oil, released into the marine environment, pose a well-documented threat to marine life mainly due to the toxic nature of some of the constituents, such as polycyclic aromatic hydrocarbons (PAHs) (Sanni et al. 2017). Beyond the harmful consequences for higher level organisms, HCs also exhibit toxic effects on smaller eukaryotes, microeukaryotes, archaea and bacteria (Brussard et al. 2016). Even a short-term oil exposure can lead to fast ecological changes, such as shifts in heterotrophic bacterial and algal composition, and loss of zooplankton predators. The toxic effect of oil contamination lead to different outcomes in terms of microbial community composition. Microorganisms, which normally dominate pristine seawater, such as members of the SAR11 clade, decrease in abundance, while secondary consumers that take advantage of organic matter released from dead organisms increase in abundance (Chronopoulou et al. 2015). Potentially, all these layers of response provide targets for bioindicator selection.

As HCs have been present in the marine environment thanks to natural sources, such as seeps, a wide range of microorganisms including hydrocarbonoclastic bacteria (HCB), yeast, fungi and algae evolved and are well-known to degrade HCs (Head et al. 2006, Das and Chandran 2011). Among these, HCB and obligate HCB (OHCB, utilizing solely HCs for growth) remain the most extensively studied and are considered the most active players in petroleum biodegradation in the marine environment (Hazen et al. 2016). Because their metabolism is dependent on HCs, OHCB are often undetectable or present at very low levels in pristine marine environments and only bloom in response to HC incidence (Yakimov and Golyshin, 2007). While HCB are represented across many different bacterial phyla (Prince et al. 2010), OHCB are limited to 8 known genera within the Proteobacteria phylum: they consist of genera within Gammaproteobacteria: *Alcanivorax*, *Cycloclasticus*, *Marinobacter*, *Neptumonas*, *Oleiphilus*, *Oleispira*, *Thalassolituus*, and Alphaproteobacteria: *Thalassospira* (Berry and Gutierrez, 2017). Most notably, the increase in relative abundance of *Alcanivorax* and *Cycloclasticus* has been reported globally in response to oil (McGenity et al. 2012). However, in cold environments *Oleispira* is often the dominant obligate alkane degrader, while in temperate conditions, *Thalassolituus* may replace *Alcanivorax*.

Although phylogenetically very diverse, HCB and OHCB use very similar enzymes to carry out HC biotransformation reactions. Depending on the length of the substrate in the case of alkanes, several different enzymes are involved in their hydroxylation. The initial biotransformation steps of volatile short-chain alkanes (C1-C4) are carried out by enzymes related to methane monooxygenases, medium-chain (C5-C17) alkanes are activated by AlkB monooxygenases and P450 monooxygenases (van Beilen and Funhoff 2007), while the first steps in the degradation of long-chain alkanes (>C18) are carried out by flavin-binding monooxygenase (AlmA) and long chain alkane monooxygenase (LadA) enzymes (Wang and Shao 2012, Wang and Shao 2013). The initial biotransformation of mono- and polyaromatic compounds is mediated primarily by dioxygenases (Rieske non-haem iron oxygenases, flavoprotein monooxygenases and soluble diiron oxygenases) and/or CoA ligases that convert them to central intermediates such as catechol, protocatechuate, and gentisate. These intermediates are subsequently cleaved by intra-diol or extra-diol dioxygenases such as catechol 1,2-dioxygenase (1,2-CAT), protocatechuate 3,4-dioxygenases (3,4-PCA), and gentisate 1,2-dioxygenase (1,2-GDO) enzymes into TCA cycle intermediates (Duarte et al. 2014).

Microbial response to oil pollution at sea depends on various environmental factors, such as nutrient concentrations, temperature, oxygen availability, number and type of HCB and OHCB present etc., as well as the type of oil. Temperature has a strong influence on the dominant genera, as was shown by Liu et al. (2017). They found that colder incubation temperature (4°C) favored *Cycloclasticus*, *Pseudoalteromonas*, *Sulfitobacter*, and *Reinekea*, in comparison to warmer incubations (24°C), where *Oleibacter*, *Thalassobius*, *Phaeobacter*, and *Roseobacter* were enriched. The temporal dynamic of HC biodegradation: short-chain alkanes, long-chain alkanes, small aromatics and then PAHs, seems to be consistent across several environmental variables such as temperature and nutrient availability. Since biodegradation of petroleum HCs is a very complex process and HCB and OHCB populations can vary between related sites, understanding which organisms are the most important players in a specific environment can be challenging. Location and environmental conditions both influence *in situ* microbial community composition and function, resulting in variable functional gene composition (Bargiela et al. 2016).

Overall, the biodegradation of petroleum compounds in the environment is directly linked to the genomic capabilities of microorganisms present and the high relative abundance of specific organisms or genes reflect the occurrence of their specific or preferred HC substrates. Moreover, since aliphatic compounds are more readily biotransformed than aromatics, the abundance or expression of different catabolic genes involved in their degradation can be used as an indicator for the weathering status of the oil pollution (Hu et al. 2017). Monitoring phylogenetic or functional nucleic acid markers of marine HC-degrading bacteria therefore offers a valid approach for tracking HC pollution events or biodegradation of specific compounds.

14.3 MOLECULAR MARKERS OF OIL CONTAMINATION

Both prokaryotic and eukaryotic microorganisms that tend to respond to oil presence can be used when selecting bioindicators of oil pollution in the sea. The development of high-throughput sequencing technologies alongside advanced bioinformatics and statistical analysis, supports a comprehensive identification of bioindicator taxa (Lanzén et al. 2016). Moreover, thanks to metagenomics, even sequences with unknown function or with functions not directly related to HC biodegradation can be identified as molecular markers through *in silico* analysis (Wu et al. 2013). In the following sections, we will focus only on oil-related bacterial phylogenetic and functional marker genes.

14.3.1 Phylogenetic Markers

Molecular methods for identification and classification of bacteria are mostly based on ribosomal gene sequences, i.e., 16S rRNA (Balch et al. 1977, Woese and Fox 1977, Větrovský and Baldrian 2013). Other genes, such as *gyrB* (DNA gyrase subunit B), *rpoC1* (polymerase gamma subunit), *rpoD1* (principal sigma factor of *E. coli* sigma [70] type for DNA-dependent RNA polymerase), *rbcL* (ribulose 1,5-bisphosphate carboxylase/oxygenase, rubisco), chloroplast genes as *matK* (encoding maturase), and *ndhF* (encoding subunit F of NADP dehydrogenase) have also been proposed (Krawczyk and Sawicki 2013, Peeters and Willems 2011, Seo and Yokota 2003, Wu et al. 2013). For the identification of fungi and cyanobacteria, internal transcribed spacers (ITSs) of ribosomal RNA are more suitable (Iwen et al. 2002, Boyer et al. 2001, Janse et al. 2003) and in many cases, eukaryotic microorganisms are best classified based on the cytochrome c oxidase I mitochondrial gene (Strüder-Kypke and Lynn 2010). The main advantage of using the SSU rRNA gene is the large amount of sequences available through specific databases; SILVA, RDP, and GreenGenes. Some argue that the disadvantage of using SSU rRNA is the presence of multiple copies (from 1 to 15 or more copies) of the rRNA operon (Klappenbach et al. 2000) but that may be perceived as a favourable property to enhance detection capability. Moreover, 16S copy number variance is often insignificant when quantified on a log scale.

Quantitative assessment of the abundance of genera including *Alcanivorax*, *Colwellia*, *Marinobacter*, *Cycloclasticus*, and *Alteromonas* is often employed in studies assessing the impact of oil exposure on marine microbial communities (Gutierrez et al. 2013). The widely used PhyloChip contains probes for *Oleispira*, *Thalassomonas*, *Pseudomonas*, *Oceanospirillales*, *Alteromonadales*, *Arcobacter*, *Thiotrichales*, *Flavobacteriales*, *Rhodobacteriales*, and *Sulfitobacter* as well (Hu et al. 2017). SSU rRNA oligonucleotide probes specific to cold water HCB; *Colwellia* and *Oleispira*, were also developed by Królicka et al. (2014) and tested in a mesocosm experiment. Results of this experiment showed a clear dose-response increase of *Oleispira antarctia* and DWH *Oceanospirillales* gene copy numbers already after one day of incubation with oil (Baussant et al. 2016). Additionally, qPCR assays designed for a *GyrB* of *Cycloclasticus pugetii* were found to be a good indicator of PAH contamination.

Selecting the right indicator species remains challenging. Recent results indicate that focusing solely on OHCB may not be appropriate for detecting acute and fresh oil spill events, rather targeting ubiquitously present generalists (HCB or other, mainly biosurfactant producers) may be more suitable (Królicka et al. *In press*). Moreover, besides focusing on absolute changes in abundance of microorganisms of interest, relative changes in the abundance of two (or more) taxa can also be used to decipher an oil pollution event. Smith et al. (2015) found that an enrichment of *Oceanospirillaceae* and a simultaneous depletion of *Pelagibacteraceae* clearly differentiated pristine from contaminated sites. Such ratios could be helpful in deciphering the age of a pollution event based on the apparent memory of bacterial communities from previous exposures.

Phylogenetic markers are widely used and can provide useful information on microbial structural changes resulting from oil exposure. However, another approach is to use gene markers that are directly related to organism function related to the oil degradation process.

14.3.2 Functional Markers

It is well known that phylogenetically closely related species may have very different metabolic (functional) capabilities, making the link between bacterial phylogeny and specific biodegradation function difficult. Many pathways for the degradation of HCs are biochemically conserved and the genes encoding enzymes in these pathways can also be similar within phylogenetically diverse organisms (Buchan et al. 2001). This phenomenon is known as functional redundancy. HCB and OHCB may be difficult to detect through a 16S rRNA gene-based approach as each taxon might be present in low abundance. Hence targeting the collective pool of their functional genes with markers may be advantageous as it can provide a set of higher copy number targets allowing for an easier detection.

The detection of oil contamination in the environment using functional gene markers is based on targeting enzymes that play key roles in degradation of aliphatic or aromatic HCs. The genes encoding for *alkB* and cytochrome P450 monooxygenase are commonly used as functional markers for aerobic alkane degradation (Lu et al. 2012, Wang et al. 2010a and b, Wang et al. 2014, Wasmund et al. 2009, Zhang et al. 2017). *AlkB*, *P450* and *almA* have been found to be widespread within marine microbiomes (Viggor et al. 2013, Wang and Shao 2012). Although most of these genes can be found in "pristine" waters, PCR or microarray-based assays have shown that *alkB* genes were more prevalent in oil-contaminated seawater.

Dioxygenase-encoding genes are commonly used as molecular marker of aromatic HC biodegradation due to the presence of conserved motifs allowing primer design targeting distantly related sequences (Iwai et al. 2011). *RHD* genes (e.g., *nah*, *phn*) have been frequently used as biomarkers in contaminated sites (Lozada et al. 2008, Sauret et al. 2016, Wang et al. 2014, Zhang et al. 2017).

For PCR primer and microarray probe design, nucleotide and protein sequences of functional genes for the microbial degradation of aromatic or aliphatic HCs can be retrieved from public databases such as HyDeg (Scoma et al. 2015), AromaDeg (Duarte et al. 2014), Fungene (Fish et al. 2013), and BioSurfDB (Oliveira et al. 2015). Several software packages that use alignments of nucleotide sequences such as CEMAsuite (Lane et al. 2015) and Primaclade (Gadberry et al. 2005) or highly divergent proteins like j-CODEHOP (Boyce et al. 2009) have been developed to facilitate the primer design. However, designing broad-specificity primers is not an easy task. AlkB protein is characterized by high sequence variation and sub-clusters are present rather than a fully conserved region across the whole alignment (Scoma et al. 2015). Using a single primer pair targeting one of the sub-clusters is not sufficient to cover the diversity within the gene family, with the risk of underestimation of the gene abundance. Jurelevicius et al. (2013) proposed to use several pairs of *alkB*-targeting primers to improve the coverage of alkane degraders belonging to different phylogenetic groups. In contrary to the *alkB* genes, one primer set can be used to cover all sub-clusters of the Rieske nonheme iron dioxygenase genes (Iwai et al. 2011).

An excellent overview of challenges related to functional gene detection in environmental samples can be found in Vilachez-Vargas et al. (2013). Although the application of functional markers to detect oil presence has largely focused on land environments and sediments, the success of such approaches indicates the potential to adapt these markers for genosensing applications in seawater.

14.4 GENOSENSING TECHNIQUES

In genosensors *per se* oligonucleotides act as biological recognition elements for the targeted analysis of DNA and RNA from biological material, and the interaction between the nucleic acid-target analyte and the oligo probe(s) is then reported by an integrated transducer. Nevertheless, as mentioned in the introduction, we consider a broader definition and include whole-cell based and amplification-based techniques in this discussion, as they have shown potential for *in situ* applications. Bioreporter systems, such as those based on genetically modified bacteria, and other techniques such as the portable bioelectronic sensing system (BESSY) (Zhou et al. 2017) are not discussed.

The classification of molecular techniques for the detection of marine bacteria presented in Figure 14.1 is neither absolute nor exhaustive. In their review, Medlin and Orozco (2017) divided these techniques first into amplification-based and hybridization-based, then further separated hybridization-based methods into whole-cell based and cell-free. Fluorescence *in situ* hybridization (FISH), developed almost three decades ago (Amann et al. 1990), enables direct observation and quantification of selected microorganisms under epifluorescence microscopy based on differential staining of nucleic acids with fluorescent dyes. FISH was used for detection and enumeration of HC-degrading microorganisms in onshore (Jjemba et al. 2006) and marine environments (Gerdes et al. 2005). However, to the best of our knowledge, FISH has not yet been implemented for environmental monitoring.

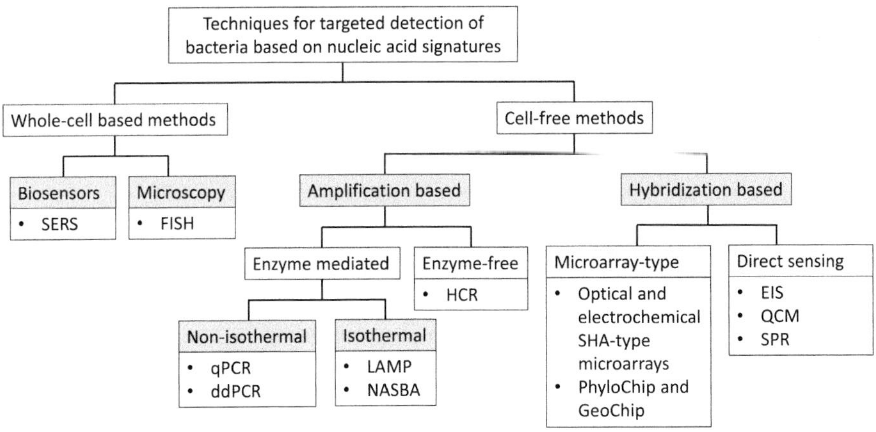

FIGURE 14.1 Overview of molecular techniques for nucleic acid-based detection. SERS—surface enhanced Raman spectroscopy, FISH—fluorescence *in situ* hybridization, PCR—polymerase chain reaction, LAMP—loop-mediated isothermal amplification, NASBA—nucleic acid sequence-based amplification, HCR—hybridization chain reaction, EIS—electrochemical impedance spectroscopy, QCM—quartz crystal microbalance, SPR—surface plasmon resonance.

A potential advantage of integrating FISH on an autonomous platform would be to gain real-time insight into ongoing processes (via labelled probes targeting mRNA) without the need for nucleic acid extraction. The second example of a whole-cell based technique noted on Figure 14.1 is the non-invasive surface enhanced Raman spectroscopy (SERS) (Wang et al. 2017). This technique has so far been considered only as a potential chemical analyser for PAHs in the deep ocean (Zhang et al. 2012) despite its potential as a bacterial identification method and a tool for microbial ecology (Germond et al. 2017). Miniaturized underwater Raman spectrometers (e.g., HORIBA's underwater construct: http://www.horiba.com/scientific/products/raman-spectroscopy/in-situ-raman-spectroscopy-in-the-worlds-oceans/) coupled to microfluidic devices capturing bacterial cells could be used to detect the spectra of HC degrading microorganisms, without the need for nucleic acid extraction. The following sections will discuss cell-free techniques, more common for use on genosensors.

14.4.1 Amplification-Based Techniques

14.4.1.1 Thermal Cycling (PCR)-Based Amplification

Real-time or quantitative PCR (qPCR) utilizes a standard end-point PCR reaction where the increase in PCR product can be monitored in "real time" during each cycle of the PCR amplification by fluorescence detection. Detection is based on the binding of the SYBR Green dye to double-stranded PCR products in a sequence independent manner, or based on specific hybridization of a fluorescently labelled TaqMan probe to the PCR product. Target quantification is achieved from a standard curve generated from amplification of known copy numbers of template DNA (Smith and Osborn 2009). qPCR techniques have mainly been used for quantifying phylogenetic markers of HC degrading genera and functional genes associated with HC biodegradation in soils and sediments (Kostka et al. 2011). Only a handful of studies assessing HC biodegradation in seawater employed this technique (Gutierrez et al. 2013, Mishamandani et al. 2014, Wang et al. 2014). qPCR was successfully demonstrated for gene detection and quantification in real-time under water (Preston et al. 2011). Design of qPCR primers is, however, challenging as high efficiency and accurate quantitative estimates can only be obtained when the primers contain little degeneracy and amplify short fragment sizes (<250 bp) (Meynet et al. 2015). Other challenges include the need for preparing standard curves, which can be difficult to establish and also time consuming and costly. A technique that circumvents this issue is droplet digital PCR (ddPCR). In ddPCR, individual DNA or RNA molecules are partitioned into droplets, resulting in one or zero molecules in most droplets (Hindson et al. 2013). After end-point amplification using fluorescent-labeled PCR reagents (SYBR Green primer-based assays or TaqMan probe-based assays), fluorescence is detected in the droplets that contain PCR products. Each droplet is then categorized into positive (containing amplified target DNA) or negative reaction (no amplified target DNA). Statistical correction is applied to account for random target distribution in each droplet. This technique does not require a calibration curve therefore provides for a significantly more robust workflow compared to qPCR. ddPCR will likely replace qPCR methods in the near future due to its superior sensitivity, reduced susceptibility to inhibition, and increased reproducibility (Taylor et al. 2017).

14.4.1.2 Isothermal Amplification

Implementing the thermal cycling necessary for qPCR or ddPCR applications introduces some complexity for *in situ* genosensors. Isothermal nucleic acid amplification offers an alternative circumventing such challenges. Several isothermal methods are showing great promises: strand displacement amplification (SDA), recombinase polymerase amplification (RPA), and helicase dependent amplification (HDA) (Fischbach et al. 2017). Loop-mediated isothermal amplification (LAMP), and nucleic acid sequence-based amplification (NASBA) have been shown to reach detection limits and sensitivities similar to PCR. Unlike RT-PCR (reverse transcription-PCR), NASBA does not require the reverse transcription step (Hønsvall and Robertson 2017a), hence it can amplify the mRNA of several target organisms directly, with low detection limits (Loukas et al. 2016 and 2017, Dimov et al. 2008, Hønsvall and Robertson 2017b).

14.4.2 Detection without Amplification

Most methods described in this section follow a so-called sandwich hybridization assay (SHA) approach. Like qPCR using sequence-specific TaqMan probes, SHA also involves two hybridization events, where two probes "sandwich" the target DNA or RNA. (Medlin and Orozco 2017). One of the probes (capture probe) can be immobilized on a membrane, electrode, or microtiter plate. This will bind a specific target to the surface. The second probe (signal probe) is typically used to produce a fluorescent or electrochemical signal while the second hybridisation step takes place. SHA approaches have the advantages of being highly specific and not requiring extensive rRNA purification prior to detection. An illustration of SHA is provided in Figure 14.2 for the case of an SPR-based detection of target analyte (the SPR technique is described later in the chapter).

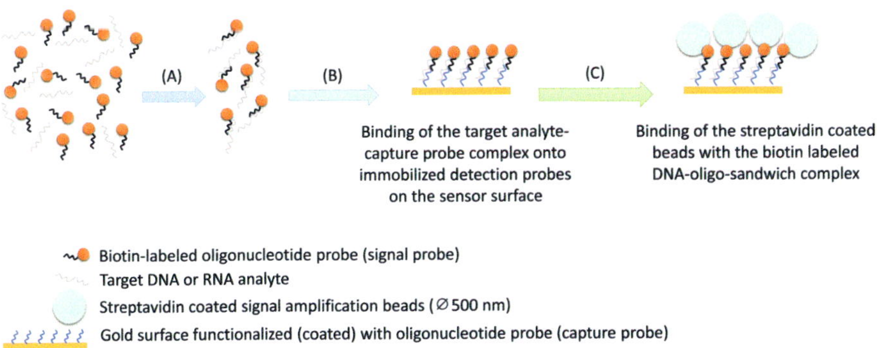

FIGURE 14.2 Illustration of a surface plasmon resonance (SPR) detection-based sandwich hybridization assay (SHA) using magnetic beads for signal amplification. The arrows show: (A) hybridization of the target nucleic acid with a signal probe, (B) immobilization (hybridization) of the analyte-oligo complex onto the SPR sensor surface, and (C) binding of the signal amplification beads onto the immobilized sandwich complex.

14.4.2.1 Microarray Techniques

Microarrays are fabricated on solid surfaces, such as glass or silicon thin-film cells functionalized with DNA probes. The typical microarray workflow consists of several steps: nucleic acid extraction, labelling, hybridization, and microarray analysis by imaging. Several microarrays were developed to help understand functional changes in response to environmental perturbations (Bouskill et al. 2011, Brodie et al. 2006, He et al. 2007, Lu et al. 2012, Wawrik et al. 2012, Zhou et al. 2013). High-density microarrays, such as the PhyloChip, allow community composition analysis by using phylogenetic probes targeting SSU rRNA exclusively (DeAngelis et al. 2011). PhyloChip G3 demonstrated a dramatic shift in microbial communities with mainly *Oceanospirillales* and 15 distinct taxa (classified as Gammaproteobacteria) being significantly enriched due to dispersed HC plume following the Deepwater Horizon (DWH) spill in the Gulf of Mexico (Hazen et al. 2010, Lu et al. 2012). In terms of functional microarrays, the most comprehensive commercially available one to date is the GeoChip 5.0 (Glomics, USA, 2014) (http://www.glomics.com/gch-tech.html).

14.4.2.2 Direct Sensing Techniques

To date, a wide range of direct detection (biosensor-on-chip) techniques have been developed, and many different types of signal transduction mechanisms have been used (Palchetti and Mascini 2008, Chen et al. 2017 [and references therein]). The two approaches cited as most promising for environmental applications are electrochemical and optical sensing. Basic concepts underlying electrochemical nucleic acids biosensors, and their recent advances are discussed in a recent review by Ferapontova (2017). Optical genosensors, also referred to as nanophotonic label-free genosensors (Nguyen et al. 2015, Carrascosa et al. 2016) have been reviewed by Chocarro-Ruiz et al. (2017). Among these, surface plasmon resonance (SPR) based techniques are particularly interesting. These rely on the change in refractive index near thin gold nanolayers following binding or hybridization between probe and target (Schasfoort 2017). SPR's potential in environmental monitoring has been recognized, mainly for the detection of small molecules and toxins (Soelberg and Furlong 2012). One example of a compact SPR-based sensor uses *Spreeta* chip technology (Soelberg et al. 2004). This instrument is promising for detection of oil-related bacteria based on 16S rRNA probes. Other low-cost and portable SPR units have been developed in the past few years with the view of them fitting into autonomous sampling and processing devices (Brulé et al. 2017, Kim 2006, Mauriz 2006, Zhang et al. 2017). A compact SPRi sensor unit is also under development in the JERICO-NEXT project (http://www.jerico-ri.eu/). Detection limits for DNA targets typically reached with SPR are on the order of 100 nM, which can be significantly improved, down to 1 pM, with a one-step selective purification of target analytes from a complex matrix using oligonucleotide coated magnetic nanoparticles (50–500 nm in diameter) (Soelberg et al. 2009). When using quantum dots instead of magnetic nanoparticles, extremely low detection limits in the sub-femtomole range have been achieved with an SPR imaging (SPRi) system on fragmented 16S rRNA targets (Foudeh et al. 2014). This is at least 100 times lower than the detection limit achieved by a chemiluminescence based assay for *Vibrio* species (Da-Silva et al. 2017). One major disadvantage of SPR is low throughput. With the improvements in nanofabrication technologies, the number of target analytes can be increased by creating multiple sensing regions on each chip, similarly to SPRi technology, which in the future should allow higher throughput for SPR (Chinowsky et al. 2007).

14.5 GENOSENSING PLATFORMS AND EXAMPLES OF APPLICATION

The major components that need to be integrated for full automatization and fluid handling on genosensing platforms include: (1) a sampling module—collection of seawater samples and filtration, (2) a sample processing module—nucleic acid extraction and purification, (3) a molecular analytical module, and (4) processing and communication of data to users or decision-makers. However, most of the existing systems perform a limited number of tasks, such as the Microbial Sampler-Submersible Incubation Device (MS-SID), which only collects samples for nucleic acid analysis (Ottesen 2016). The number of *in situ* devices under development, with extensive capability from sample collection to nucleic acids extraction and more, is increasing and many prototypes have been launched for autonomous real-time *in situ* surveillance of the European seas (see, e.g., projects: DEVOTES, BRAAVOO, MARIABOX, SenseOcean, NeXOS, Common Sense, EnviGuard, JERICO Next, SEA-on-a-CHIP, etc.; details available at http://cordis.europa.eu/projects/home_en.html).

Two advanced and versatile platforms capable of performing all *in situ* steps from collection to analysis and communication are the Environmental Sample Processor (ESP) and the Autonomous Microbial Genosensor (AMG) (McQuillan and Robidart 2017). In the AMG, NASBA is used to amplify and detect specific mRNA targets such as the *rbcL* gene of the red tide forming dinoflagellate *Karenia brevis* (Casper 2007, Paul et al. 2007). The ESP is currently using qPCR and SHA for nucleic acid target analysis, whereas ddPCR will replace qPCR in the future development (Cao et al. 2016). Several applications of the ESP have been reported in recent years and this fully automatized instrument is integrating further development to move from a moored to a long-range AUV-based platform. The current device has been used in several deployable configurations such as moored, free-drifting or on pier, and for detection of various marine targets and processes, such as faecal indicators and harmful algae (Yamahara et al. 2015), seasonal dynamic patterns of microbial populations (Robidart et al. 2012) and bacterial gene regulation of dimethylsulfoniopropionate (Varaljay et al. 2015). ESP was also successfully used to identify methane monooxygenase (*pmoA*) genes in methane-rich seawater in the deep ocean (Ussler et al. 2013). Protocols mimicking ESP's were also developed and tested for detection of HC degraders in a laboratory setup and showed good correlation with bench RT-qPCR assays, particularly for some oil-related 16S rRNA genes such as *Oleispira* and *Marinomonas* (Królicka et al. 2014).

14.6 CHALLENGES AND FUTURE OUTLOOK

Future marine environmental monitoring will most likely include the use of intelligent autonomous systems operating interconnectedly on large-scales (Bohan et al. 2017, Molin and Benoit-Bird 2016). Several autonomous underwater platforms equipped with a payload of sensors are already in use by oil and gas operators for monitoring the physio-chemical status of seawater and sediment, and for detecting oil leaks by means of chemical or visual sensors. Biosensors are needed but their development for *in situ* underwater application is still rare because of technical challenges to measure biological event “on-line”. Genosensors are promising biosensors for real-time *in situ* monitoring. Many molecular techniques have shown exciting potential for being integrated into autonomous underwater systems, such as qPCR, SHA, ELISA on ESP and NASBA on AMG. These examples are encouraging for further research into gene-based assay development and operational application of genosesensors on automatised systems for environmental surveillance in offshore marine operations.

Some of the challenges linked to using genosensors for monitoring oil pollution include the selection of relevant molecular target markers, and purpose built, robust and cost-efficient sensor instrumentation (Medlin and Orozco 2017). For HCB and OHCB-based oil pollution monitoring, determining threshold values to reliably distinguish contaminated from clean sites, and acute versus chronic pollution will be crucial for decision-making.

Selecting the best marker(s) poses one of the greatest challenges, as indicator species may vary between sites and seasons, due to the influence of environmental variables (temperature, nutrient availability, oxygen content, organic load etc.). Large-scale sequencing studies can become the fundament for deciphering correlations between environmental variables and bioindicator taxa/sequences. Advanced data analysis tools, e.g., machine learning techniques (Smith et al. 2015), will clearly play an important role in marker selection.

Advances in sensor technology, especially in the field of label-free optical sensing are promising. Detection limits of SPR assays for pathogen detection have already achieved picomole levels relevant to the marine environment. Genosensors with multiplexing capacity are expected to detect low levels as well as a wide range of specific targets in the near future. When it comes to assay specificity and robustness, novel nucleic acid analogs, such as morpholino probes, and other novel types of recognition elements for nucleic acids such as aptamers, represent further potential for improvement.

Despite many challenges, genosensing appears to be a promising tool for monitoring in the marine environment, particularly in the perspective of further exploitation of marine resources and industrialization.

REFERENCES

Amann, R.I., Binder, B.J., Olson, R.J., et al. 1990. Combination of 16S rRNA-targeted oligonucleotide probes with flow cytometry for analyzing mixed microbial populations. *Applied Environmental Microbiology* 56:1919–25.

Balch, W.E., Magrum, L.J., Fox, G.E., Wolfe, R.S., and Woese, C.R. 1977. An ancient divergence among the bacteria. *Journal of Molecular Evolution* 9:305–11.

Bargiela, R., Yakimov, M.M., Golyshin, P.N., and Ferrer, M. 2016. Distribution of hydrocarbon degradation pathways in the sea. In: *Consequences of microbial interactions with hydrocarbons, oils and lipids: Production of fuels and chemicals, Handbook of Hydrocarbon and Lipid Microbiology*, Lee, S.Y. (ed). Springer International Publishing, Heidelberg, Germany, DOI 10.1007/978-3-319-31421-1_383-1.

Baussant, T., Królicka, A., Preston, C., et al. 2016. Can we use microbes to tell us about the presence of oil in water? Chapter 12, pp 153–166, In: *Arctic Approach—The Northern Area Program*, ed. Sand, G., Dahle, S., and Misund, A. Fagbokforlaget, Bergen, Norway.

Berry, D., and Gutierrez, T. 2017. Evaluating the Detection of Hydrocarbon-Degrading Bacteria in 16S rRNA Gene Sequencing Surveys. *Frontiers in Microbiology*. doi: 10.3389/fmicb.2017.00896.

Bohan, D.A., Vacher, C., Tamaddoni-Nezhad, A., et al. 2017. Next-Generation Global Biomonitoring: Large-scale, Automated Reconstruction of Ecological Networks. *Trends in Ecology and Evolution*. 32(7):477–487.

Bouskill, N.J., Eveillard, D., O'Mullan, G., et al. 2011. Seasonal and annual reoccurrence in betaproteobacterial ammonia-oxidizing bacterial population structure. *Environmental microbiology* 13:872–86.

Boyce, R., Chilana, P., and Rose, T.M. 2009. iCODEHOP: a new interactive program for designing COnsensus-DEgenerate Hybrid Oligonucleotide Primers from multiply aligned protein sequences. *Nucleic Acids Research* 37(Web Server issue): W222–228.

Boyer, S.L., Flechtner, V.R., and Johansen, J.R. 2001. Is the 16S-23S rRNA internal transcribed spacer region a good tool for use in molecular systematics and population genetics? A case study in cyanobacteria. *Molecular Biology and Evolution* 18:1057–69.

Brodie, E.L., Desantis, T.Z., Joyner, D.C., et al. 2006. Application of a high-density oligonucleotide microarray approach to study bacterial population dynamics during uranium reduction and reoxidation. *Applied Environmental Microbiology* 72:6288–98.

Brulé, T., Granger, G., Bukar, N., et al. 2017. A field-deployed surface plasmon resonance (SPR) sensor for RDX quantification in environmental waters. *Analyst* 142:2161–68.

Brussard, C.P.D., Paperzak, L., Beggah, S., et al. 2016. Immediate ecotoxicological effects of short-lived oil spills on marine biota. *Nature Communications* 7:11206.

Buchan, A., Neidle E. L., and Moran M.A. 2001. Diversity of the ring-cleaving dioxygenase gene pcaH in a salt marsh bacterial community. *Applied Environmental Microbiology* 67(12):5801–09.

Cao, Y., Griffith, J.F., and Weisberg, S.B. 2016. The Next-Generation PCR-Based Quantification Method for Ambient Waters: Digital PCR. In: *Marine Genomics: Methods and Protocols, Methods in Molecular Biology*, Sarah J. Bourlat (ed.). New York: Springer Science+Business Media. Vol. 1452: DOI 10.1007/978-1-4939-3774-5_7.

Carrascosa, L.G., Huertas, C.S., and Lechuga, L.M. 2016. Prospects of optical biosensors for emerging label-free RNA analysis. *Trends in Analytical Chemistry* 80:177–189.

Casper, E.T., Patterson, S.S., Bhanushali, P., et al. 2007. A handheld NASBA analyzer for the field detection and quantification of *Karenia brevis*. *Harmful Algae* 6:112–118.

Chen, S., and Shamsi, M.H. 2017. Biosensors-on-chip: a topical review. *Journal of Micromechanics and Microengineering* 27. doi: 10.1088/1361-6439/aa7117.

Chinowsky, T.M., Soelberg, S.D., Baker, P., et al. 2007. Portable 24-analyte surface plasmon resonance instruments for rapid, versatile biodetection. *Biosensors and Bioelectronics* 22:2268–75.

Chocarro-Ruiz, B., Fernandez-Gevela, A., Herranz, S., and Lechuga, L.M. 2017. Nanophotonic label-free biosensors for environmental monitoring. *Current Opinion in Biotechnology* 45:175–83.

Chronopoulou, P.-M., Sanni, G.O., et al. 2015. Generalist hydrocarbon-degrading bacterial communities in the oil-polluted water column of the North Sea. *Microbial Biotechnology* 8(3):434–447.

Das, N., and Chandran, P. 2010. Microbial Degradation of Petroleum Hydrocarbon Contaminants: An Overview. *Biotechnology Research International* doi: 10.4061/2011/941810.

Da-Silva, E., Barthelmebs, L., and Baudart, J. 2017. Development of a PCR-free DNA based assay for the specific detection of *Vibrio* species in environmental samples by targeting the 16S rRNA. *Environmental Science and Pollution Research* 24:5690–5700.

DeAngelis, K.M., Wu, C.H., Beller, H.R., et al. 2011. PCR Amplification-Independent Methods for Detection of Microbial Communities by the High-Density Microarray PhyloChip. *Applied and Environmental Microbiology* 77:6313–22.

Dimov, I.K., Garcia-Cordero, J.L., O'Grady, J., et al. 2008. Integrated microfluidic tmRNA purification and real-time NASBA device for molecular diagnostics. *Lab a Chip* 8(12): 2071–2078.

Duarte, M., Jauregui, R., Vilchez-Vargas, R., Junca, H., and Pieper, D. H. 2014. AromaDeg, a novel database for phylogenomics of aerobic bacterial degradation of aromatics. *Database (Oxford).* doi: 10.1093/database/bau118. Print 2014.

Farre, M., Kantiani, L., Perez, S., and Barcelo, D. 2009. Sensors and biosensors in support of EU Directives. *Trac-Trends in Analytical Chemistry* 28:170–85.

Ferapontova, E.E. 2017. Basic concepts and recent advances in electrochemical analysis of nucleic acids. *Current Opinion in Electrochemistry.* https://doi.org/10.1016/j.coelec.2017.09.026

Fischbach, J., Frohme, M., and Glökler, J. 2017. Hinge-initiated Primer-dependent Amplification of Nucleic Acids (HIP)—A New Versatile Isothermal Amplification Method. *Scientific Reports* 7. doi:10.1038/s41598-017-08067-x.

Fish, J.A., Chai, B., Wang, Q., et al. 2013. FunGene: the functional gene pipeline and repository. *Frontiers in Microbiology* 4:291.

Foudeh, A.M., Daoud, J.T., Faucher, S.P., et al. 2014. Sub-femtomole detection of 16S rRNA from *Legionella pneumophila* using surface plasmon resonance imaging. *Biosensors and Bioelectronics* 52:129–135.

Gadberry, M.D., Malcomber, S.T., Doust, A.N., and Kellogg, E.A. 2005. Primaclade—a flexible tool to find conserved PCR primers across multiple species. *Bioinformatics* 21(7):1263–64.

Gerdes, B., Brinkmeyer, R., Dieckmann, G., and Helmke, E. 2005. Influence of crude oil on changes of bacterial communities in Arctic sea-ice. *FEMS Microbiology Ecology* 55:129–39.

Germond, A., Kumar, V., Ichimura, T., Moreau, J., Furusawa, C., Fujita, H., Watanabe, T.M. 2017. Raman spectroscopy as a tool for ecology and evolution. *Journal of the Royal Society Interface* 14(131):20170174.

Gutierrez, T., Singleton, D. R., Berry, D., et al. 2013. Hydrocarbon-degrading bacteria enriched by the Deepwater Horizon oil spill identified by cultivation and DNA-SIP. *The ISME Journal* 7:2091–2104.

Hazen, T.C., Dubinsky, E.A., DeSantis, T.Z., et al. 2010. Deep-sea oil plume enriches indigenous oil-degrading bacteria. *Science* 330:204–8.

Hazen, T.C., Prince, R.C., and Mahmoudi, N. 2016. Marine oil biodegradation. *Environmental Science and Technology* 50(5):2121–2129.

He, Z., Gentry, T.J., Schadt, C.W., et al. 2007. GeoChip: a comprehensive microarray for investigating biogeochemical, ecological and environmental processes. *The ISME Journal* 1:67–77.

Head, I.M., Jones, D.M., and Röling, W.F.M. 2006. Marine microorganisms make a meal of oil. *Nature Reviews Microbiology* 4:173–82.

Hindson, C.M., Chevillet, J.R., Briggs, H.A., et al. 2013. Absolute quantification by droplet digital PCR versus analog real-time PCR. *Nature Methods* 10(10):1003–05.

Hønsvall, B.K., and Robertson, L.J. 2017a. Will NASBA make the leap? *Water Research* 109:389–97.

Hønsvall, B.K., and Robertson, L.J. 2017b. Real-time nucleic acid sequence-based amplification (NASBA) assay targeting MIC1 for detection of *Cryptosporidium parvum* and *Cryptosporidium hominis* oocysts. *Experimental Parasitology* 172:61–67.

Hu, P., Dubinsky E.A., Probst, A.J., et al. 2017. Simulation of Deepwater Horizon oil plume reveals substrate specialization within a complex community of hydrocarbon degraders. *Proceedings of the National Academy of Sciences of the United States of America* 114(28):7432–37.

Iwai, S., Johnson, T. A., Chai, B., Hashsham, S.A., and Tiedje, J.M. 2011. Comparison of the specificities and efficacies of primers for aromatic dioxygenase gene analysis of environmental samples. *Applied Environmental Microbiology* 77(11):3551–57.

Iwen, P.C., Hinrichs, S.H., and Rupp, M.E. 2002. Utilization of the internal transcribed spacer regions as molecular targets to detect and identify human fungal pathogens. *Medical Mycology* 40:87–109.

Janse, I., Meima, M., Kardinaal, W.E., and Zwart, G. 2003. High-resolution differentiation of Cyanobacteria by using rRNA-internal transcribed spacer denaturing gradient gel electrophoresis. *Applied Environmental Microbiology* 69:6634–43.

Jjemba, P.K., Kinkle, B.K., and Shann, J.R. 2006. In-situ enumeration and probing of pyrene-degrading soil bacteria. *FEMS Microbiology Ecology* 55:287–98.

Jurelevicius, D., Alvarez, V. M., Peixoto, R., Rosado, A. S., and Seldin, L. 2013. The Use of a Combination of alkB Primers to Better Characterize the Distribution of Alkane-Degrading Bacteria. *PLoS One* 8(6):e66565.

Justino, C.I.L., Freitas, A.C., Duarte, A.C., and Santos, T.A.P.R. 2015a. Sensors and biosensors for monitoring marine contaminants. *Trends in Environmental Analytical Chemistry*. doi: 10.1016/j.teac.2015.02.00.

Justino, C.I.L., Freitas, A.C., Pereira, R., Duarte, A.C., and Santos, T. 2015b. Recent developments in recognition elements for chemical sensors and biosensors. *Trac-Trends in Analytical Chemistry* 68:2–17. doi: 10.1016/j.trac.2015.03.006.

Kim, S.J., Gobi, K.V., Harada, R., Shankaran, D.R., and Miura, N. 2006. Miniaturized portable surface plasmon resonance immunosensor applicable for on-site detection of low-molecular-weight analytes. *Sensors and Actuators B-Chemical* 115:349–56. doi: 10.1016/j.snb.2005.09.025.

Klappenbach, J.A., Dunbar, J.M., and Schmidt, T.M. 2000. rRNA Operon Copy Number Reflects Ecological Strategies of Bacteria. *Applied and Environmental Microbiology* 66(4):1328–1333.

Kostka, J.E., Prakash, O., Overholt, W.E., et al. 2011. Hydrocarbon-Degrading Bacteria and the Bacterial Community Response in Gulf of Mexico Beach Sands Impacted by the Deepwater Horizon Oil Spill. *Applied Environmental Microbiology* 77(22):7962–74.

Krawczyk, K., and Sawicki, J. 2013. The Uneven Rate of the Molecular Evolution of Gene Sequences of DNA-Dependent RNA Polymerase I of the Genus Lamium L. *International Journal of Molecular Sciences* 14:11376–91.

Kroger, S., and Law, R.J. 2005. Biosensors for marine applications - We all need the sea, but does the sea need biosensors? *Biosensors & Bioelectronics* 20:1903–13. doi: 10.1016/j.bios.2004.08.036.

Kroger, S., Parker, E.R., Metcalfe, J.D., et al. 2009. Sensors for observing ecosystem status. *Ocean Science* 5 523–35.

Królicka, A., Boccadoro, C., Maeland, M., et al. 2014. Detection of Oil Leaks by Quantifying Hydrocarbonoclastic Bacteria in Cold Marine Environments Using the Environmental Sample Processor. *Arctic and Marine Oilspill Program conference paper* 791–807.

Królicka, A., Boccadoro, C., Nilsen, M.M., and Baussant, T. In press. Capturing early changes in the marine bacterial community as a result of crude oil pollution in a mesocosm experiment. *Microbes and Environments*.

Lane, C.E., Hulgan, D., O'Quinn, K., and Benton, M.G. 2015. CEMAsuite: open source degenerate PCR primer design. *Bioinformatics* 31(22):3688–90.

Lanzén, A., Lekang, K., Jonassesn, I., Thompson, E.M., and Troedsson, C. 2016. High-throughput metabarcoding of eukaryotic diversity for environmental monitoring of offshore oil-drilling activities. *Molecular Ecology* 25(17):4392–4406.

Liu, J., Bacosa, H.P., and Liu, Z. 2017. Potential Environmental Factors Affecting Oil-Degrading Bacterial Populations in Deep and Surface Waters of the Northern Gulf of Mexico. *Frontiers in Microbiology* 7:2131. doi: 10.3389/fmicb.2016.02131.

Loukas, C.-M. 2016. Lab-on-a-chip technology for *in situ* molecular analysis of marine microorganisms. University of Southampton, Ocean & Earth Science, Doctoral Thesis, p. 141.

Loukas C.-M., McQuillan, J.S., Laouenan, F., et al. 2017. Detection and quantification of the toxic microalgae *Karenia brevis* using lab on a chip mRNA sequence-based amplification. *Journal of Microbiological Methods* 139:189–95.

Lozada, M., Riva Mercadal, J.P., Guerrero, L.D., et al. 2008. Novel aromatic ring-hydroxylating dioxygenase genes from coastal marine sediments of Patagonia. *BMC Microbiology* 8:50.

Lu, Z., Deng, Y., Van Nostrand, J.D., et al. 2012. Microbial gene functions enriched in the Deepwater Horizon deep-sea oil plume. *The ISME Journal* 6:451–60.

Mauriz, E., Calle, A., Montoya, A., and Lechuga, L.M. 2006. Determination of environmental organic pollutants with a portable optical immunosensor. *Talanta* 69:359–364. doi: 10.1016/j.talanta.2005.09.049.

McGenity, T.J., Folwell, B.D., McKew, B.A., and Sanni, G.O. 2012. Marine crude-oil biodegradation: a central role for interspecies interactions. *Aquatic Biosystems* 8:10. doi:10.1186/2046-9063-8-10.

McQuillan J.S., and Robidart J.C. 2017. Molecular-biological sensing in aquatic environments: recent developments and emerging capabilities. *Current Opinion in Biotechnology* 45:43–50.

Medlin, L.K. 2016. Mini Review: Molecular techniques for identification and characterization of marine biodiversity. *Annals of Marine Biology and Research* 3(2):1015.

Medlin, L.K., and Orozco, J. 2017. Molecular Techniques for the Detection of Organisms in Aquatic Environments, with Emphasis on Harmful Algal Bloom Species. *Sensors (Basel, Switzerland)* 17:1184.

Meynet, P., Head, I.M., Werner, D., Davenport, R.J. 2015. Re-evaluation of dioxygenase gene phylogeny for the development and validation of a quantitative assay for environmental aromatic hydrocarbon degraders. *FEMS Microbiology Ecology* 91(6). doi: 10.1093/femsec/fiv049.

Mills, G., and Fones, G. 2012. A review of *in situ* methods and sensors for monitoring the marine environment. *Sensor Review* 32:17–28. doi:10.1108/02602281211197116.

Mishamanandi, S., Guiterrez, T., and Aitken, M.D. 2014. DNA-based stable isotope probing coupled with cultivation methods implicates *Methylophaga* in hydrocarbon degradation. *Frontiers in Microbiology* 5:76.

Molin, M.A., and Benoit-Bird, K. 2016. Sensor Fusion and Autonomy as a Powerful Combination for Biological Assessment in the Marine Environment. *Robotics* 5(1):4. doi:10.3390/robotics5010004.

Nguyen, H.H., Park J., Kang S., and Kim M. 2015. Surface plasmon resonance: A versatile technique for biosensor applications. *Sensors* 15:10481–510.

Niu, H., Adams, S., Lee, K., Husain, T., and Bose, N. 2009. Applications of Autonomous Underwater Vehicles in Offshore Petroleum Industry Environmental Effects Monitoring. *Journal of Canadian Petroleum Technology* 48:12–16.

Oliveira, J.S., Araujo, W., Lopes Sales, A.I., et al. 2015. BioSurfDB: knowledge and algorithms to support biosurfactants and biodegradation studies. *Database (Oxford)*. doi: 10.1093/database/bav033.

Ottesen, E.A. 2016. Probing the living ocean with ecogenomic sensors. *Current Opinion in Microbiology* 31:132–9.

Palchetti, I., and Mascini, M. 2008. Nucleic acid biosensors for environmental pollution monitoring. *Analyst* 133:846–54.

Paul, J., Scholin, C., Van Den Engh, G., and Perry, M.J. 2007. *In Situ* Instrumentation. *Oceanography* 20:70–8.

Puillat, I., Farcy, P., Durand, D., et al. 2016. Progress in marine science supported by European joint coastal observation systems: The JERICO-RI research infrastructure. *Journal of Marine Systems* 162:1–3. doi: 10.1016/j.jmarsys.2016.06.004.

Peeters, K., and Willems, A. 2011. The gyrB gene is a useful phylogenetic marker for exploring the diversity of Flavobacterium strains isolated from terrestrial and aquatic habitats in Antarctica. *FEMS Microbiology Letters* 321:130–140.

Preston, C.M., Harris, A., Ryan, J.P., et al. 2011. Underwater application of quantitative PCR on an ocean mooring. *PLoS One* 6(8):e22522.

Prince, R.C., Gramain, A., and McGenity, T. J. 2010. Prokaryotic Hydrocarbon Degraders. In: *Handbook of Hydrocarbon and Lipid Microbiology*, Springer Protocols Handbooks. Ed: Timmis, K.N. Springer-Verlag Berlin, Heidelberg, Germany. 1669–1692.

Robidart, J.C., Preston, C.M., Paerl R.W., et al. 2012. Seasonal Synechococcus and Thaumarchaeal population dynamics examined with high resolution with remote *in situ* instrumentation. *The ISME Journal* 6:513–23.

Rodriguez-Mozaz, S., de Alda, M.J.L., and Barcelo, D. 2006. Biosensors as useful tools for environmental analysis and monitoring. *Analytical and Bioanalytical Chemistry* 386:1025–41.

Sanni, S., Björkblom, C., et al. 2017. I: Biomarker quantification in fish exposed to crude oil as input to species sensitivity distributions and threshold values for environmental monitoring. *Marine Environmental Research* 125:10–24.

Sauret, C., Tedetti, M., Guique, C., et al. 2016. Influence of PAHs among other coastal environmental variables on total and PAH-degrading bacterial communities. *Environmental Science and Pollution Research International* 23(5):4242–56.

Schasfoort, R.B.M. 2017. Handbook of Surface Plasmon Resonance: Edition 2. *Royal Society of Chemistry, London, UK*. ISBN 978-1-78262-730-2.

Scholin, C., Doucette, G., Jensen, S., et al. 2009. Remote Detection of Marine Microbes, Small Invertebrates, Harmful Algae, and Biotoxins Using the Environmental Sample Processor (Esp). *Oceanography* 22:158–67.

Scholin C.A. 2010. What are "ecogenomic sensors"? A review and thoughts for the future. *Ocean Science* 6:51–60.

Scoma, A., Hernandez-Sanabrina, E., Lacoere, T., et al. 2015. Primers: Bacterial genes encoding enzymes for aerobic hydrocarbon degradation. In: *Hydrocarbon and Lipid Microbiology* Protocols, Springer Protocols Handbooks. Ed: McGenity, T. J., Timmis, K., and Nogales, B. Springer-Verlag Berlin, Heidelberg, Germany. 23–37.

Seo, P.S., and Yokota, A. 2003. The phylogenetic relationships of cyanobacteria inferred from 16S rRNA, gyrB, rpoC1 and rpoD1 gene sequences. *The Journal of General and Applied Microbiology* 49:191–203.

Siontorou, C.G., Georgopoulos, K.N., and Nalantzi, M.-M.E. 2017. Designing biosensor networks for the environmental risk assessment of aquatic systems. *Critical Reviews in Environmental Science and Technology* 47(1):40–63.

Smith, C.J., and Osborn, A.M. 2009. Advantages and limitations of quantitative PCR (Q-PCR)-based approaches in microbial ecology. *FEMS Microbiology Ecology* 67(1):6–20.

Smith, M.B., Rocha, A.M., Smillie, C.S., et al. 2015. Natural Bacterial Communities Serve as Quantitative Geochemical Biosensors. *mBio* 6(3):e00326–15.

Soelberg, S.D., and Furlong, C.E. 2012. Biosensors and Bioassays for Ecological Risk Monitoring and Assessment. In: *Environmental Toxicology: Selected Entries from the Encyclopedia of Sustainability Science and Technology*, ed: E. A. Laws. Springer Science+Business Media LLC, New York, USA. doi 10.1007/978-1-4614-5764-0_6.

Soelberg, S.D., Chinowsky, T., Geiss, G., et al. 2004. A portable surface plasmon resonance sensor system for real-time monitoring of small to large analytes. *Journal of Industrial and Microbiology and Biotechnology* 32:669–74.

Soelberg, S.D., Stevens, R.C., Limaye, A.P., and Furlong, C.E. 2009. Surface plasmon resonance detection using antibody-linked magnetic nanoparticles for analyte capture, purification, concentration, and signal amplification. *Analytical Chemistry* 81:2357–63.

Strüder-Kypke, M.C., and Lynn, D.H. 2010. Comparative analysis of the mitochondrial cytochrome c oxidase subunit I (COI) gene in ciliates (Alveolata, Ciliophora) and evaluation of its suitability as a biodiversity marker. *Systematics and Biodiversity* 8:131–48.

Taylor, S.C., Laperriere, G., and Germain, H. 2017. Droplet Digital PCR versus qPCR for gene expression analysis with low abundant targets: from variable nonsense to publication quality data. *Scientific Reports* 7(1):2409.

Tecon, R., and van der Meer, J.R. 2008. Bacterial biosensors for measuring availability of environmental pollutants. *Sensors* 8:4062–80.

Ussler, W. 3rd, Preston, C., Tavormina, P., et al. 2013. Autonomous application of quantitative PCR in the deep sea: *in situ* surveys of aerobic methanotrophs using the deep-sea environmental sample processor. *Environmental Science and Technology* 47(16):9339–46.

van Beilen, J.B., and Funhoff, E.G. 2007. Alkane hydroxylases involved in microbial alkane degradation. *Applied Microbiology and Biotechnology* 74(1):13–21.

Varaljay, V.A., Robidart, J., Preston, C.M., et al. 2015. Single-taxon field measurements of bacterial gene regulation controlling DMSP fate. *The ISME Journal* 9:1677–86.

Větrovský, T., and Baldrian, P. 2013. The Variability of the 16S rRNA Gene in Bacterial Genomes and Its Consequences for Bacterial Community Analyses. *PLoS ONE* 8, e57923.

Viggor, S., Juhanson, J., Joesaar, M., et al. 2013. Dynamic changes in the structure of microbial communities in Baltic Sea coastal seawater microcosms modified by crude oil, shale oil or diesel fuel. *Microbiological Research* 168(7):415–27.

Vilchez-Vargas, R., Geffers, R., Suarez-Diez, M., et al. 2013. Analysis of the microbial gene landscape and transcriptome for aromatic pollutants and alkane degradation using a novel internally calibrated microarrays system. *Environmental Microbiology* 15(4):1016–39.

Wang, J. 2000. From DNA biosensors to gene chips. *Nucleic Acids Research* 28:3011–16.

Wang, L., Wang, W., Lai, Q., and Shao, Z. 2010a. Gene diversity of CYP153A and AlkB alkane hydroxylases in oil-degrading bacteria isolated from the Atlantic Ocean. *Environmental Microbiology* 12(5):1230–42.

Wang, W., Wang, L., and Shao, Z. 2010b. Diversity and abundance of oil-degrading bacteria and alkane hydroxylase (*alkB*) genes in the subtropical seawater of Xiamen Island. *Microbial Ecology* 60(2):429–39.

Wang, W., and Shao, Z. 2012. Diversity of flavin-binding monooxygenase genes (almA) in marine bacteria capable of degradation long-chain alkanes. *FEMS Microbiology Ecology* 80(3):523–33.

Wang, W., and Shao, Z. 2013. Enzymes and genes involved in aerobic alkane degradation. *Frontiers in Microbiology* 4:116.

Wang, W., Zhang, R., Shan, D., and Shao, Z. 2014. Indigenous oil-degrading bacteria in crude oil-contaminated seawater of the Yellow sea, China. *Applied Microbiology and Biotechnology* 98(16):7253–69.

Wang, W., and Shao, Z. 2012. Diversity of flavin-binding monooxygenase genes (*alm*A) in marine bacteria capable of degradation long-chain alkanes. *FEMS Microbiology Ecology* 80(3):523–533.

Wang, C., Madiyar, F., Yu, C., and Li, J. 2017. Detection of extremely low concentration waterborne pathogen using a multiplexing self-referencing SERS microfluidic biosensor. *Journal of Biological Engineering* 11:9. doi: 10.1186/s13036-017-0051-x.

Wasmund, K., Burns, K.A., Kurtböke, I., and Bourne, D.G. 2009. Novel Alkane Hydroxylase Gene (*alk*B) Diversity in Sediments Associated with Hydrocarbon Seeps in the Timor Sea, Australia. *Applied and Environmental Microbiology* 75(23):7391–7398.

Woese, C.R., and Fox, G.E. 1977. Phylogenetic structure of the prokaryotic domain: the primary kingdoms. *Proceedings of the National Academy of Sciences of the United States of America* 74:5088–90.

Wu, D., Jospin, G., and Eisen, J. A. 2013. Systematic Identification of Gene Families for Use as "Markers" for Phylogenetic and Phylogeny-Driven Ecological Studies of Bacteria and Archaea and Their Major Subgroups. *PLoS ONE* 8, e77033.

Yakimov, M.M., Timmis, K.N., and Golyshin, P.N. 2007. Obligate oil-degrading marine bacteria. *Current Opinion in Biotechnology* 18(3):257–266.

Yamahara K.M., Demir-Hilton, E., Preston, C.M., et al. 2015. Simultaneous monitoring of faecal indicators and harmful algae using an *in-situ* autonomous sensor. *Letters in Applied Microbiology* 61:130–8.

Zhang X.-I., Liu Y., Fan T., et al. 2017. Design and Performance of a Portable and Multichannel SPR Device. *Sensors*. doi:10.3390/s17061435.

Zhang, K., Sun, Y., Cui, Z., et al. 2017. Periodically spilled-oil input as a trigger to stimulate the development of hydrocarbon-degrading consortia in a beach ecosystem. *Scientific Reports* 7(1):12446.

Zhang, X., Kirkwood, W.J., Walz, P.M., Peltzer, E. T., and Brewer, P. G. 2017. A Review of Advances in Deep-Ocean Raman Spectroscopy. *Applied Spectroscopy* 66(3):237–249.

Zhou, A., He, Z., Qin, Y., et al. 2013. StressChip as a high-throughput tool for assessing microbial community responses to environmental stresses. *Environmental Science & Technology* 47:9841–49.

Zhou, A.Y., Baruch, M., Ajo-Franklin, C.M., and Maharbiz, M.M. 2017. A portable bioelectronic sensing system (BESSY) for environmental deployment incorporating differential microbial sensing in miniaturized reactors. *PLoS ONE* 12(9):e0184994.

Part VI

Microbial Exploration and Enhanced Oil Recovery

15 Application of Nitrate-Reducing Bacteria in Oil Reservoirs for Enhanced Oil Recovery

Navreet Suri and Gerrit Voordouw

CONTENTS

15.1 INTRODUCTION

Conventional oil recovery technologies in use provide limited oil extraction leaving up to two-thirds of the original oil in oil reservoirs. To meet global oil dependent energy needs, attention has to be focused on developing cost-effective enhanced oil recovery (EOR) technologies to extract maximum oil from active and abandoned oilfields. Microbially enhanced oil recovery (MEOR), is a potentially low-priced technology that uses microbial cells and their metabolites. It has been studied since 1926 as an alternative to conventional EOR technologies (Lazar et al. 2007; Beckman 1926). Lab studies and field trials have shown that it is not only feasible but also provides added advantages of being eco-friendly and requiring only simple microbial growth substrates (Lazar et al. 2007). Nitrate, a commonly used souring control agent in oil reservoirs, is one of the substrates tested to enhance the production of desired microbial products and/or activity to aid in increasing oil production (Lazar et al. 2007; Youssef et al. 2009; Safdel et al. 2017). Upon injection, nitrate stimulates the growth of nitrate-reducing bacteria (NRB), which reduces it to di-nitrogen (denitrification) or ammonia, using inorganic and organic electron donors (Hubert and Voordouw 2007). The sulfide-oxidizing NRB (soNRB) and heterotrophic NRB (hNRB) couple reduction of nitrate to oxidation of sulfide and oil organics, respectively.

15.2 DIVERSITY OF NITRATE-REDUCING BACTERIA

Different nitrate-reducing bacteria (NRB), either indigenous or introduced, during water flooding operations are associated with oil reservoirs. The soNRB and hNRB were found to be microbial community members of the mesothermic Medicine Hat Glauconitic C (MHGC) field in Alberta, Canada, which has been subjected to continuous injection of nitrate to remove sulfide. Epsilonproteobacterial soNRB identified by 16S rRNA gene analysis included *Arcobacter*, *Sulfurimonas,* and *Sulfurospirillum* species (Cornish-Shartau et al. 2010). Two mesophilic soNRB strains CVO and FWKO_B, identified as belonging to the genus *Sulfurimonas* (formerly *Thiomicrospira*) and *Arcobacter*, respectively, were also previously isolated from produced brine from the Coleville oilfield in Saskatchewan, Canada (Gevertz et al. 2000). These genera were also reported in the studies on high temperature North Sea oil reservoirs impacted by nitrate injection (Bodtker et al. 2009; Gittel et al. 2009). Strain FWKO_B uses sulfide, hydrogen, or formate while strain CVO oxidizes sulfide and elemental sulfur for reduction of nitrate to nitrite. Strain CVO can further reduce nitrite formed to N_2 (Gevertz et al. 2000; Hubert and Voordouw 2007).

The majority of the mesophilic NRB belong to the phylum *Proteobacteria* while under thermophilic conditions many of these fall under the phylum *Firmicutes* (Fida et al. 2016). Analysis of nitrate reductase genes in metabolic gene clone libraries showed that NRB in a long-term water-flooded low temperature petroleum reservoir in China mainly belonged to *Pseudomonas, Azospirillium, Bradyrhizobium, Thauera, Magnetospirillum, Azoarcus, Rhodobacter,* and *Sinorhizobium* (Gao et al. 2015). The hNRB identified in MHGC waters included *Dechlorospirillum, Azoarcus, Diaphorobacter, Thauera, Moritella, Pseudomonas, and Thermomonas* (Cornish-Shartau et al. 2010; Agrawal et al. 2012). The genera *Azoarcus, Diaphorobacter, Pseudomonas,* and *Thauera* are known to reduce nitrate to N_2 using hydrocarbons as electron donors (Foght 2008; Mbadinga et al. 2011). Among these genera, *Thauera/Azoarcus* within the class *Betaproteobacteria* using alkylbenzenes for nitrate reduction have been extensively studied. Several of these were originally classified as *Pseudomonas* (Spormann and Widdel 2000; Chakraborty and Coates 2004). *Pseudomonas* and *Thauera* can also grow on water-soluble substrates like acetate, glucose, and molasses under denitrifying conditions (Gassara et al. 2017).

Among the alkylbenzenes, the use of toluene by hNRB has been most comprehensively investigated. *Thauera aromatica* strains K172 and T1 and *Azoarcus tolulyticus* Tol4 isolated under denitrifying conditions metabolized toluene as an organic substrate (Evans et al. 1991; Anders et al. 1995; Zhou et al. 1995). Several other known toluene-oxidizing *Thauera* or *Azoarcus* species are also capable of using xylenes and other alkylbenzenes. Metabolic studies conducted with *Azoarcus* sp. strain T, formerly known as *Pseudomonas* sp. strain T, demonstrated oxidation of toluene and *m*-xylene under dentrifying conditions (Seyfried et al. 1994; Krieger et al. 1999).

A. tolulyticus strain Td15 used toluene and *m*-xylene (Fries et al. 1994) while *Azoarcus* sp. strain EB1 used ethylbenzene for nitrate reduction (Ball et al. 1996). The preferential utilization of toluene by hNRB of the genus *Thauera* enriched from the MHGC field has also been shown in lab studies (Lambo et al. 2008; Suri et al. 2017). However, *Thauera* strain TK001 and strain N2 isolated from the same oil reservoir used organic acids such as acetate, propionate, butyrate, and lactate for nitrate reduction. Strain TK001 did not use alkylbenzenes for denitrification (Hubert and Voordouw 2007; Fida et al. 2016).

Anaerobic oxidation of alkylbenzenes in crude oil coupled to the reduction of nitrate was first demonstrated with pure isolates that included strains ToN1, mXyN1, and EbN1. These strains metabolized toluene, *m*-xylene and ethylbenzene in oil for nitrate reduction (Rabus and Widdel 1995 and 1996). Other hydrocarbons (C_5 to C_{12} alkanes) were used by hNRB affiliated with *Betaproteobacteria* only after depletion of alkylbenzenes in the oil. The *n*-hexane appeared to be the preferred alkane utilized by dominant *Azoarcus* and *Thauera* species in these enrichment cultures (Rabus et al. 1999). An *Azoarcus* sp. capable of using heptane as electron donor has been enriched from the MHGC field (Kryachko and Voordouw 2014). However, utilization of alkanes was not observed in the incubations with significant proportions of *Thauera* species even in the presence of excess nitrate (Lambo et al. 2008; Agrawal et al. 2012; Suri et al. 2017).

Two denitrifying *Betaproteobacteria* strains OcN1 and HxN1 oxidize C_6-C_{12} *n*-alkanes as electron donors for nitrate reduction (Ehrenreich et al. 2000). The metabolism of strain HxN1 has been characterized in detail and it has been shown to co-metabolize short chain (C_4-C_5) and cyclic alkanes (cyclopentane and methylcyclopentane) in addition to utilization of C_6-C_8 *n*-alkanes as substrates (Wilkes et al. 2003; Grundmann et al. 2008). The phylogenetic analysis of 16S rRNA sequences showed that strain HxN1 clustered with alkylbenzene oxidizing *Azoarcus* sp. strains such as PbN1 and mXyN1. Strain OcN1 was closely related to benzene utilizing *Dechloromonas* species (Mbadinga et al. 2011). In contrast, anaerobic oxidation of long chain alkanes (C_{15}-C_{18}) is catalyzed by Gammaproteobacterial *Pseudomonas balearica* strain BerOc6 for nitrate reduction (Grossi et al. 2008).

Analysis of nitrate-reducing microbial communities in produced waters from three different temperature (37°, 58°, and 60°C) oil reservoirs in China showed the presence of the classes *Betaproteobacteria* and *Gammaproteobacteria*. Microorganisms closely related to *Thauera* sp. MZ1T, *Azoarcus* sp. BH72, and *Pseudomonas stutzeri* A1501 were enriched in the incubations of these produced waters with a mixture of C_{15}-C_{20} alkanes and nitrate (Feng et al. 2011). The presence of these classes was also detected in a low temperature (30°C) oil reservoir in Canada. However, utilization of alkylbenzenes, particularly toluene, was favoured over alkanes by the nitrate-reducing *Thauera* and *Pseudomonas* sp. detected in the produced waters from this field (Agrawal et al. 2012; Suri et al. 2017).

Overall, these studies have indicated that NRB are phylogenetically diverse and have the ability to use organic acids as well as a wide variety of oil hydrocarbons as substrates. The intended use of these hNRB in oil reservoir applications such as souring control or MEOR is therefore dependent on their distinct capabilities to reduce nitrate using a particular electron donor (Suri et al. 2017).

15.3 NRB-INDUCED MECHANISMS OF MEOR

The potential MEOR roles of NRB activities and their metabolic products through different mechanisms are discussed individually in the following sections. However, the impact on oil recovery is most likely a synergistic contribution of all these mechanisms.

15.3.1 Biomass and/or Biopolymer Formation

The concept of using biomass and/or biopolymers to selectively plug highly permeable oil-depleted zones has received much attention in MEOR studies (Gray et al. 2008; Youssef et al. 2009; Safdel et al. 2017). The injected nitrate and/or other nutrients penetrate deep into the reservoir and stimulate introduced or indigenous NRB growth. Accumulation of biomass and secreted exopolysaccharides (EPS) and formation of biofilms in preferential high permeable regions diverts the floodwater towards oil-rich zones, thus improving oil production (Sen 2008; Youssef et al. 2009; Siegert et al. 2013; Patel et al. 2015). A diversion of water flow is less likely to occur in homogeneous permeability regions, as the pores are evenly flooded with water. In contrast, plugging would be more noticeable in heterogeneous regions where high permeability zones would receive more bacteria and nutrients. However, the ability of injected bacteria to survive and produce the desired products within the reservoir is a key factor in the success of this MEOR approach.

Laboratory studies have shown significant reduction in permeability in sandstone cores and other reservoir model systems upon stimulation of denitrifying bacteria (Raiders et al. 1989; Hitzman et al. 2004; Youssef et al. 2009). The injection of nitrate-amended medium stimulated the growth of resident bacteria and resulted in a more than 90% decrease in the permeability of sandstone cores. The residual oil recoveries ranged from 8% to 35% (Raiders et al. 1989). The decline in oil production was slowed in oilfields injected with nitrate, nitrite, and other inorganic nutrients (Hitzman et al. 2004; Brown et al. 2002). It was suggested through reservoir modeling analysis that reservoirs with less than 6% of the pore volume in high permeability zones would offer the best prospects for bacterial plugging. This implies that plugging offers good potential for fractured reservoirs (Gray et al. 2008).

Plugging with biopolymers has been found to be more effective than biomass plugging. Polymers dissolved in water increases its viscosity. Environmental scanning electron microscopy (ESEM) analysis showed that the bacteria grew and produced biopolymer in the sand pack. Production of biopolymer by bacteria also increased the viscosity of a cell-free supernatant to 120 mPa · s at 28°C (Harish et al. 2009). Note that 1 mPa · s is the same as 1 centipoise (1cP) and that the viscosity of water at 20°C is 1 cP. The viscosity (η) and permeability (K) are inversely related according to the Darcy equation (see Equation 15.1).

$$q = -(K * A * \Delta P)/(\eta * L) \tag{15.1}$$

where q is the flow rate, A is the cross-section area, ΔP is the pressure gradient, and L is the length of the sand pack column.

Biopolymers produced by bacteria can also be directly injected into oil reservoirs for chemically enhanced oil recovery (CEOR). Species of *Pseudomonas*, *Thauera,* and

Azoarcus can produce polyhydroxyalkanoates (PHA) using various substrates (Saharan et al. 2014). Oil recovery increased by an additional 30% with the injection of water-dissolved biopolymer as compared to water injection (Rangel et al. 2012). Although direct injection of biopolymers is a proven EOR method, their loss upon injection due to adsorption to reservoir rock and degradation under reservoir conditions are factors that may prevent successful application (Buller and Vossoughi 1990; Huh et al. 1990).

15.3.2 Biosurfactant Production

Biosurfactants are amphiphilic molecules produced by microorganisms extracellularly or on their cell surfaces as a part of cell membranes during exponential or stationary phases of their growth. Biosurfactants accumulate between two fluid phases of different polarities. Production of biosurfactants by bacteria can aid MEOR by reducing interfacial tension (IFT), increasing oil emulsification, and detaching oil from the rock pores (Banat et al. 2000; Sen 2008; Youssef et al. 2009). The IFT between hydrocarbons and water is typically of the order of 30–40 mN/m. A biosurfactant that decreases IFT to below 0.4 mN/m is considered to be effective (Gray et al. 2008). The biosurfactants are classified into five categories. Among these, lipopeptides and glycolipids (e.g., rhamnolipids) are the most studied for MEOR as these can lower hydrocarbon-water IFT below 0.1 mN/m. *Bacillus* and *Pseudomonas* species are known to produce lipopeptides and glycolipids, respectively (Youssef et al. 2009; Safdel et al. 2017). The production of biosurfactants by *Pseudomonas aeruginosa, Bacillus subtilis,* and *Rhodococccus erythropolis* decreased surface tension from 72 to <30 mN/m, which resulted in residual oil recovery of 7% to 14% from a sandstone core flooding system (Xia et al. 2011).

Multiple substrates can induce production of biosurfactants by microorganisms (Jain et al. 1992; Moran et al. 2000; Nordmann and Janssen 2002). Microbial cells play a significant role in the surface interaction between oil and water, which can promote oil emulsification. Nitrate-mediated increased heavy oil recovery was found to be associated with increased oil emulsification, likely due to the surface activity of NRB (Gassara et al. 2015 and 2017). The formation of oil-water emulsions was found to depend on the concentration of biomass (Kaster et al. 2012). A decrease in interfacial tension from 20.8 to 14.5 mN/m gave more oil production from columns fed with nitrate (Silva et al. 2014). No changes in the surface activity were observed upon injection of nutrients including glucose, sodium nitrate, and trace metals into the reservoir well even though presence of biosurfactant producers was detected in the study (Youssef et al. 2007).

Biosurfactants are not products of a microbial cell's main energy metabolism. Strictly anaerobic microorganisms produce lower amounts of biosurfactants than aerobes and the former mostly dominate in oil reservoirs. However, biosurfactant production can increase under facultative conditions, for example, upon nitrate reduction by NRB (Domingues et al. 2017). The minimum concentrations of lipopeptides required to mobilize entrapped oil is 10 mg/L. A field study has demonstrated average production of 90 mg/L of lipopeptides in the production fluids of wells injected with two bacterial strains and nutrients including glucose, sodium nitrate, and trace metals. CO_2, acetate, lactate, ethanol, and 2,3-butanediol were the main fermentation products. The data provided for modeling MEOR processes *in situ*, including growth rates, carbon balances, biosurfactant rates, and biosurfactant yields demonstrated the feasibility of biosurfactant-mediated oil recovery (Youssef et al. 2007).

15.3.3 Gas Formation

Nitrate reducers produce N_2 and CO_2 gases that could increase reservoir pressure, but reports of increased pressures in oil reservoirs through nitrate stimulation are limited (Youssef et al. 2009; Siegert et al. 2013). An additional 3% of oil recovery from oil-saturated sand pack flooding experiment was predicted to result from re-pressurization by N_2 formation after incubation with denitrifying bacteria. There was an increment of 2000 mbar in pressure (Nuryadi et al. 2011); note that 1000 mbar = 1 bar = 0.987 atmosphere (atm). Formation of gases including N_2 and CO_2 by an enrichment culture containing *Pseudomonas* sp. gave an increase of 1500 mbar in a well stimulation experiment conducted in an oilfield in Russia (Belyaev et al. 2004). A much lower increment of 20 mbar in pressure within sand packs was observed upon production of N_2 and CO_2 gases by another enrichment culture dominated by *Thauera* and *Pseudomonas* sp. The higher additional oil production of 16% in these studies indicated the operation of mechanisms other than gas formation in the observed oil recovery (Gassara et al. 2015). Most importantly, small pressure increases are insufficient for re-pressurization and it seems unlikely to maintain these pressure increases throughout the reservoir for oil displacement (Bryant and Lockhart 2002; Youssef et al. 2009).

Gases produced in sufficient amounts particularly at high pressure can result in swelling of crude oil upon dissolution and reduce its viscosity. The solubility of gases (CO_2 and N_2) in crude oil increases with pressure and with API gravity while it decreases with temperature. Hence, the degree of viscosity reduction that can be achieved is more likely in case of light oils than heavy oils (Alvarado and Monrique 2010; Anne Perera et al. 2016). A higher solubility of CO_2 than of N_2 in crude oil at a pressure of 10000 mbar was indicated in lab studies that involved injection of CO_2 and N_2 mixtures (Farias and Watson 2007). The injection of CO_2 in an oilfield in China resulted in oil swelling of around 47% and viscosity reduction of 63.2%. It was expected to recover an additional 5% to 10% of original oil in place (OOIP; Peng 2010). The injection of N_2 has also been proposed to increase oil recovery in other reservoirs (Alvarado and Monrique 2010). Information on the impact of microbially produced CO_2 and N_2 on oil production is also available in the literature (Siegert et al. 2013; Patel et al. 2015; Safdel et al. 2017).

15.4 NITRATE-MEDIATED MEOR

The feasibilities of MEOR processes that involve the use of nitrate to aid in increasing oil production have been tested not only in lab studies but also in oilfields (Lazar et al. 2007; Youssef et al. 2009; Safdel et al. 2017). Nitrate is highly soluble in water. Maximizing biomass formation for selective plugging by supplying nitrate as an electron acceptor is one of the mechanisms evaluated in laboratory studies for improving oil production (Maudgalya et al. 2007; Youssef et al. 2009). Injection of nitrate together with water-soluble electron donors such as molasses, glucose, or acetate into the oil-saturated model systems improved production of oil. hNRB biomass, gases (N_2 and CO_2) and other microbial by-products were most likely involved in additional oil production of up to 0.18 PV (36%) of residual oil (see Table 15.1; Gassara et al. 2017).

Use of nitrate in MEOR field trials conducted worldwide has been reviewed by several authors (Lazar et al. 2007; Maudgalya et al. 2007; Safdel et al. 2017). A total of 407 field trials were carried out that included 314 in sandstone reservoirs and 89 in carbonate reservoirs, according to Maudgalya et al., 2007. Based on the primary MEOR mechanisms (see the previous section) and the type of microorganisms involved, these trials were classified into six categories. According to this classification, the trials that involved the stimulation of NRB *in situ* showed diversion of flood water towards unswept oil reservoir zones through permeability modifications. Overall, many trials conducted using nitrate injection were successful, indicating its high potential for MEOR applications in oil reservoirs (Maudgalya et al. 2007; Youssef et al. 2007; Safdel et al. 2017).

Apart from stimulation of NRB growth and activity in conjunction with molasses and/or other water-soluble substrates, nitrate can also support the growth of oil-oxidizing hNRB in oilfields (Youssef et al. 2009; Voordouw 2011). These bacteria prefer to use alkylbenzenes, especially toluene, as an electron donor (Lambo et al. 2008; Agrawal et al. 2012; Suri et al. 2017). Nitrate reduction increases with the increased toluene content of oil resulting in more biomass accumulation that could potentially enhance oil production (Suri et al. 2017). Lab studies using sand pack bioreactors have shown increased production of heavy MHGC oil through nitrate injection (Kryachko et al. 2014). Based on these observations, a hydrocarbon- and nitrate-mediated MEOR (HN-MEOR) technology was proposed (Gassara et al. 2015).

15.4.1 Hydrocarbon- and Nitrate-Mediated MEOR

The technology involves the sequential injection of water-dissolved light hydrocarbons followed by injection of nitrate and other nutrients (such as phosphate) with or without an hNRB consortium. The nitrate-reducing ability of hNRB that use low molecular weight hydrocarbons might be limited by the low concentrations of these oil components in many heavy oils such as MHGC oil. The concentrations of the alkylbenzenes (e.g., toluene and ethylbenzene) are 1–2 mM in MHGC oil. Toluene and ethylbenzene have water solubility of 5.6 mM (0.52 g/1000 mL) and 1.5 mM (0.15 g/1000 mL), respectively, at 20°C. Hence, the concentrations of these light hydrocarbons in oil can be increased by their aqueous injection into oil saturated model systems. Toluene and ethylbenzene transfer to the oil phase to improve nitrate reduction capacity of indigenous or introduced hNRB using oil along the water flow channels as an electron donor.

15.4.1.1 Increasing Alkylbenzene Concentrations in HN-MEOR

The distribution of injected alkylbenzenes (e.g., toluene) between oil and aqueous phases is governed by the partition coefficient $\pi = C_o/C_w$, where C_o and C_w are the equilibrium concentrations of toluene in the oil and water phases, respectively. Numerical values for π have been estimated to be in the range of 1000 to 10,000 indicating that water injected toluene would transfer to the oil phase (Gassara et al. 2015).

Water flooded sand pack model columns containing 0.5 pore volumes (PV) of residual MHGC oil were injected with 2 PV of aqueous toluene (C_w = 3 mM). Less than 1 mM toluene was detected in the aqueous phases of effluents collected during injection, indicating retention of toluene in the columns. Prior to toluene injection the measured toluene concentration in MHGC oil was 2.2 ± 0.3 mM (N = 2). Following injection the volume weighted C_0 increased to 2.7 ± 1.0 mM (N = 5), representing increases of 1.5 mM. The toluene concentration was highest in the oil from the near injection region indicating that hNRB activity would be highest in this region upon introduction of nitrate (Gassara et al. 2015; Suri et al. 2017). The inclusion of nitrate has also been previously shown to stimulate the activity of hNRB and soNRB in the near injection wellbore region (NIWR) in lab studies (Callbeck et al. 2011; Agrawal et al. 2012)

15.4.1.2 Oil Production through HN-MEOR

Different concentrations of alkylbenzenes (1.4, 2.8, 5.6, or 11.2 mM) and nitrate (10, 20, 40, or 80 mM) were tested to stimulate hNRB growth and activity at a temperature of 22–25°C and a pressure of 1 atm or 27.2 atm in model oil containing columns for improving oil recovery. The oil production increased with increased alkylbenzene and nitrate concentrations. Columns with or without nitrate only produced up to 0.02 PV of residual oil while columns with low concentrations of toluene (1.4, 2.8 or 5.6 mM) and nitrate (10, 20 or 40 mM) produced up to 0.04 PV of residual oil. On the other hand, columns injected with high toluene (11.2 mM) and nitrate (80 mM) produced up to 0.11 PV of residual oil. Similarly, the injection of high ethylbenzene (9.5 mM) and nitrate (80 mM) into MHGC oil containing model columns gave up to 0.08 PV of residual oil (see Table 15.1 and Figure 15.1).

The oil production was associated with increased nitrate reduction through oxidation of toluene or ethylbenzene by hNRB within the columns. Note that because injected aqueous toluene or ethylbenzene is transferred from the aqueous phase into a viscous oil phase where diffusion is slow, the concentrations in volume elements lining aqueous

TABLE 15.1
The Maximum Percentage (%) Residual Oil Production from MHGC Oil Containing Model Columns Operated at a Temperature of 22–25°C and an Atmospheric Pressure. The Columns Were Injected with hNRB Consortiums Enriched in Batch Cultures Containing 2% (v/v) MHGC Oil with Additional Specific Electron Donors and 80 mM Nitrate. These Were Inoculated with Produced Water from MHGC Field.

Substrates		Dominant Taxon in hNRB Consortiums (%)		Maximum Residual Oil Production (%)
Electron donor	Electron acceptor	*Thauera*	*Pseudomonas*	
Acetate[1]	Nitrate	35	8	18
Glucose[1]	Nitrate	3	80	36
Molasses[1]	Nitrate	11	38	17
Toluene[1]	Nitrate	95	1	23
Ethylbenzene	Nitrate	86	8	18

[1] *From Gassara et al. 2015 and 2017*

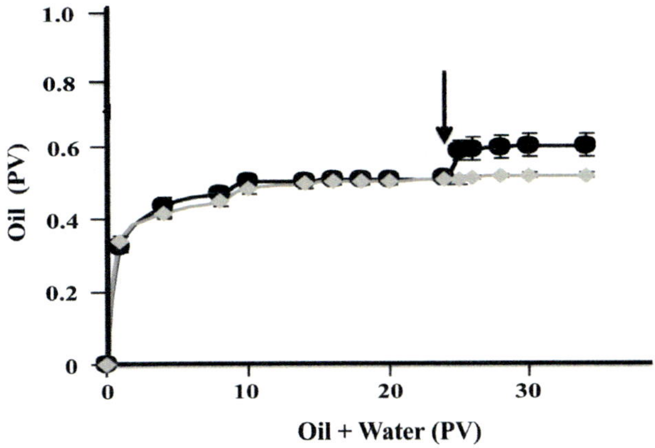

FIGURE 15.1 Oil production from sand pack model columns without (♦) or with (•) additional toluene in the oil phase and nitrate in the aqueous phase. The columns were injected with nutrient medium at the rate of 1 PV/day. The hNRB consortium was injected into columns at the indicated (↓) time and columns were incubated at room temperature (22–25 °C) for two weeks. Each point is an average oil production observed from the replicate columns (adapted from Gassara et al. 2015).

channels may be higher than the calculated average C_0 based on the amount of injected aqueous toluene or ethylbenzene. Hence, the hNRB growth resulting in biomass accumulation would be highest in these aqueous channels suggesting a role of selective plugging mechanism for oil recovery. Oil production correlated positively with the amount of hNRB biomass formed in the columns. The reduction of nitrate and oxidation of toluene by hNRB also led to the formation of N_2 and CO_2 gases that helped to push oil out of the columns, thus increasing water-flood independent oil recovery (see Figure 15.2). Formation of stable oil-water emulsions in the effluents collected from the columns

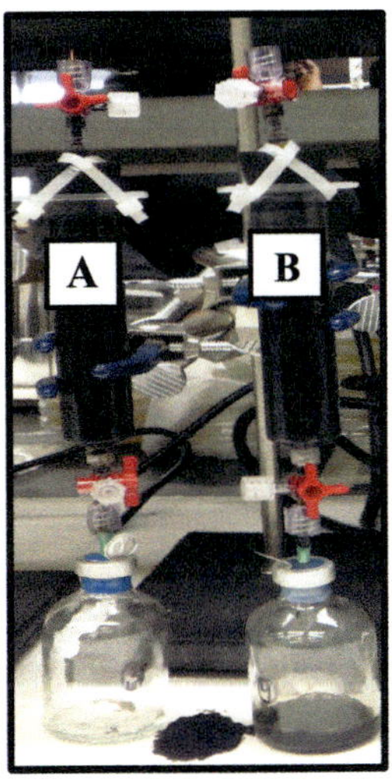

FIGURE 15.2 Oil production during incubation from columns injected with aqueous ethylbenzene and an hNRB consortium. The concentrations of nitrate in the aqueous phases of these columns were (A) 0 mM and (B) 80 mM. The top ends of the columns were kept closed while the bottom ends were open and connected to anaerobic vials through syringe needles to collect oil produced through formation of gases (N_2 and CO_2).

indicated the formation of microbial products such as biosurfactants. Thus, the additional oil production observed from the columns injected with high alkylbenzene and nitrate concentrations was likely due to the effect of multiple mechanisms.

15.4.1.3 Key Players of HN-MEOR

Microbial community analysis of 16S rRNA amplicons of hNRB consortia used in HN-MEOR studies has shown a predominance of phylum *Proteobacteria.* Within this phylum, *Thauera* of the class *Betaproteobacteria* and *Pseudomonas* of the class *Gammaproteobacteria* were the two most abundant genera. The fractions of the bacterial taxa *Thauera* and *Pseudomonas* were 0.03% and 12.3%, respectively, in produced water from the MHGC oilfield. The proportion of *Thauera* increased up to 95% while that of *Pseudomonas* decreased up to 1.5% upon its incubation with MHGC oil with additional toluene in the presence of nitrate (see Table 15.1). Similar observations were made for the hNRB consortia grown on ethylbenzene and nitrate. However, in the presence of water-dissolved oil insoluble electron donors such as molasses, sucrose, glucose and acetate, the proportions of *Thauera* (up to 11%) were much lower than *Pseudomonas* (up to 80%; see Table 15.1; Gassara et al. 2015). This indicates that *Thauera* is involved in nitrate reduction using alkylbenzenes as the electron donors. *Thauera* is a known denitrifying bacterium that uses alkylbenzenes or volatile fatty acids (VFA) as electron donors (Chakraborty and Coates 2004; Hubert and Voordouw 2007; Agrawal et al. 2012).

15.4.1.4 Advantages of HN-MEOR

MEOR studies in the past five decades have focused on using readily fermentable substrates such as molasses (Youssef et al. 2007; Maudgalya et al. 2007; Safdel et al. 2017). The success rate of field trials that used molasses was considerable. Some of these also involved the injection of nitrate together with molasses to promote NRB growth and activity (Lazar et al. 2007; Patel et al. 2015; Safdel et al. 2017). On the other hand, studies on the use of crude oil as a substrate by microrganisms to promote oil production have also been reported in the literature (Sen 2008; Gudina et al. 2012; Kryachko and Voordouw 2014).

Molasses is inexpensive and is highly soluble in water, which makes it easy to inject. The use of highly water-soluble substrates seems to be advantageous for MEOR applications. However, it should also be considered that water injection into oil reservoirs could easily remove these substrates, hence limiting their availability for microbial consumption. In contrast, the use of oil-soluble substrates such as alkylbenzenes in HN-MEOR could be advantageous in this respect, as these could bind to oil preventing them from being lost during continued water injection. HN-MEOR has the potential to be a cost-effective technology because the cost of industrial grade toluene is $1/L and the other substrate, nitrate is already in use in many oilfields to decrease souring (Gassara et al. 2015).

15.5 CONCLUSIONS

Nitrate-mediated MEOR studies have shown that oil production can be improved by stimulating hNRB activity in the presence of nitrate and water or oil-soluble electron donors. Model column studies have shown that oil production using HN-MEOR can be increased up to 23% of residual oil. The combined formation of hNRB biomass, gases (N_2 and CO_2) and other microbial products seems to be involved in enhancing oil recovery. The possibility of sequential injection of alkylbenzenes and their retention within the oil-containing zones for availability to hNRB is a unique aspect of this technology. It has the potential to be economically beneficial. Further studies should define the exact mechanisms of additional oil production. Field trials are also needed to test the applicability of this technology.

ACKNOWLEDGMENTS

This work was supported by an NSERC Industrial Research Chair award to Dr. Gerrit Voordouw, which was also supported by Baker Hughes, BP America Production Co, Computer Modeling Group Ltd., ConocoPhillips, Intertek, Dow Microbial Control, Enbridge Inc., Enerplus Corporation, Oil Search Limited, Shell Global Solutions International, Suncor Energy Inc., and Yara Norge as well as by Alberta Innovates. We are thankful to Petroleum Microbiology Research Group (PMRG) members for support provided.

REFERENCES

Agrawal, A., H.S. Park, S. Nathoo, et al. 2012. Toluene depletion in produced oil contributes to souring control in a field subjected to nitrate injection. *Environ. Sci. Technol.* 46(2): 1285–1292.

Alvarado, V., and E. Monrique. 2010. Enhanced oil recovery: an update review. *Energies.* 3: 1529–1575.

Anders, H.J., A. Kaetzke, P. Kaempfer, et al. 1995. Taxonomic position of aromatic-degrading denitrifying pseudomonad strains K 172 and KB 740 and their description as new members of the genera *Thauera*, as *Thauera aromatica* sp. nov., and *Azoarcus*, as *Azoarcus evansii* sp. nov., respectively, members of the beta subclass of the Proteobacteria. *Int. J. Syst. Bacteriol.* 45: 327–333.

Anne Perera, M.S., R.P. Gamage, T.D. Rathnaweera, et al. 2016. A Review of CO_2-Enhanced Oil Recovery with a Simulated Sensitivity Analysis. *Energies.* 9: 481.

Ball, H.A., H.A. Johnson, M. Reinhard, et al. 1996. Initial reactions in anaerobic ethylbenzene oxidation by a denitrifying bacterium, strain EB1. *J. Bacteriol.* 178: 5755–5761.

Banat, I.M., R.S. Makkar, and S.S. Cameotra. 2000. Potential commercial applications of microbial surfactants. *Appl. Microbiol. Biotechnol.* 53: 495–508.

Beckman, J.W. 1926. The action of bacteria on mineral oil. *Indus. Eng. Chem. News.* 4: 3.

Belyaev, S.S., I.A. Borzenkov, T.N. Nazina, et al. 2004. Use of microorganisms in the biotechnology for the enhancement of oil recovery. *Microbiol.* 73(5): 590–598.

Bodtker, G., K. Lysnes, T. Torsvik, et al. 2009. Microbial analysis of backflowed injection water from a nitrate-treated North Sea oil reservoir. *J. Ind. Microbiol. Biotechnol.* 36: 439–450.

Brown, L.R., A.A. Vadie, and J.O. Stephens. 2002. Slowing production decline and extending the economic life of an oilfield: New MEOR technology. *Soc. Pet. Eng. Reservoir Eval. Eng.* 5: 33–41.

Bryant, S.L., and T.P. Lockhart. 2002. Reservoir engineering analysis of microbial enhanced oil recovery, SPE 63229 (SPE 79719). SPE Annual Technical Conference and Exhibition, Dallas, TX, 1 4 October 2000, SPE.

Buller, C.S., and S. Vossoughi. 1990. Subterranean preliminary modification by using microbial polysaccharide polymers. US patent office. Patent No. 4,941533.

Callbeck, C.M., X. Dong, I. Chatterjee, et al. 2011. Microbial community succession in a bioreactor modeling a souring low temperature oil reservoir subjected to nitrate injection. *Appl. Microbiol. Biotechnol.* 91: 799–810.

Chakraborty, R., and J.D. Coates. 2004. Anaerobic degradation of monoaromatic hydrocarbons. *Appl. Microbiol. Biotechnol.* 64: 437–446.

Cornish-Shartau S.L., M. Yurkiw, S. Lin, et al. 2010. Ammonium concentrations in produced waters from a mesothermic oilfield subjected to nitrate injection decrease through formation of denitrifying biomass and anammox activity. *Appl. Environ. Microbiol.* 76: 4977–4987.

Domingues, P.M., A. Almeida, L.S. Leal, et al. 2017. Bacterial production of biosurfactants under microaerobic and anaerobic conditions. *Rev Environ Sci Biotechnol.* 16: 239–272.

Ehrenreich, P., A. Behrends, J. Harder, et al. 2000. Anaerobic oxidation of alkanes by newly isolated denitrifying bacteria. *Arch. Microbiol.* 173: 58–64.

Evans, P.J., D.T. Mang, K.S. Kim, et al. 1991. Anaerobic degradation of toluene by a denitrifying bacterium. *Appl. Environ. Microbiol.* 57: 11390–1145.

Farias, M.J., and R.W. Watson. 2007. Interaction of Nitrogen/CO_2 Mixtures with Crude Oil. Report by: The Pennsylvania State University, University Park, PA 16802.

Feng, W., J. Liu, J. Gu, et al. 2011. Nitrate-reducing community in production water of three oil reservoirs and their responses to different carbon sources revealed by nitrate-reductase encoding gene (napA). *Inter. Biodeterio. Biodegrad.* 65: 1081–1086.

Fida, T.T., C. Chen, G. Okpala, et al. 2016. Implications of limited thermophilicity of nitrite reduction for control of sulfide. *Appl. Environ. Microbiol.* 82 (14): 4190–4199.

Foght, J. 2008. Anaerobic Biodegradation of Aromatic Hydrocarbons: Pathways and Prospects. *J. Mol. Microbiol. Biotechnol.* 15: 93–120.

Fries, M.R., J. Zhou, J. Chee-Sanford, et al. 1994. Isolation, characterization, and distribution of denitrifying toluene degraders from a variety of habitats. *Appl. Environ. Microbiol.* 60: 2802–2810.

Gao, P., H. Tian, G. Li, et al. 2015. Microbial diversity and abundance in the Xinjiang Luliang long-term water-flooding petroleum reservoir. *Microbiol. Open.* 4(2): 332–342.

Gassara, F., N. Suri, and G. Voordouw. 2017. Nitrate-mediated microbially enhanced oil recovery (N-MEOR) from model upflow bioreactors. *J. Hazard. Mater.* 324: 94–99.

Gassara, F., N., Suri, P., Stanislav, et al. 2015. Microbially enhanced oil recovery by sequential injection of light hydrocarbon and nitrate in low- and high-pressure bioreactors. *Environ. Sci. Technol.* 49: 12594–12601.

Gevertz, D., A.J. Telang, G. Voordouw, et al. 2000. Isolation and characterization of strains CVO and FWKO B, two novel nitrate-reducing, sulfide-oxidizing bacteria isolated from oilfield brine. *Appl. Environ. Microbiol.* 66: 2491–2501.

Gittel, A., K.B. Sorensen, T.L. Skovhus, et al. 2009. Prokaryotic community structure and activity of sulfate reducers in production water from high-temperature oil reservoirs with and without nitrate treatment. *Appl. Environ. Microbiol.* 75: 7086–7096.

Gray, M.R., A. Yeung, J.M. Foght, et al. 2008. Potential Microbial enhanced oil recovery processes: a critical analysis. In *SPE Annual Technical Conference and Exhibition,* SPE114676; Denver, CO: SPE Publications.

Grossi, V., C. Cravo-Laureau, R. Guyoneaud, et al. 2008. Metabolism of n-alkanes and n-alkenes by anaerobic bacteria: a summary. *Org. Geochem.* 39: 1197–1203.

Grundmann, O., A. Behrends, R. Rabus, et al. 2008. Genes encoding the candidate enzyme for anaerobic activation of n-alkanes in the denitrifying bacterium, strain HxN1. *Environ. Microbiol.* 10(2): 376–385.

Gudina, E.J., J.F.B. Pereira, L.R. Rodrigues, et al. 2012. Isolation and study of microorganisms from oil samples for application in Microbial Enhanced Oil Recovery. *Inter. Biodeterior. and Biodegrad.* 68: 56–64.

Harish, S., K. Hingurao, A. Desai, et al. 2009. Selective Plugging Strategy-Based Microbial-Enhanced Oil Recovery Using Bacillus licheniformis TT33. *J. Microbiol. Biotechnol.* 19(10): 1230–1237.

Hitzman, D.O., M. Dennis, and D.C. Hitzman. 2004. "Recent Successes: MEOR Using Synergistic H_2S Prevention and Increased Oil Recovery Systems." SPE 89453. Proceedings of Fourteenth Symposium on Improved Oil Recovery. Society of Petroleum Engineers, Richardson, TX.

Hubert, C., and G. Voordouw. 2007. Oilfield souring control by nitrate-reducing *Sulfospirillium* spp. That outcompete sulfate-reducing bacteria for organic electron donors. *Appl. Environ.Microbiol.* 73: 2644–2652.

Huh, C., E.A. Lange, and W.J. Cannella. 1990. "Polymer Retention in Porous Media." SPE 20235. Proceedings of SPE/DOE Enhanced Oil Recovery Symposium. Society of Petroleum Engineers, Richardson, TX.

Jain, D.K., H. Lee, and J.T. Trevors. 1992. Effect of addition of *Pseudomonas aeruginosa* UG2 inocula or biosurfactants on biodegradation of selected hydrocarbons in soil. *J. Ind. Microbiol.* 10: 87–93.

Kaster, K.M., A. Hiorth, G. Kjeilen-Eilertsen, et al. 2012. Mechanisms Involved in Microbially Enhanced Oil Recovery. *Transp. Porous Med.* 91:59–79.

Krieger, C.J., H.R. Beller, M. Reinhard, et al. 1999. Initial reactions in anaerobic oxidation of *m*-xylene in the denitrifying bacterium *Azoarcus* sp. strain T. *J. Bacteriol.* 181: 6403–6410.

Kryachko, Y., and G. Voordouw. 2014. Microbially enhanced oil recovery from miniature model columns through stimulation of indigenous microflora with nitrate. *Int. Biodeterior. Biodegrad.* 96: 135–143.

Lambo, A.J., K.I. Noke, S.R. Larter, et al. 2008. Competitive, microbially mediated reduction of nitrate with sulphide and aromatic oil components in a low temperature, Western Canadian oil reservoir. *Environ. Sci. Technol.* 42(23): 8941–46.

Lazar, I., I.G. Petrisor, and T.F. Yen. 2007. Microbial enhanced oil recovery (MEOR). *Pet. Sci. Technol.* 25: 1353–1366.

Maudgalya, S., R.M. Knapp, and M. McInerney. 2007. Microbially enhanced oil recovery technologies: a review of the past, present and future. In Proceedings of SPE Production and Operations Symposium, Oklahoma City, OK.

Mbadinga, S.M., L. Wang, L. Zhou, et al. 2011. Microbial communities involved in anaerobic degradation of alkanes. *Inter. Biodeterior. & Biodegrad.* 65: 1–13.

Moran, A.C., N. Olivera, M. Commendatore, et al. 2000. Enhancement of hydrocarbon waste biodegradation by addition of a biosurfactant from *Bacillus subtilis* O9. *Biodegrad.* 11: 65–71.

Noordman, W.H., and D.B. Janssen. 2002. Rhamnolipid Stimulates Uptake of Hydrophobic Compounds by *Pseudomonas aeruginosa. Applied and Environmental Microbiology.* 68(9): 4502–08.

Nuryadi, A., A. Kishita, N. Watanabe, et al. 2011. EOR simulation by *in situ* nitrogen production via denitrifying bacteria and performance improvement by nitrogen alternating surfactant injection. In: SPE Asia Pacific Oil and Gas Conference and Exhibition, Jakarta, Indonesia, SPE 147823. Society of Petroleum Engineers.

Patel, J., S. Borgohain, M. Kumar, et al. 2015. Recent developments in microbial enhanced oil recovery. *Renewable and Sustainable Energy Reviews.* 52: 1539–1558.

Peng, B. 2010. CO_2 Storage and Enhanced Oil Recovery in Jilin Oilfield; China Australia CO_2 Geological Storage Workshop, January 19–21, 2010, Canberra, Australia.

Rabus, R., and F. Widdel. 1995. Anaerobic degradation of ethylbenzene and other aromatic hydrocarbons by new denitrifying bacteria. *Arch. Microbiol.* 163: 96–103.

Rabus, R., and F. Widdel. 1996. Utilization of Alkylbenzenes during Anaerobic Growth of Pure Cultures of Denitrifying Bacteria on Crude Oil. *Appl. and Environ. Microbiol.* 62 (4), 1238–1241.

Rabus, R., H. Wilkes, A. Schramm. et al. 1999. Anaerobic utilization of alkylbenzenes and n-alkanes from crude oil in an enrichment culture of denitrifying bacteria affiliating with the β-subclass of Proteobacteria. *Environ. Microbiol.* 1(2); 145–57.

Raiders, R.A., R.M. Knapp, and M.J. McInerney. 1989. Microbial selective plugging and enhanced oil recovery. *J. Ind. Microbiol.* 4: 215–230.

Rangel, I.R., L.R. Thompson, R.G. Pereira, et al. 2012. Experimental investigation of the enhanced oil recovery process using a polymeric solution. *J. of the Braz. Soc. Of Mech. Sci and Eng.* XXXIV (3): 285–293.

Safdel, M., M.A. Anbaz, A. Daryasafar, et al. 2017. Microbial enhanced oil recovery, a critical review on worldwide implemented field trials in different countries. *Renewable and Sustainable Energy Reviews.* 74: 159–172.

Saharan, B.S., A. Grewal, and P. Kumar. 2014. Biotechnological Production of Polyhydroxyalkanoates: A Review on Trends and Latest Developments. *Chinese J. of Biol. Article* ID 802984.

Sen, R. (2008) Biotechnology in petroleum recovery: the microbial EOR. *Prog. Energy Combust. Sci.* 34: 714–724.

Seyfried, B., G. Glod, R. Schocher, et al. 1994. Initial reactions in the anaerobic oxidation of toluene and *m*-xylene by denitrifying bacteria. *Appl. Environ. Microbiol.* 60: 4047–4052.

Siegert, M., J. Sitte, A. Galushko, et al. 2013. Starting Up Microbial Enhanced Oil Recovery. *Adv. Biochem. Eng./Biotechnol.* 142: 1–94.

Silva, M.L.B., H.M. Soares, A. Furigo, et al. 2014. Effects of Nitrate Injection on Microbial Enhanced Oil Recovery and Oilfield Reservoir Souring. *Appl. Biochem. Biotechnol.* 174:1810–1821.

Spormann, A.M., and F. Widdel. 2000. Metabolism of alkylbenzenes, alkanes, and other hydrocarbons in anaerobic bacteria. *Biodegradation.* 11: 85–105.

Suri, N., J. Voordouw, and G. Voordouw. 2017. The Effectiveness of Nitrate-Mediated Control of the Oilfield Sulfur Cycle Depends on the Toluene Content of the Oil. *Front. Microbiol.* 8: 956.

Voordouw G. (2011) Production-related petroleum microbiology: progress and prospects. *Curr. Opin. Biotechnol.* 22: 401–405.

Wilkes, H., R. Rabus, T. Fischer, et al. 2002. Anaerobic degradation of *n*-alkanes in a denitrifying bacterium: further degradation of the initial intermediate (1-methylpentyl)succinate via C-skeleton rearrangement. *Arch. Microbiol.* 177: 235–243.

Xia, W., H. Dong, L. Yu, et al. 2011. Comparative study of biosurfactant produced by microorganisms isolated from formation water of petroleum reservoir. *Colloids and Surfaces A: Physicochem. Eng. Aspects.* 392: 124–130.

Youssef, N., M.S. Elshahed, and M.J. McInerney. 2009. Microbial processes in oilfields: culprits, problems and opportunities. *Adv. Appl. Microbiol.* 66: 141–251.

Zhou, J., M.R. Fries, J.C. Chee-Sanford, et al. 1995. Phylogenetic analyses of a new group of denitrifiers capable of anaerobic growth on toluene and description of *Azoarcus tolulyticus* sp. Nov. *Int. J. Syst. Bacteriol.* 45: 500–506.

16 Microbial Exploration Techniques: An Offshore Case Study

Juan de Dios Miranda, José Miguel Seoane, Ángela Esteban, and Enrique Espí

CONTENTS

16.1 INTRODUCTION

Hydrocarbon exploration is an expensive, high-risk activity where state-of-the-art technology is used in order to reduce associated costs. Considering that drilling success rate is never above 25% and that drilling a well on-shore has an approximate cost of \$25 M (offshore, costs can raise from \$100 M up to \$600 M), companies belonging to the Oil and Gas sector have always shown a deep interest in the development of new technologies to reduce exploration risks and its associated costs.

Among the different hydrocarbon prospecting technologies available, surface geochemical petroleum exploration techniques are commonly used to detect migrated surface hydrocarbons and their alteration products (Miqueletto et al. 2011). During prospection campaigns, hundreds of piston cores are extracted from surface sediments and sampled to identify and quantify different geochemical indicators that could suggest hydrocarbon presence underneath, after which they are discarded. As hypothesized in

Hubert and Judd (2010), an additional analysis of the bacterial DNA contained within these samples could potentially be used to reduce risks in the discovery of new oil and gas reservoirs. Although the usage of biological indicators for the prospection of hydrocarbons has already been addressed in the literature (C. Hubert et al. 2009; C. Hubert and Judd 2010; C. R. J. Hubert et al. 2012; Ashby et al. 2013; Baum et al. 2008; Miqueletto et al. 2011; Rasheed et al. 2014; Rasheed, Patil, and Dayal 2013) with promising results, we believe that DNA sequencing technologies, if merged together with new advancements in big data, machine learning and supercomputing, could lead to the development of a new set of tools that may improve significantly our abilities to detect hydrocarbon related anomalies both in soil and in marine sediments samples. This chapter summarizes our approach and results on a real proof of concept performed on 20 piston cores during an exploration campaign performed by Repsol S.A.

The proposed workflow could be divided in four main tasks:

1. Extract DNA information from piston core samples.
2. Perform DNA data analysis to identify biological indicators of the presence/absence of oil and gas accumulation in the subsurface.
3. Build a machine learning model able to calculate the probability of a sample containing hydrocarbons based on its genetic signature.
4. Using this information, generate a probability map of hydrocarbon presence as the final output.

16.2 BRINGING MICROBIOMICS TO OIL AND GAS EXPLORATION

16.2.1 Microbial Exploration Techniques Application Off-Shore: A Motivating Case Study

Microbial Exploration Techniques (MET) applied to non-conventional onshore plays have shown a positive correlation between the presence of certain taxonomic groups and hydrocarbon deposits. However, in despite of the high cost of drilling offshore wells, there is a lack of studies demonstrating the validity of MET in offshore areas. In such sense, we had access to 20 samples coming from the piston cores of a standard geochemical offshore exploration campaign. Samples corresponded to piston core sections of 5 meters depth and were selected to include a wide gradient of geochemical characteristics. This gradient was examined using two different methodologies (Ramette 2007): exploratory analysis along the gradient and hypotheses testing between groups of samples (as discussed below). Groups to be compared in the present pilot were clustered using geochemical data as an input and Gower distance (the product of each variable by its range) as distance measure (see Figure 16.1). Basically, interstitial light hydrocarbon (C_1-C_5) data, Total Scanning Fluorescence (TSF) related variables, and data obtained from hydrocarbon analysis extracted by hexane (including the unresolved complex mixture, UCM) were used for such cluster.

As a result, piston cores were clustered in four groups as shown in Figure 16.1: a single sample group (1) that contains PC35, a piston core that geochemical analysis classified as containing both thermogenic hydrocarbons and gases; a second group (2) that contains PC24 and PC61, classified as containing thermogenic gases; a third

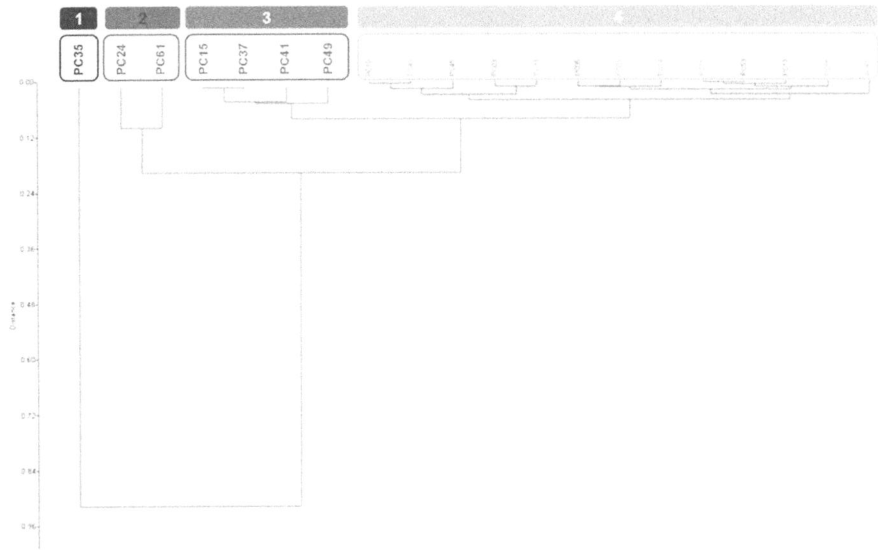

FIGURE 16.1 Paired group cluster analysis based on Gower distance measure showing piston cores geochemical groups from offshore pilot campaign.

group (3) (PC15, PC37, PC41, PC49) that contains a mix of piston cores marginally qualified to contain hydrocarbon, and piston cores qualified as containing thermogenic gases; and a fourth group (4) of piston cores disqualified.

16.2.2 Sampling Technique

Offshore prospection campaigns usually involve the uptake of subsurface seafloor samples called piston cores. The samples are collected using a piston corer, a long, heavy tube plunged into the seafloor to extract samples of mud sediment. The core rig used in this pilot was assembled to be 10 meters using core barrel sections that are 7 cm inside diameter. Piston coring offers several advantages over gravity coring: greater penetration depths, better core recovery, and higher quality (less disturbed) samples. Using piston coring to sample depths up to 6 meters significantly reduces intracore variability due to bioturbation, loss by diffusion of gases, and mixing of natural hydrocarbon seepage or pollution in the top meter of seafloor sediments with deeper sections of the core. In particular, the quality of information gained from measurement of sediment gases increases dramatically in sections collected deeper than about 2.5 meters. Sediment gas concentrations can change several orders of magnitude through the 2- to 5-meter depth range in marine sediments, greatly enhancing the type and quality of measurements that can be made on the samples.

Microbial metagenomic analysis only requires a few grams of a core sample (1 to 10 g), but there are a number of specificities that require consideration prior or during sample preparation such as sample extraction and handling, core sample subsampling, sample transport, and sample preservation and storage

16.2.3 Sample Extraction

For the current case of study, 20 sea bottom sediment samples were obtained from different locations situated between 1500 and 3500 meters below the sea level. Samples corresponded to piston core sections of 5 meters depth. Like in any environmental microbiology study, one of the main challenges was the obtention, handling, and shipping of uncontaminated samples from the area(s) under investigation. Although the impacts of microbial contamination are irrelevant for geochemical or geological analysis, the high sensitivity of DNA sequencing technologies makes these samples very sensitive to foreign DNA coming from external microorganisms as, for example, those inhabiting the operator's skin. Therefore, even minimal microbial contamination should be prevented by thoroughly cleaning sampling lines in between sample uptake, using sterile sampling equipment and bottles, using gloves during sample handling, sealing bottles containing soil material, etc. To this end, specific protocols have been described in detail by Tsesmetzis and collaborators (Tsesmetzis et al. 2017).

As soon as each acceptable core was retrieved on-deck, sample processing started. Considerable effort was put on minimizing the time between core retrieval, sample processing, and storage at –80°C. For each accepted core, the core sections taken for further processing were extracted and the top and bottom 1 cm of sediment were discarded. The core material, usually around 500 cm, was then sub-sampled according to the specified analytical requirements.

16.2.4 Core Sample Subsampling and Shipping

Microbial cell abundance in sub-seafloor sediments is known to differ by approximately five orders of magnitude at different site locations, depths, and geophysical-geochemical conditions (Kallmeyer et al. 2012). Therefore, it is very important to establish a subsampling plan prior to sample extraction, so the obtained material corresponds to equivalent sediment depths and microbial diversity anomalies can be detected by comparison between samples.

For this study, core sections were cut into consecutive short whole-round segments with a hacksaw and stored onboard at –20°C. Since seafloor cores usually contain very little biomass, obtaining sufficient amount of DNA for downstream processing is often a challenging process. We chose to work within 4 to 5 meters of sediment depth because the top 20 most abundant taxonomic classes have been shown to be well represented within this layer (Petro et al. 2017). To this end, 10- to 30-gram subsamples from the 4 to 5 meters sediments meter section were extracted in duplicate, placed in sealed plastic bottles under sterile conditions, and stored at –80°C. Samples were subsequently transferred to onshore sequencing facilities under cryogenic conditions.

16.2.5 Microbial Community Characterization of Piston Cores Samples

In order to detect hydrocarbon-related anomalies in the microbial diversity of the bacterial populations inhabiting the sediments in our prospection sites, we performed an amplicon-based taxonomic characterization of our samples. Unlike capillary

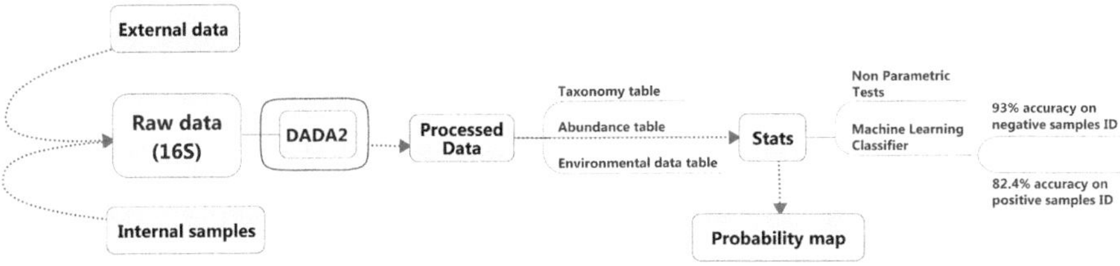

FIGURE 16.2 Graphical abstract. We generated a database based on public data obtained from amplicon studies on marine sediments worldwide. We processed these, together with data generated from 20 marine sediments samples extracted by Repsol S.A. using the DADA2 workflow. Three data matrixes were obtained as outputs, collecting abundance, taxonomy, and environmental data, respectively. Several non-parametric tests were applied to identify new biomarkers and, using our database, we trained a machine learning model to automatically identify hydrocarbon positive samples from Repsol cores. These analyses allowed us to generate a probability map of hydrocarbon presence based solely on the samples microbial signature.

sequencing or PCR-based approaches, next-generation sequencing (NGS) is a culture-free method that enables analysis of the entire microbial community within a sample. With the ability to combine many samples in a sequencing run, microbiologists can use NGS-based 16S rRNA sequencing as a cost-effective technique to identify strains that may not be found using other methods. However, 16S rRNA will not enable analysis of the "entire microbial community" (e.g., it will not pick up fungi or microeukaryotes or even Archaea if using bacterial specific primers). Also no primer set is truly universal in any particular microbial domain.

The workflow summarized in Figure 16.2 has been proven to be successful in 16S rRNA gene amplification for Illumina sequencing platform.

16.2.5.1 DNA Extraction and Quality Control

DNA-based microbial characterization methods are strongly dependent on the amount and quality of DNA extracted from samples. As we mentioned before, marine sediments do not contain high amounts of biomass and therefore achieving the required DNA extraction yields and quality levels are common bottlenecks in amplicon sequencing projects. This is because DNA extraction yields will determine the final concentration of DNA obtained from each sample and hence, the sequencing protocol that may be applied. Current standard methods in amplicon sequencing, based on the Illumina platform, require a minimal concentration of 4 nM of DNA. Such small DNA amounts are extremely sensitive to contamination from exogenous DNA introduced while manipulating the samples, when in contact with non-sterile instruments or by reagents. Therefore, it is crucial to operate under sterile conditions and maintain the working environment as clean as possible.

In addition, there are certain molecules naturally occurring in soil samples, such as clays and humic acids, that are known to reduce DNA quality and extraction yields. For this reason, there is a new family of protocols and extraction kits specifically designed for samples known to contain DNA extraction inhibitors (Kennedy et al. 2014).

Since the samples extracted from our exploration block were rich in clay, DNA extraction was performed in triplicate for each sample using an optimized protocol for the obtention of high-purity DNA under difficult conditions. Briefly, triplicates of 250 mg of sediment subsamples were passed through a "bead-beating" mechanical cell lysis pre-treatment after which an enzymatic lysis using the PowerSoil DNA isolation kit (MO BIO, USA) was performed. Lastly, a final step of DNA cleaning and concentration after DNA extraction with the High Pure PCR Template Preparation Kit (Roche) was done. DNA concentration and quality were measured using Nanodrop (Thermo Fisher, USA), confirming the absence of inhibitors and the obtention of sufficient DNA amounts for sequencing.

16.2.5.2 Gene Amplification and Sequencing Techniques

Modern environmental studies aiming to characterize microbial taxonomic diversity are almost completely reliant on NGS because of their higher throughput, reliability, and cost-effectiveness when compared with previous approaches.

Currently, there are two main different NGS approaches that are used in microbial diversity analysis, amplicons, and shotgun sequencing. The first entails sequencing all amplified products from a single gene, typically the 16S rRNA gene. This is an excellent phylogenetic marker because it contains nine well-characterized hypervariable regions (HVRs) whose presence can be used to assign taxonomic categories as long as the bacterial group they belong to is known for science and is included in the standard taxonomy databases created to this end (Caporaso et al. 2011). These include the Greengenes database (DeSantis et al. 2006), the Ribosomal Database Project (RDP) (Cole et al. 2014), and SILVA (Quast et al. 2013). However, it also presents several limitations such as known biases when resolving a significant fraction of the diversity in a community (due to errors inherent to PCR amplification, unequal weight of different loci when resolving taxa, sequencing errors leading to incorrectly assembled amplicons, etc.) and the inability to provide information on the functional diversity of the community (Sharpton 2014).

The second approach, shotgun metagenomics, seeks to avoid such limitations by, instead of targeting a specific genomic locus for amplification, shearing all DNA into small fragments and independently sequenced (Quince et al. 2017). This results in DNA sequences (called reads) that align to several genomic locations for the myriad genomes present in the sample, including non-microorganisms. Some of these reads are then sampled from taxonomically informative genomic loci (e.g., 16S rRNA), and others will be sampled from sequences coding for biological functions. As a result, shotgun metagenomics data may be used to explore the two main aspects of a microbial community simultaneously: who is there and what can they do?

While powerful, shotgun metagenomics also show some important limitations. Shotgun datasets are large and complex, needing vast amounts of computational resources for informatics analysis and access to sequencing and computational facilities is needed, which make them more expensive. All metagenomics computational tools rely to some extent on available genomes, and these are biased toward model organisms, pathogens, and easily cultivable bacteria. In addition, functional profiling is also dependent on the availability of experimentally validated annotations, which are usually missing for most genes present in a natural microbial community. Furthermore, shotgun metagenomics is also affected by multiple experimental error sources such as sequencing noise, contaminant DNA, or taxonomically complicated microorganisms and plasmids. Finally, the "live or dead" may always be considered when performing functional characterization of a microbial community. DNA may persist in the environment long after the death of the host cell, generating reads on sequences that are no longer active and hence producing misleading results (Quince et al. 2017). For this proof of concept, and considering the pros and cons exposed above, we chose to perform amplicon sequencing only. Therefore, the following methods describe in this chapter will refer exclusively to amplicon sequencing methodologies.

After DNA extraction, the region(s) of interest must be amplified and prepared for sequencing. Usually this task is achieved through polymerase chain reaction (PCR), a very straightforward and cost-effective approach but which may be a source of biases to the sample at different levels (Schirmer et al. 2015). The vast majority of PCR methods amplify a specific region of a DNA strand (the DNA target) in a two-step reaction. First, the two strands of the DNA double helix are physically separated at a high temperature in a process called DNA melting. In the second step, the temperature is lowered and the two DNA strands become templates for the enzymes DNA polymerases to selectively amplify the target DNA. The selectivity of PCR results from the usage of primers (short strands of RNA or DNA that serve as a starting point for DNA synthesis), that are complementary to the sequence around the DNA region targeted for amplification under specific thermal cycling conditions. As a consequence, a DNA target that does not present complementarity to the selected primers will not be amplified, producing a false negative for its correspondent organism in the microbiome profile. In order to overcome these limitations, we have used two different sets of primers for both bacteria and archaea (Klindworth et al. 2013, S-D-Bact-0341-b-S-17, CCTACGGGNGGCWGCAG, S-D-Bact-0785-a-A-21, GACTACHVGGGTA-TCTAATCC and Apprill et al. 2015, 806RB 5′-GGA CTA CNV GGG TWT CTA AT-3′ and 515F (5′-GTG CCA GCM GCC GCG GTA A-3′). Apprill and co-workers implemented a PCR product to detect bacterio-plankton profiles in marine environmental samples that has shown good results dealing with Archaea diversity (Cordoñer, F. personal communication). In addition, during PCR amplification partially amplified molecules might break and bind unspecifically to templates from other lineages, serving as primers for new, chimeric sequences referred to as chimeras. This arises due to sequence similarity between amplicons and the nature of the chain reaction. Factors that have been linked to increased frequency of chimeras include both laboratory settings, such as fast thermocycling during PCR and intrinsic characteristics of the sample, such as richness and diversity (Hugerth and Andersson 2017).

Finally, besides sampling and library preparation, the choice of the sequencing platform to be used will depend on the sample type and the questions to be answered. In general, we will aim towards high-throughput short-read platforms when more quantitate information is preferred over taxonomic resolution. Among the various sequencing technologies available, Illumina (San Diego, CA, USA) is by far the most widely used platform due to its high cost-effectiveness and higher throughput.

On the opposite, longer reads length would be preferred when the goal is the obtention of an accurate phylogenetic classification (Singer et al. 2016). In addition to the short read technologies used in this work, long read approaches such as

PacBio (Schloss et al. 2016) and Oxford Nanopore MinION (Kerkhof et al. 2017) are increasingly in use. Although both PacBio and Oxford Nanopore generate longer reads compared to short read Illumina or Ion sequencing, the higher error rate of both sequencers (10–15% vs. 5%) remain an issue that needs to be addressed.

In order to perform microbial profiling on the 20 samples used here as a case study, DNA obtained and purified in the previous step was sequenced in the MiSeq platform (Illumina, San Diego, CA, USA), configuration 300 cycles paired reads. In brief, the first amplification step, primers were designed containing: (1) a universal linker sequence allowing amplicons for incorporation indexes and sequencing primers by Nextera XT Index kit (ILLUMINA); and (2) the 16S rRNA gene universal primers (see above). In the second and last amplification indexes were included. Libraries were quantified by fluorometry using Quant-iT™ PicoGreen™ dsDNA Assay Kit (Thermo Fisher) and pooled prior to sequencing on the MiSeq platform. The size and quantity of the pool were assessed on the Bioanalyzer 2100 (Agilent) and with the Library Quantification Kit for Illumina (Kapa Biosciences), respectively. PhiX Control library (v3) (Illumina) was combined with the amplicon library (expected at 20%). Sequencing data were available within approximately 56 hours. Image analysis, base calling, and data quality assessment were performed on the MiSeq instrument.

16.2.5.3 DNA-Seq Analysis Workflow

Along the next section, we will introduce the reader to the Dada2 pipeline, an open-source R package which we have used to analyze our sequences. Some of the methods used here, such as data quality-check or filtering, are generic but others, such as the denoising algorithm Dada2 (Callahan et al. 2016), use a different strategy with respect to more standard amplicon data analysis workflows such as Sickle (Joshi and Fass 2011), QIIME (Caporaso et al. 2011), and MOTHUR (Schloss et al. 2016). We chose to use DADA2 workflow for this proof of concept because it is built as a stand-alone R package, excellently well documented, able to scale up and use multiple processors, and has easy and straightforward integration with powerful statistical data analysis R packages such as Phyloseq (McMurdie and Holmes 2013) and CARET (Max et al. 2017).

16.2.5.4 Data Quality Check

The initial sequencing data processing step is filtering based on read quality scores and the presence of the expected primer and adapter sequences, and the removal of these non-biological sequences (filtering and trimming, respectively). Hence, in order to correctly parametrize this step, a previous exploration of the raw data is required. FastQC (Andrews 2014) and MultiQC (Ewels et al. 2016) are the most commonly used tools to address this need.

16.2.5.5 Filtering and Trimming

This step is specific to amplicon sequencing. DNA sequences (i.e., reads) with a low-quality score (Phred score < 20) or short length lower than the cutoff value are filtered out, and primer sequences are removed. Common stand-alone tools designed to achieve this task are Usearch (R C Edgar 2010) and FLASH (Magoč and Salzberg 2011), but it is also supported by most amplicon sequencing analysis workflows such as MOTHUR, Qiime, or DADA2 (Schloss et al. 2009; Caporaso et al. 2010; Callahan et al. 2016). More recently, Chen and co-workers have developed AfterQC (Chen et al. 2017), a highly automated quality control and data filtering tool that greatly facilitates these steps. In addition, it can also correct wrong bases in overlapping regions and detect and visualize sequencing bubbles, which are a common source of sequencing errors.

Filtering and trimming was performed in our samples before denoising. The DADA2 *fastqPairedFilter* function was used to remove sequences with Ns or more than two expected errors, trim the first 10 nt and the last 10 nt (forward reads) or 10–50 nt (reverse reads), depending on the quality profile of the sample data and a Q threshold default value of 2 as a special quality score indicating the end of good quality sequencing in Illumina 1.8+.

16.2.5.6 Denoising

In addition to quality filtering, errors in Illumina-based amplicon sequencing are currently addressed by clustering sequences that differ by less than a fixed dissimilarity threshold (typically 3%) in constructions called closed-reference operational taxonomic units (OTUs) (Schloss et al. 2009; Caporaso et al. 2010; Edgar 2010). In this way, the rate at which sequencing errors are incorrectly considered as taxonomic diversity is reduced but at the expense of losing taxonomic resolution and the inability to resolve fine-scale variation (Rosen et al. 2012). Fine-scale resolution has been shown to be very informative when working with methane natural seepages, providing insight on ecological niches, temporal dynamics, and population structure (Roalkvam et al. 2012; Niu et al. 2017; Winkel et al. 2018). In addition, closed-reference OTUs are extremely dependent on the set of reference sequences used in their assignment. Therefore, closed-reference OTUs assigned against different reference databases are not comparable. For keeping OTUs from different experiments comparable over time, reference databases should remain static or closed-reference OTU tables should be regenerated whenever a new or an upgraded reference database is adopted (Callahan, Mcmurdie, and Holmes 2017).

In order to address these limitations, new algorithms have been recently proposed to resolve amplicon sequence variants (ASVs) from amplicon data without the need of building OTUs dependent on an arbitrary similarity threshold. These approaches are based in the statistical modeling of the error rates inherent to the Illumina platform on the basis of, in part, the expectation that biological sequences are more likely to be repeatedly observed than are error-containing sequences (Callahan et al. 2016). Once the error rate can be controlled, amplicon sequence variants (ASVs) can be resolved down to the level of single-nucleotide differences over the sequenced gene region. Hence, they have also been referred as exact-sequence methods (Callahan, Mcmurdie, and Holmes 2017b). As a consequence, ASV inference can only be performed in a sample but, unlike de novo OTUs, ASVs can be considered as permanent labels since they represent the DNA sequence of the assayed organism, independently of the context within which data were analyzed. Thus, ASVs inferred independently from different studies or different samples can be validly compared (Callahan, Mcmurdie, and Holmes 2017). Because ASVs are consistent labels derived from the data alone, ASVs remain unvarying into the indefinite future.

This is especially relevant to achieve our long-term goals, since we intend to use data from different exploratory campaigns to build statistical models that would help us to classify future samples based in the information obtained in previous campaigns. The most commonly used tools for exact-sequence methods are DADA2 (Callahan et al. 2016), now also available in QIIME2 (Caporaso et al. 2010), UNOISE2 (Robert C Edgar 2016), offered in USEARCH (R C Edgar 2010), and Deblur (Amir et al. 2017), built in QIIME2 (Caporaso et al. 2010).

Using the *mergePairs* function within the DADA2 workflow, the forward and reverse reads were independently denoised and then merged. Chimeras were removed from the denoised output of DADA2 by the *isBimeraDenovo* function, obtaining a list of output sequences (ASVs) and associated abundances analog to the OTU table typically obtained as an output from other more popular workflows such as QIIME or MOTHUR.

16.2.5.7 Taxonomic Annotation

Once ASVs have been resolved, it is common at this point to perform taxonomic classification. The DADA2 package provides a native implementation of the RDP's naive Bayesian classifier for this purpose. The *assignTaxonomy* function takes a set of sequences and a training set of taxonomically classified sequences, and outputs the taxonomic assignments with a confidence interval given by the user (Callahan et al. 2016).

16.2.6 Amplicon Data Analysis

16.2.6.1 Preprocessing

Outputs from most amplicon sequencing analysis workflows are usually summarized as a contingency table of read counts per OTU or ASV per sample. This information is usually wrapped together with the corresponding sample and information metadata through the Biological Observation Matrix format BIOM (McDonald et al. 2012), a standard format compatible with most of the existing bioinformatics tools for amplicon data analysis. Often in environmental samples, where both diversity within and between samples is very high, the number of possible variables (taxa) is huge but a majority of the "observations" (counts) in the table are 0, frequently greater that 90% (Paulson et al. 2013). This is meant to represent that a given "observation" is not detected in the corresponding sample (McDonald et al. 2012). The fraction of the table that has non-zero values is defined as the "density," in contrast to "sparse" representations, where most of the elements are zeros. As the collection of samples becomes more diverse, these tables become even sparser and their size (both on disk and in memory) becomes a considerable barrier to performing meta-analyses (McDonald et al. 2012). This is particularly problematic in oilfield microbiology, where a long tail of taxa belongs to the rare biosphere.

However, the interpretation of counts equal to 0 has further statistical implications if we consider they may represent the true absence of an observation but also its presence below the detection limit. Sparsity implies that the counts of rare taxa are uncertain, since they are at the limit of sequencing detection ability when there are many sequences per sample (i.e., large library size) and are undetectable when there are few sequences per sample (Paulson et al. 2013). In addition, experimental limitations such as different yields between sequencing runs or unequal representation of samples in pooled sequencing libraries may lead to different detection limits between samples within the same study (Hugerth and Andersson 2017). Therefore, the microbial community in each biological sample may be represented by very different numbers of sequences, reflecting differential efficiency of the sequencing process rather than true biological variation. Finally, since a sample only represents a small fraction of the original environment, the total number of counts is not a direct measurement of the absolute number of microorganisms present in the environment but compositional data, which have been defined historically as random vectors with strictly positive components whose sum is constant (e.g., 100, one, a million). More recently, the term covers all those vectors representing parts of a whole which only carry relative information, thus including not only parts per unit or percentages, but also molar compositions, abundance of different species etc. Since compositional data are constrained within a given sampling space, often referred to as the simplex (i.e., sum to 1), standard statistical techniques lose their applicability. Uneven sampling depth, sparsity, and the need of extracting conclusions at the ecosystem level from taxa relatives abundances obtained through sampling represent serious challenges for interpreting data from microbial survey studies (Weiss et al. 2017).

In order to mitigate the undesired bias these statistical challenges may introduce on the results, several approaches have been proposed. Most frequently used amplicon data analysis workflows, such as MOTHUR (Schloss et al. 2009), Qiime (Caporaso et al. 2010), or Phyloseq (McMurdie and Holmes 2013) suggest filtering taxa based on different criteria such as having less than a given number or percentage of observations, being present in fewer than a given number of samples, etc. There is currently a lack of consensus on methods or protocols specifying the criteria on which one could decide how or where to set these cutoffs, as it greatly depends on the nature of samples, the questions being asked or the number of samples available. However, bias due to DNA extraction and PCR amplification protocols have been shown to be much greater than the effects of sequencing and data analysis (Brooks et al. 2015).

16.2.6.2 Normalization

In addition to filtering taxa, most datasets are often normalized prior to downstream analysis. Normalization is the set of methods aiming to transform a given dataset by adjusting values measured on different scales to a notionally common scale in order to allow a non-biased comparison between statistics from different measurements. As mentioned before, sparsity is an inherent property in microbiome datasets and sampling depth is often uneven, mostly due to biases related with differences in sampling, library preparation, and sequencing. Therefore, normalization is always required before downstream data analysis and hence it has become an increasingly important topic within the field in an effort to standardize methods and make analysis results more comparable.

Recent advancements in normalization methods for metagenomic data have been extensively discussed by McMurdie and Holmes (2014) and Weiss and collaborators (Weiss et al. 2017). Between the different options available, dividing the counts by the total count of the sample is by far the most popular one, since it is intuitive and little computational power is required. However, this treatment removes the independence between observations, since an increase in the relative abundance of one

OTU induces a perceived reduction in all others. A similar approach is using relative abundances, which should be less sensitive (Bullard et al. 2010).

One very common alternative to relative abundances is rarefying, a normalization method aiming to remove the bias introduced by uneven sampling depths by random down-sampling of every sample to the smallest sample size of the cohort (Hugerth and Andersson 2017). Although this procedure is a recommended approach in widely used microbiomic data analysis workflows such as Qiime (Caporaso et al. 2010), McMurdie and Holmes (McMurdie and Holmes 2014) explored how downsampling entails data waste and loss of statistical power for downstream analysis, questioning this practice. In addition, they showed how using relative abundances cannot control for overdispersion and, since it breaks data independence, increases the rate of false positives.

Instead, McMurdie and Holmes (2014) have proposed using methods initially designed for mRNA and marker gene sequencing data analysis, such as DESeq2 (Love, Huber, and Anders 2014), edgeR (Robinson, McCarthy, and Smyth 2010), and metagenomeSeq (Paulson et al. 2013). These methodologies control the dispersion of the data through appropriate statistical models, minimizing error rates. The Qiime team has extended this analysis recently (Weiss et al. 2017), showing that the high rate of false discoveries attributed to rarefaction by McMurdie and Holmes was due to an artifact on the simulation. Instead, they found that rarefaction leads to lower sensitivities (higher false negatives rate), due to the rejection of important amounts of data and the nonparametric nature of the Wilcoxon rank-sum. The severity of the power decrease caused by rarifying depends upon how much data have been thrown away and how many samples were collected. This problem has been known for a long time, leading to the general guideline of rarefying to the highest depth possible (Weiss et al. 2017). Their results also discarded DESeq2 as a useful normalization technique prior to most standard downstream data analysis methods in environmental microbial ecology, except weighted UniFrac (see below). This was particularly relevant for very low library sizes (under approximately 1000 sequences per sample) or if presence/absence metrics like unweighted UniFrac were used. As explained by the authors, these negative results are due to the fact that both DESeq and edgeR-TMM normalization methods work under the assumptions of a constant abundance of a majority of species and of a balance of increased/decreased abundance for those species that do change. These are requirements that don't apply to most microbial environments, which are extremely variable in microbial composition, making them less suitable for normalizing data from environmental samples. However, these methods performed better when used for differential abundance testing and are now also available in QIIME version 2.0 (Caporaso et al. 2010).

16.3 DOWNSTREAM MICROBIOMIC DATA ANALYSIS METHODS

16.3.1 Comparing Bacterial Communities between Samples (β-diversity)

The term "beta diversity" was firstly introduced by Whittaker (Whittaker 1960), and defined as "the extent of change in community composition, or degree of community differentiation, in relation to a complex-gradient of environment, or a pattern of environments." Simplifying, beta-diversity can be understood as a metric that addresses the degree of similarity between populations belonging to different samples.

There is a plethora of mathematical indices of community similarity available. As an example, Mothur (Schloss et al. 2009) provides 37 different tests while Phyloseq (McMurdie and Holmes 2013) has implemented up to 46 measures. Due to the strong interdependence between distance metrics and normalization, which can increase or reduce the relative distance between two samples, the choice of beta-diversity index can provide apparently very different results for the same dataset, making it advisable to perform several different analyses on the same dataset prior to extracting conclusions.

Most available beta diversity indexes could be grouped within one of these three approaches: taxonomy-based ordination methods, phylogeny-based ordination methods, and interaction networks analysis. Traditional taxonomy-based methods, such as Jaccard (Jaccard 1901) or Bray-Curtis (J. Roger Bray and John T. Curtis 1957), focus on taxa co-occurrence among samples, quantified directly from taxa abundance data. Bray–Curtis and Jaccard indices are rank-order similar, but the Jaccard index is metric, and probably should be preferred instead of the default Bray-Curtis, which is semimetric. More recently, phylogeny-based methods have been developed based on common evolutionary traits between communities instead of shared taxa (Swenson 2011). Between these, Unifrac (Lozupone and Knight 2005) is perhaps the most widely used by microbial ecologists to address sample dissimilarities. Unifrac measures the phylogenetic distance between sets of taxa in a phylogenetic tree as the fraction of the branch length of the tree that leads to descendants from either one environment or the other, but not both. Lozupone and Knight (2005) developed this approach as a measurement of the total amount of evolution that is unique to each sample, under the assumption that adaptive traits to one environment would be deleterious in the other. Therefore, if two environments are similar, few adaptations would be unique to only one community. Consequently, most nodes in a phylogenetic tree would have descendants from both communities, and much of the branch length in the tree would be shared. In contrast, if two environments are so different that most of the microorganisms adapted to one could not survive in the other, then the lineages in each community would be distinct, and most of the branch length in the tree would lead to descendants from only one of the two communities. Therefore, traditional and phylogenetic metrics may provide complementary insights into community composition when combined, allowing a deeper understanding of the system under study.

16.3.2 Biomarkers Discovery through Statistics

Amplicon data analysis outputs can usually be divided into three data types: a matrix containing the taxonomy associated to each detected ASV's or OTU, a matrix containing microorganisms counts, with sample names in rows and measured variables (biological taxa, OTUs, etc.) for those samples in columns, and a "metadata" matrix that in our case contains all kinds of geological, geochemical, and geographical variables available for every sample. Using these data we created a new categorical variable called "Hydrocarbon Presence" representing the existence or absence of hydrocarbons for every sample based on their geochemistry, which was later used in the analysis as a piston core classification metric. In the next section we explore the most useful statistical methods that can be used to identify microbial anomalies and new biomarkers associated with hydrocarbon presence. We do not intend to make an exhaustive revision since recent detailed reviews are available (Paliy and Shankar 2017; Hugerth and Andersson 2017; Ramette 2007) but suggest which approaches would fit the best to our goals. Most of the techniques explored here could be classified as "multivariate statistical techniques." This is because, when working with microbial communities and/or metagenomics obtained by high throughput sequencing methods, one of the main challenges is to condense all the information (coming from numerous OTUs, taxa, etc., and/or samples) into two- or three-dimensional spaces (Hugerth and Andersson 2017a). On the other hand one should not over-interpret the data obtained by these techniques by relying on unjustified causality (Ramette 2007), mostly because these techniques may only suggest cause or factors (that would be necessary to validate), and because the synthetic variables, axes, or clusters derived from these analysis do not necessarily have a biological or ecological meaning in nature (Ramette 2007).

Exploratory approaches are a set of statistical tools that aim to help us to visually understand patterns in large datasets: the basic idea is that similar objects (or samples) would be found near each other, while dissimilar samples would be depicted farther apart from each other. The most known on this "category" is Principal Components Analysis (PCA; Hugerth and Andersson 2017): new synthetic variables (principal components, that are linear combinations of the original variables) are calculated, accounting for as much of the original data variance as possible (Ramette 2007). Data are represented in a new system of dimensions (we would have many dimensions as variables are included in the analysis, but normally the first two or three dimensions are depicted). The first principal component is by definition placed in the direction representing the largest variation of the data. The second component is placed in the direction orthogonal to this that explains the largest amount of the remaining variation, and so on. The first few components often explain a large amount of the variation, allowing a visual inspection of the distance between samples in two- or three-dimensional space (Hugerth and Andersson 2017).

However, since PCA is based on linear methods that don't perform well on non-continuous datasets, its application in ecology is limited since species normally display nonlinear responses to environmental gradients (Ramette 2007). In other words, euclidean distance is rarely appropriate for microbial data, since pairs of samples with many common counts of 0 will be depicted at a shorter distance (Hugerth and Andersson 2017). Bray-Curtis dissimilarity, in contrast, only considers taxa present in at least one of the samples being compared (Hugerth and Andersson 2017).

Alternatively, Principal Coordinate Analysis (PCoA) should be used instead of PCA. Both algorithms share the objective of explaining most of the variance of the original dataset but the second can work with any dissimilarity measure (such as Bray-Curtis dissimilarity) without the constraint of the euclidean distances (Hugerth and Andersson 2017). PCoA has gained recent popularity in microbial ecology (Paliy and Shankar 2017) due to its ability to use phylogenetic distance (e.g., UniFrac distance, see Section 16.2.3.1). However, since PCoA employs the distance matrix as an input, it does not provide a direct link between the principal coordinate axes (components) and the original variables, so the interpretation of variable contribution could be more difficult and complex (Ramette 2007).

Another interesting exploratory approach is Correspondence Analysis (CA). This test aims to provide an answer to the common question of whether certain species have some ecological preferences or occur at specific sites (Ramette 2007). This method maximizes the correspondence between species and samples rather than maximizing the percentage of variance explained by the ordination (Ramette 2007; Hugerth and Andersson 2017). This technique is recommended when species display unimodal relationships with environmental gradients, which is shown when a maximum of abundances (or a peak of presence) is found when the optimal conditions for this species are met (Ramette 2007).

Contrary to these techniques, in NMDS (Nonmetric Multidimensional Scaling) the number of dimensions (n) to which your data must be reduced is selected a priori; a matrix of object dissimilarities is then calculated (so a distance metric should be chosen as in PCoA) and ranked. Hence, this algorithm intends to find a simplified n-dimensional ordination space in which these ranked distances are preserved (Ramette 2007; Hugerth and Andersson 2017; Paliy and Shankar 2017). The distances in the original configuration are compared by regression with the calculated distances based on a stress function, which indicates how different the ranks on the new configuration are from the ranks in the original distance matrix (Ramette 2007). Since the result is based on the random original positions of the objects in the ordination space, a permutation step is included to minimize the stress function, and no unique solution is produced (Ramette 2007; Paliy and Shankar 2017). Ordination distances among objects do not correspond to original distances in NMDS, but the proximity between objects correspond to their similarity (Ramette 2007). Although NMDS is more computer intensive than PCoA (Ramette 2007; Paliy and Shankar 2017), it often produces better results by compressing distance relationships among objects into fewer dimensions (Ramette 2007). In datasets with many overlapping gradients, NMDS will often produce a clearer overview of data distribution than other techniques that don't need to limit the number of dimensions (Hugerth and Andersson 2017; Paliy and Shankar 2017).

Microbiomics could be defined as the study of the totality of microorganisms in particular environments, such as the human gut or an oil well. Hence, many environmental microbiomics studies aim to identify the best mathematical relationship among the species data matrix and the environmental matrix. Although some of the exploratory methods described above, such as NMDS, could be used to make such comparisons, variables are added to the existing ordination by linear regression of their values onto the existing ordination axes (Ramette 2007). Alternatively, we recommend the use of "interpretative methods" (Paliy and Shankar 2017), where only the variation in the species data matrix that can be explained by the environmental data is displayed and analyzed. As an output, we obtain an ordination plot that includes samples, species, and environmental variables so that relationships among them could be easily derived (Ramette 2007). Between the last, also referred to as direct gradient analysis (or constrained/canonical analyses), Canonical Correspondence Analysis (CCA) fits very well our objective of finding new biomarkers significantly related to different "environmental" types of samples. The goal of CCA is to find significant relationships between two sets of variables: species data (dependent variables) and the environmental data (independent variables). Multiple linear regression is applied to explain variation between these two sets of data" after which an iterative procedure is performed to find the best ordination (Ramette 2007). Particularly, CCA assumes unimodal species-environment relationships, which has been demonstrated to be robust enough even when species display bimodal responses, unequal ranges, etc. (Ramette 2007). Therefore, CCA could be seen as a constrained version of CA, where ordination axes are linear combinations of environmental variables (Ramette 2007; Paliy and Shankar 2017). As a main drawback, CCA seems to be highly sensitive to odd species from low diversity samples.

Having in mind this variety of techniques, it is worth trying different approaches and keeping not only those that explain the largest proportion of dataset variation, but also those ordinations with some underlying biological mechanisms well-disposed for further research (Hugerth and Andersson 2017). Keep in mind that none of the techniques discussed here can distinguish correlation from causation, except perhaps in intervention studies, or experiments under controlled conditions. More importantly, clusters and gradients produced along artificial axes do not necessarily correspond to any underlying biological effect (Hugerth and Andersson 2017).

16.3.2.1 Hypothesis Testing

In addition to visually inspecting the data, it is important to test statistically the plausibility of the proposed hypotheses (Hugerth and Andersson 2017). In order to obtain a measure of confidence for multivariate statistical analysis, since no data distribution is available before running any tests, parametric testing (based on the comparisons between the output and the distribution model expected, and the probability—p-value—of getting the output distribution by chance) cannot be applied. Alternatively, other methods based on bootstrap, jackknife, and permutation tests are available (see Paliy and Shankar 2017 for more details). Choosing the most appropriate test for our data will strongly depend on the type of question we are trying to answer. For example, if we aim to identify which microbial communities are different between predefined groups or clusters, ANOSIM and PERMANOVA could be useful techniques (Hugerth and Andersson 2017a; Paliy and Shankar 2017). When asking if the changes of the community between samples reflect any other contextual/environmental parameter, the Mantel test becomes a robust approach. This test compares two distance matrices that were calculated for the same set of objects but that were based on sets of variables that are independent (e.g., a species dissimilarity matrix and an environmental dissimilarity matrix or site distance matrix for the same samples). It calculates the correlation between values in the corresponding positions of two matrices and assesses significance by permutation (Ramette 2007; Paliy and Shankar 2017).

Finally, and fitting with our main objective of revealing biological indicators or biomarkers, most readers might be familiar with ANOVA (Analysis of Variance). However, most microbiomics data follow a non-normal distribution and hence ANOVA requirements are not met. Therefore, non-parametric versions of this analysis are required: Mann-Whitney U-test or Kruskal-Wallis test are suitable approaches for comparing two or more than two classes or groups, respectively. At this point, one could be tempted to test if any of the OTU/taxa identified during an amplicon analysis could be significantly associated with any classes or groups by repeating a Mann-Whitney test as many times as needed (hundreds or even thousands of times). Although this approach is correct for an individual taxon, when testing for multiple species the probability of finding significant differences between classes would be artificially augmented just by chance. There are some methods that deal with this problem setting a new (and more restrictive) significant p-value cut-off. As an example, Bonferroni correction set a new p-value cut-off at α/n, with n=number of variables or species to be tested. This correction is considered too conservative but new corrections over this one are available, and the complexity of this problem is too wide to be solved here.

It is important to keep in mind that running a very high number of tests will reduce the p-value cut-off, which may lead to overlooking significant differences between classes or groups (Noble 2009). This effect is particularly relevant for small sample numbers, which is a common issue in microbiomics studies.

In order to avoid these issues, we have applied an alternative method as a previous step: Similarity Percentages Breakdown (SIMPER). This test measures the contribution of individual taxa to a chosen dissimilarity distance (i.e., Bray-Curtis) between sample groups (Hugerth and Andersson 2017). It is not unusual that a subset of taxa might show significant differences between samples even when no significant differences were detected when using the whole community composition. This is an important issue in the development of diagnostic tools, environmental surveillance strategies, and/or the selection of biomarkers (Hugerth and Andersson 2017). We have used SIMPER to filter the taxa (those with higher contributions, and higher mean differences between groups) to be tested for differences (by Mann-Whitney U test). Using this approach, the number of comparison tests was reduced from several hundreds to less than 10 taxa, avoiding unwanted reduction effects on the p-value cut-off.

Complementarily, we have used the R package "IndicSpecies" (De Cáceres and Legendre 2009) to identify those taxa that could be considered to be "indicators" or "markers" of a class or group of samples using the *IndVal* index. *IndVal* is the product of two components (A and B). Component A (specificity or positive predictive value) is the probability that the surveyed object belongs to the target group given the fact that the species has been found. Component B (fidelity or sensitivity) is the probability of finding the species in objects belonging to the class group. *IndVal* for each species and group is calculated and then the group corresponding with the highest association value is selected. The statistical significance is tested by permutation. In this way, *IndVal* allows estimating the co-occurrence of taxa and samples. In our experience, this approach has demonstrated to be very useful since it offers well organized information and has helped us to verify the SIMPER + multiple testing approach. Results were very similar except for some scarce taxa only present in one group of samples, which IndSpecies identified as bioindicators but were not selected by our SIMPER protocol (they were scarce species only present in one group of samples, so the difference between group mean abundance was really low).

Although the previous approach yielded consistent results, we acknowledge that "manual" and supervised steps are required, which are time consuming and may hinder test reproducibility. Alternatively, recent statistical tools have been developed that allow performing this kind of test automatically when sample size is high enough. These methodologies control the dispersion of the data through appropriate statistical models, minimizing error rates. Many of these methods were initially developed for RNA-sequencing data but can also be used for microbiome studies (Hugerth and Andersson 2017). Among the most popular we can highlight DESeq2 (Love, Huber, and Anders 2014) and edgeR (Robinson, McCarthy, and Smyth 2010); both work under the same assumptions: both depend on the negative binomial distribution and assume that taxa with similar abundance should have similar dispersions. As a rule of thumb, these conditions would be invalidated when the observed variance is at least twofold the mean variance (Hugerth and Andersson 2017). SAMSeq (Li and Tibshirani 2013) was released as an alternative method to the previous ones: it conducts the Mann–Whitney U test on multiple resampling of the data to account for different sequencing depths, giving fewer false-positives than the negative-binomial tests but showing lower sensitivity for small sample sizes (Hugerth and Andersson 2017). Another alternative would be metagenomeSeq (Paulson et al. 2013), which uses models developed explicitly for taxa count data. All these packages and others can be easily used in R in combination with other microbiomics tools through the package Phyloseq (Hugerth and Andersson 2017).

Finally, the linear discriminant analysis effect size method LEfSe (Segata et al. 2011) is a powerful test that follows a different methodology. It couples statistical significance (standard statistical tests) with biological consistency (previous biological knowledge in its search for markers) and effect size estimation (Segata et al. 2011; Hugerth and Andersson 2017). First, OTU differentially abundant between classes are identified by a first round of feature selection through Kruskal-Wallis' sum-rank test; then, pairwise Wilcoxon's tests are used to discard OTU whose differential abundance isn't consistent across sub-classes (this step should remove spurious correlations; Hugerth and Andersson 2017a). Finally, it uses linear discriminant analysis (LDA) to estimate the effect size of each differentially abundant OTU (an important step in biomarker discovery; Hugerth and Andersson 2017a), so OTUs with abundance significantly different are ranked according to its effect size (Segata et al. 2011).

Using the statistical toolbox describe above (SIMPER followed by differences testing by the Mann-Whitney U test and complementarily "IndicSpecies" R package), we have identified a set of microbial taxonomical groups (for the purpose of this project, order taxonomical resolution has demonstrated to be enough showing a high predictive capacity) that allow us to discriminate between the previous groups of piston cores (see above and Figure 16.1). We have found three taxa (P1, P2, and P3) that showed greater mean relative abundances in the piston core qualified to contain thermogenic hydrocarbons (both liquids and gases) than in the piston cores disqualified (see Figure 16.3a). These three taxa also showed greater mean relative abundances (but the differences between mean relative abundances of the positive and negative groups were smaller) in the piston cores marginally qualified containing thermogenic hydrocarbons (oil and/or gas) than in the negative piston cores or disqualified (see Figure 16.3c). Only two (P2 and P3) out of these three taxa showed greater mean relative abundances in the piston cores qualified to contain thermogenic gases (see Figure 16.3b), and no differences were found in the third taxon when comparing with piston cores disqualified (see Figure 16.3b). Similarly, two groups of taxa (N1 and N2) were identified to have greater mean relative abundances in the piston cores

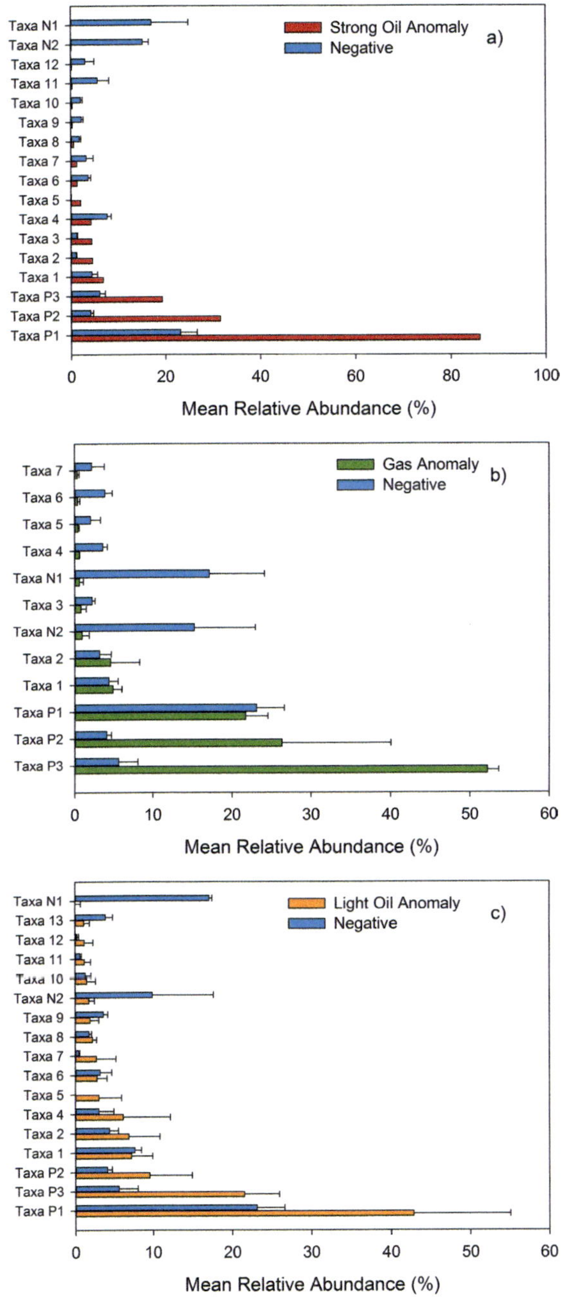

FIGURE 16.3 Taxa mean relative abundance (%) comparisons between a) piston cores qualified to contain thermogenic hydrocarbons (both oil and gas) and piston cores disqualified; b) piston cores qualified to contain thermogenic gases and piston cores disqualified; and c) piston cores marginally qualified containing thermogenic hydrocarbons (liquids and/or gases) and negative piston cores or disqualified.

disqualified than in the rest of piston cores (see Figure 16.3). P1, P2, and P3 corresponded to orders known to include genus able to degrade hydrocarbons, or with specific functions from complementary pathways to carbon oxidation such as sulfur cycle (Hawley et al. 2014; Håvelsrud et al. 2011) while N1 and N2 correspond to orders that have been identified in hydrocarbon pristine sea environments (Kostka et al. 2011; Dubinsky et al. 2013).

16.3.3 Using Biomarkers to Make Predictions on New Samples

The exponential decline in the cost of next-generation sequencing technology is being accompanied by innovations in bioinformatics and by a new "rebirth" of data mining or data science (nowadays called Big Data) approaches, mainly due to a great and cost-effective development of computer power and supercomputing facilities. The recent application of machine learning to high-throughput sequence data under the human microbiome project (Yazdani et al. 2016; Caporaso et al. 2011; Knights, Costello, and Knight 2011; Statnikov et al. 2013) have shown that these techniques can be successfully applied to classify environmental samples solely based on their microbial signature (Smith et al. 2015). Based on this idea, during this offshore pilot we have also explored several machine learning approaches with the aim to generate predictive models that should be able to classify piston cores as positive or negative for hydrocarbon presence solely based on their microbial community profile. In addition, this approach also allowed us to assign probability values for every piston core, which in turn are used to generate probability maps. The ultimate goal of this approach is to obtain an additional information layer based on the microbial signature that could be easily integrated with other tools currently used in exploration campaigns. In this sense, many tools and algorithms are available (Statnikov et al. 2013). However, regarding microbiomics, random forest (RF) and support vector machines (SVM) have the wider applications (Statnikov et al. 2013). Even though SVM performs better in terms of calculation time and accuracy, SVM models are black-box classifiers, which can tag our samples but cannot offer any explanation on the taxa they used to make such decisions. However, Random Forests (RF) both classifies and also offers as an output a ranked classification of taxa in terms of importance, making it the preferred algorithm for classifying samples based on their microbial profiles (Knights, Costello, and Knight 2011; Statnikov et al. 2013). Briefly, RF are an ensemble of learning methods for classification (and also for regression) that work by constructing many decision trees at training time and giving as an output the class (i.e., the mode of the classes; classification) or mean prediction (regression) for all the individual trees generated (Paliy and Shankar 2017). They have shown to achieve a high discriminating power and accuracy of classification because RF final output consolidates decisions across hundreds/thousands of individual decision trees (Paliy and Shankar 2017).

However, the main limitation of machine learning techniques for being applied on the field is that they require a minimum of 50 samples to generate reliable models. In order to overcome this limitation, we built a database using public data available in the literature (Ruff et al. 2015; Ristova et al. 2015). To this end, we first generated a

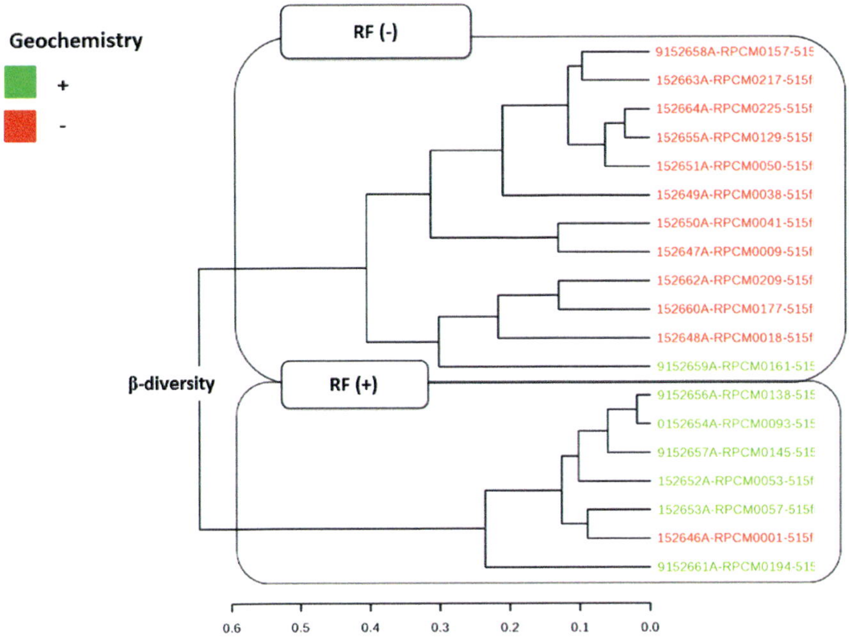

FIGURE 16.4 Comparison between geochemistry-based and machine learning-based piston cores classification. β-diversity analysis shows two well separated groups of samples based on their microbial populations phylogenetic distances. Machine learning classifier shows very good agreement with the geochemistry analysis for both negative (red) and positive samples (green) (only one disagreement respectively) and a perfect fit with the phylogenetic tree generated by the β-diversity analysis.

dataset of marine sediment samples obtained all around the globe and considered then as negative cases for hydrocarbon presence. As positives, we added samples obtained from cold seeps samples (areas of the ocean floor where hydrogen sulfide, methane, and other hydrocarbon-rich fluid seepage naturally occurs). Finally, we merged these two datasets and used this database to train a machine learning algorithm.

Once trained, the resulting model was used to classify our samples. The model classified correctly (based on geochemistry classification) 12 of the 13 samples disqualified (class error =7.7%), and 5 out of 7 samples qualified as containing hydrocarbons (oil and/or gases) by geochemical analysis. The Out of Bag error (OOB) for the model was 18.15% (see Figure 16.4), which overall suggests a good agreement between our model and the biochemistry results. The "positives" and "negatives" samples identified by the model are also consistent with the phylogenetic groups generated by the β-diversity analysis. Using this model, we have assigned a probability (based on the number of trees that assigned a kind of class to each sample) to each piston core, which once combined with geostatistics and kriging methods, have allowed us to generate a probability contour map (see Figure 16.5).

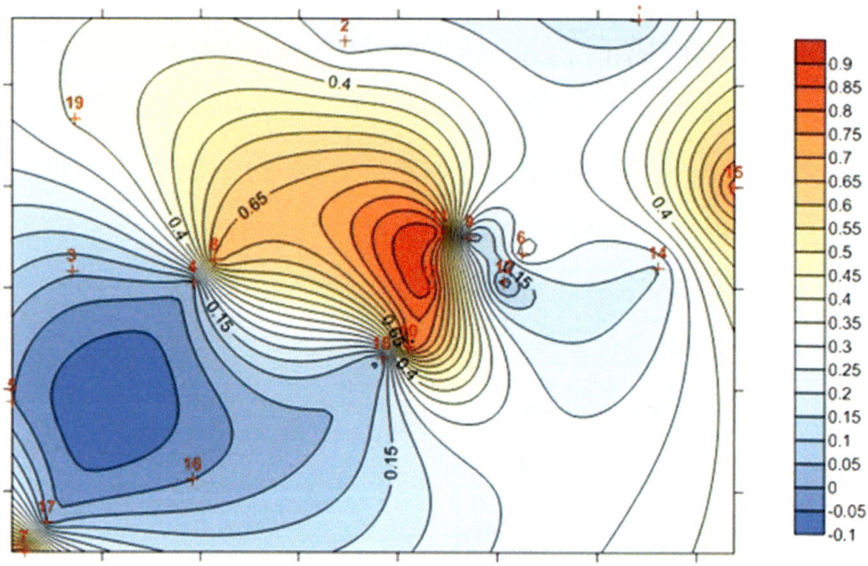

FIGURE 16.5 Contour hydrocarbon subsurface presence probability map obtained from the probability calculated for each piston core by a random forest model trained using our "cold seeps" database generated from public available data. Geographic coordinates have been hidden due to privacy issues.

16.4 CONCLUDING REMARKS

The low prices of the oil barrel, low successful discovery rates, and increasing costs of exploratory campaigns due to the increasing inaccessibility of undiscovered reservoirs, have generated the need to incorporate new cost-effective technologies into oil and gas hydrocarbon discovery strategies. The rapid development of microbiomics (the study of the totality of microorganisms in particular environments, such as the human gut or an oil well) offers now new possibilities for the analysis of piston cores routinely sampled for geochemical test, since they contain useful biological information that we currently fail to extract and interpret.

In this chapter we have shown how we successfully applied this concept to the classification of 20 piston cores obtained from an exploration campaign performed by Repsol S.A. Additionally, we have revised current methods to extract meaningful biological information, analyzed it, and used the produced data to tag samples through machine learning classification algorithms. Considering the predictive ability of the models developed here, we strongly believe that incorporating microbiomics-based techniques into current oil and gas discovery strategies could make a significant contribution to reduce uncertainty and promote data-driven decision-making in exploration campaigns.

REFERENCES

Amir, Amnon, Daniel McDonald, Jose A Navas-Molina, Evguenia Kopylova, James T Morton, Zhenjiang Zech Xu, Eric P Kightley, et al. 2017. "Deblur Rapidly Resolves Single-Nucleotide Community Sequence Patterns." Jack A Gilbert (ed.). *MSystems* 2 (2).

Andrews, S. 2014. *FastQC A Quality Control Tool for High Throughput Sequence Data. Http://www.Bioinformatics.Babraham.Ac.Uk/Projects/Fastqc/.*

Apprill, Amy, Sean McNally, Rachel Parsons, and Laura Weber. 2015. "Minor Revision to V4 Region SSU rRNA 806R Gene Primer Greatly Increases Detection of SAR11 Bacterioplankton." *Aquatic Microbial Ecology.* 75:129–137.

Bray, J Roger, and John T. Curtis. 1957. "An Ordination of the Upland Forest Communities of Southern Wisconsin." *Ecol Monogr* 27: 325–349.

Brooks, J Paul, David J Edwards, Michael D Harwich, Maria C Rivera, Jennifer M Fettweis, Myrna G Serrano, Robert A Reris, et al. 2015. "The Truth about Metagenomics: Quantifying and Counteracting Bias in 16S rRNA Studies." *BMC Microbiology* 15 (March): 66. https://doi.org/10.1186/s12866-015-0351-6.

Bullard, James H, Elizabeth Purdom, Kasper D Hansen, and Sandrine Dudoit. 2010. "Evaluation of Statistical Methods for Normalization and Differential Expression in MRNA-Seq Experiments." *BMC Bioinformatics* 11 (February): 94. https://doi.org/10.1186/1471-2105-11-94.

Cáceres, Miquel De, and Pierre Legendre. 2009. "Associations between Species and Groups of Sites: Inindices and Statistical Inference." *Ecology* 90 (12): 3566–3574. https://doi.org/10.1890/08-1823.1.

Callahan, Benjamin J, Paul J Mcmurdie, and Susan P Holmes. 2017. "Exact Sequence Variants Should Replace Operational Taxonomic Units in Marker-Gene Data Analysis" *ISME* 11 (12): 2639–43.

Callahan, Benjamin J, Paul J McMurdie, Michael J Rosen, Andrew W Han, Amy Jo A Johnson, and Susan P Holmes. 2016. "DADA2: High-Resolution Sample Inference from Illumina Amplicon Data." *Nature Methods* 13 (May): 581. http://dx.doi.org/10.1038/nmeth.3869.

Caporaso, J Gregory, Justin Kuczynski, Jesse Stombaugh, Kyle Bittinger, Frederic D Bushman, Elizabeth K Costello, Noah Fierer, et al. 2010. "QIIME Allows Analysis of High-Throughput Community Sequencing Data." *Nature Methods* 7 (April): 335. http://dx.doi.org/10.1038/nmeth.f.303.

Caporaso, J Gregory, Christian L Lauber, William A Walters, Donna Berg-Lyons, Catherine A Lozupone, Peter J Turnbaugh, Noah Fierer, and Rob Knight. 2011. "Global Patterns of 16S rRNA Diversity at a Depth of Millions of Sequences per Sample." *Proceedings of the National Academy of Sciences* 108 (Supplement 1): 4516–22. https://doi.org/10.1073/pnas.1000080107.

Chen, Shifu, Tanxiao Huang, Yanqing Zhou, Yue Han, Mingyan Xu, and Jia Gu. 2017. "AfterQC: Automatic Filtering, Trimming, Error Removing and Quality Control for Fastq Data." *BMC Bioinformatics* 18 (Suppl 3): 80. https://doi.org/10.1186/s12859-017-1469-3.

Cole, James R, Qiong Wang, Jordan A Fish, Benli Chai, Donna M McGarrell, Yanni Sun, C Titus Brown, Andrea Porras-Alfaro, Cheryl R Kuske, and James M Tiedje. 2014. "Ribosomal Database Project: Data and Tools for High Throughput rRNA Analysis." *Nucleic Acids Research* 42 (Database issue): D633–42. https://doi.org/10.1093/nar/gkt1244.

DeSantis, TZ, P Hugenholtz, N Larsen, M Rojas, EL Brodie, K Keller, T Huber, D Dalevi, P Hu, and GL Andersen. 2006. "Greengenes, a Chimera-Checked 16S rRNA Gene Database and Workbench Compatible with ARB." *Applied and Environmental Microbiology* 72 (7): 5069–72. https://doi.org/10.1128/AEM.03006-05.

Dubinsky, Eric A., Mark E Conrad, Romy Chakraborty, Markus Bill, Sharon E Borglin, James T Hollibaugh, Olivia U Mason, et al. 2013. "Succession of Hydrocarbon-Degrading Bacteria in the Aftermath of the Deepwater Horizon Oil Spill in the Gulf of Mexico." *Environmental Science and Technology.* https://doi.org/10.1021/es401676y.

Edgar, RC. 2010. "Search and Clustering Orders of Magnitude Faster than BLAST." *Bioinformatics* 26: 2460–2461.

Edgar, RC. 2016. "UNOISE2: Improved Error-Correction for Illumina 16S and ITS Amplicon Sequencing." *BioRxiv*, October. http://biorxiv.org/content/early/2016/10/15/081257.abstract.

Ewels, Philip, Måns Magnusson, Sverker Lundin, and Max Käller. 2016. "MultiQC: Summarize Analysis Results for Multiple Tools and Samples in a Single Report." *Bioinformatics* 32 (19): 3047–48. http://dx.doi.org/10.1093/bioinformatics/btw354.

Håvelsrud, Othilde, Thomas HA Haverkamp, Tom Kristensen, Kjetill S Jakobsen, and Anne Rike. 2011. "A Metagenomic Study of Methanotrophic Microorganisms in Coal Oil Point Seep Sediments." *BMC Microbiology* 11 (1): 221. https://doi.org/10.1186/1471-2180-11-221.

Hawley, Erik R, Hailan Piao, Nicole M Scott, Stephanie Malfatti, Ioanna Pagani, Marcel Huntemann, Amy Chen, et al. 2014. "Metagenomic Analysis of Microbial Consortium from Natural Crude Oil That Seeps into the Marine Ecosystem Off-shore Southern California." *Standards in Genomic Sciences* 9 (3): 1259–74. https://doi.org/10.4056/sigs.5029016.

Hubert, C, and A Judd. 2010. "Using Microorganisms as Prospecting Agents in Oil and Gas Exploration." In *Handbook of Hydrocarbon and Lipid Microbiology*, K N Timmis (ed.), 2713–26. Berlin Heidelberg: Springer-Verlag. https://doi.org/10.1007/978-3-540-77587-4.

Hugerth, Luisa W, and Anders F Andersson. 2017. "Analysing Microbial Community Composition through Amplicon Sequencing: From Sampling to Hypothesis Testing." *Frontiers in Microbiology* 8: 1561.

Jaccard, Paul. 1901. "Étude Comparative de La Distribution Florale Dans Une Portion Des Alpes et Des Jura." *Bulletin Del La Société Vaudoise Des Sciences Naturelles* 37: 547–79.

Joshi, NA, and JN Fass. 2011. "Sickle: A Sliding-Window, Adaptive, Quality-Based Trimming Tool for FastQ Files (Version 1.33). http://www.citeulike.org/user/mvermaat/article/13260426"

Kallmeyer, Jens, Robert Pockalny, Rishi Ram Adhikari, David C Smith, and Steven D'Hondt. 2012. "Global Distribution of Microbial Abundance and Biomass in Subseafloor Sediment." *Proceedings of the National Academy of Sciences of the United States of America* 109 (40): 16213–16.

Kennedy, Nicholas A, Alan W Walker, Susan H Berry, Sylvia H Duncan, Freda M Farquarson, Petra Louis, and John M Thomson. 2014. "The Impact of Different DNA Extraction Kits and Laboratories upon the Assessment of Human Gut Microbiota Composition by 16S rRNA Gene Sequencing." *PLOS ONE* 9 (2): e88982.

Kerkhof, Lee J, Kevin P Dillon, Max M Häggblom, and Lora R Mcguinness. 2017. "Profiling Bacterial Communities by MinION Sequencing of Ribosomal Operons," *Microbiome* 5:116.

Klindworth, Anna, Elmar Pruesse, Timmy Schweer, Jörg Peplies, Christian Quast, Matthias Horn, and Frank Oliver Glöckner. 2013. "Evaluation of General 16S Ribosomal RNA Gene PCR Primers for Classical and Next-Generation Sequencing-Based Diversity Studies." *Nucleic Acids Research* 41 (1): e1–e1. https://doi.org/10.1093/nar/gks808.

Knights, Dan, Elizabeth K Costello, and Rob Knight. 2011. "Supervised Classification of Human Microbiota." *FEMS Microbiology Reviews* 35 (2): 343–59. https://doi.org/10.1111/j.1574-6976.2010.00251.x.

Kostka, Joel E, Om Prakash, Will A Overholt, Stefan J Green, Gina Freyer, Andy Canion, Jonathan Delgardio, Nikita Norton, Terry C Hazen, and Markus Huettel. 2011. "Hydrocarbon-Degrading Bacteria and the Bacterial Community Response in Gulf of Mexico Beach Sands Impacted by the Deepwater Horizon Oil Spill." *Applied and Environmental Microbiology* 77 (22): 7962–74. https://doi.org/10.1128/AEM.05402-11.

Li, Jun, and Robert Tibshirani. 2013. "Finding Consistent Patterns: A Nonparametric Approach for Identifying Differential Expression in RNA-Seq Data." *Stat Methods Med Res* 22 (5): 519–36. https://doi.org/10.1177/0962280211428386.Finding.

Love, Michael I, Wolfgang Huber, and Simon Anders. 2014. "Moderated Estimation of Fold Change and Dispersion for RNA-Seq Data with DESeq2." *Genome Biology* 15 (12): 550. https://doi.org/10.1186/s13059-014-0550-8.

Lozupone, Catherine, and Rob Knight. 2005. "UniFrac: A New Phylogenetic Method for Comparing Microbial Communities." *Applied and Environmental Microbiology* 71 (12): 8228–35. https://doi.org/10.1128/AEM.71.12.8228-8235.2005.

Magoč, Tanja, and Steven L Salzberg. 2011. "FLASH: Fast Length Adjustment of Short Reads to Improve Genome Assemblies." *Bioinformatics* 27 (21): 2957–63. https://doi.org/10.1093/bioinformatics/btr507.

McDonald, Daniel, Jose C Clemente, Justin Kuczynski, Jai Ram Rideout, Jesse Stombaugh, Doug Wendel, Andreas Wilke, et al. 2012. "The Biological Observation Matrix (BIOM) Format or: How I Learned to Stop Worrying and Love the Ome-Ome." *GigaScience* 1 (July): 7. https://doi.org/10.1186/2047-217X-1-7.

McMurdie, Paul J, and Susan Holmes. 2013. "Phyloseq: An R Package for Reproducible Interactive Analysis and Graphics of Microbiome Census Data." *PLOS ONE* 8 (4): e61217.

McMurdie, Paul J, and Susan Holmes. 2014. "Waste Not, Want Not: Why Rarefying Microbiome Data Is Inadmissible." *PLOS Computational Biology* 10 (4): e1003531.

Miqueletto, Paula B, Fernando D Andreote, Armando CF Dias, Justo C Ferreira, Eugênio V Dos Santos Neto, and Valéria M de Oliveira. 2011. "Cultivation-Independent Methods Applied to the Microbial Prospection of Oil and Gas in Soil from a Sedimentary Basin in Brazil." *AMB Express* 1 (1): 35. https://doi.org/10.1186/2191-0855-1-35.

Niu, Mingyang, Xibei Fan, Guangchao Zhuang, Qianyong Liang, and Fengping Wang. 2017. "Methane-Metabolizing Microbial Communities in Sediments of the Haima Cold Seep Area, Northwest Slope of the South China Sea." *FEMS Microbiology Ecology* 93 (9): fix101-fix101. http://dx.doi.org/10.1093/femsec/fix101.

Paliy, O, and V Shankar. 2017. "Application of Multivariate Statistical Techniques in Microbial Ecology." *Molecular Biology and Evolution* 25 (5): 1032–57. https://doi.org/10.1111/mec.13536.Application.

Paulson, Joseph N, O Colin Stine, Héctor Corrada Bravo, and Mihai Pop. 2013. "Differential Abundance Analysis for Microbial Marker-Gene Surveys" *Nature Methods* 10 (12): 1200–1202. https://doi.org/10.1038/nmeth.2658.

Petro, Caitlin, Andreas Schramm, Caitlin Petro, Piotr Starnawski, Andreas Schramm, and Kasper U Kjeldsen. 2017. "Microbial Community Assembly in Marine Sediments Microbial Community Assembly in Marine Sediments," *Aquatic Microbial Ecology* 79: 177–195.

Quast, Christian, Elmar Pruesse, Pelin Yilmaz, Jan Gerken, Timmy Schweer, Pablo Yarza, Jörg Peplies, and Frank Oliver Glöckner. 2013. "The SILVA Ribosomal RNA Gene Database Project: Improved Data Processing and Web-Based Tools." *Nucleic Acids Research* 41 (Database issue): D590–96. https://doi.org/10.1093/nar/gks1219.

Quince, Christopher, Alan W Walker, Jared T Simpson, Nicholas J Loman, and Nicola Segata. 2017. "Shotgun Metagenomics, from Sampling to Analysis." *Nature Biotechnology*. 35: 833–844.

Ramette, Alban. 2007. "Multivariate Analyses in Microbial Ecology." *FEMS Microbiology Ecology* 62 (2): 142–60. https://doi.org/10.1111/j.1574-6941.2007.00375.x.

Ristova, Petra, Frank Wenzhöfer, Alban Ramette, Janine Felden, and Antje Boetius. 2015. "Spatial Scales of Bacterial Community Diversity at Cold Seeps (Eastern Mediterranean Sea)." *The ISME Journal* 9 (6): 1306–18. https://doi.org/10.1038/ismej.2014.217.

Roalkvam, Irene, Håkon Dahle, Yifeng Chen, Steffen Jørgensen, Haflidi Haflidason, and Ida Steen. 2012. "Fine-Scale Community Structure Analysis of ANME in Nyegga Sediments with High and Low Methane Flux." *Frontiers in Microbiology*. https://www.frontiersin.org/article/10.3389/fmicb.2012.00216.

Robinson, Mark D, Davis J McCarthy, and Gordon K Smyth. 2010. "EdgeR: A Bioconductor Package for Differential Expression Analysis of Digital Gene Expression Data." *Bioinformatics* 26 (1): 139–40. https://doi.org/10.1093/bioinformatics/btp616.

Rosen, Michael J, Benjamin J Callahan, Daniel S Fisher, and Susan P Holmes. 2012. "Denoising PCR-Amplified Metagenome Data." *BMC Bioinformatics* 13 (1): 283. https://doi.org/10.1186/1471-2105-13-283.

Ruff, S Emil, Jennifer F Biddle, Andreas P Teske, Katrin Knittel, Antje Boetius, and Alban Ramette. 2015. "Global Dispersion and Local Diversification of the Methane Seep Microbiome." *Proceedings of the National Academy of Sciences* 112 (13): 4015–20. https://doi.org/10.1073/pnas.1421865112.

Schirmer, Melanie, Umer Z Ijaz, Rosalinda D'Amore, Neil Hall, William T Sloan, and Christopher Quince. 2015. "Insight into Biases and Sequencing Errors for Amplicon Sequencing with the Illumina MiSeq Platform." *Nucleic Acids Research* 43 (6): e37–e37. https://doi.org/10.1093/nar/gku1341.

Schloss, Patrick D, Matthew L Jenior, Charles C Koumpouras, Sarah L Westcott, and Sarah K Highlander. 2016. "Sequencing 16S rRNA Gene Fragments Using the PacBio SMRT DNA Sequencing System." Christine Josenhans (ed.). *PeerJ* 4: e1869. https://doi.org/10.7717/peerj.1869.

Schloss, Patrick D, Sarah L Westcott, Thomas Ryabin, Justine R Hall, Martin Hartmann, Emily B Hollister, Ryan A Lesniewski, et al. 2009. "Introducing Mothur: Open-Source, Platform-Independent, Community-Supported Software for Describing and Comparing Microbial Communities." *Applied and Environmental Microbiology* 75 (23): 7537–41. https://doi.org/10.1128/AEM.01541-09.

Segata, Nicola, Jacques Izard, Levi Waldron, Dirk Gevers, Larisa Miropolsky, Wendy S Garrett, and Curtis Huttenhower. 2011. "Metagenomic Biomarker Discovery and Explanation." *Genome Biology* 12 (6): R60. https://doi.org/10.1186/gb-2011-12-6-r60.

Sharpton, Thomas J. 2014. "An Introduction to the Analysis of Shotgun Metagenomic Data." *Frontiers in Plant Science* 5 (June): 209. https://doi.org/10.3389/fpls.2014.00209.

Smith, Mark B, Andrea M Rocha, Chris S Smilliie, Scott W Olesen, Charles Paradis, Wu. Liyou, James H Campbell, et al. 2015. "Natural Bacterial Communities Serve as Quantitative Geochemical Biosensors." *MBio* 6 (3): 1–13. https://doi.org/10.1128/mBio.00326-15.

Statnikov, Alexander, Mikael Henaff, Varun Narendra, Kranti Konganti, Zhiguo Li, Liying Yang, Zhiheng Pei, Martin J Blaser, Constantin F Aliferis, and Alexander V Alekseyenko. 2013. "A Comprehensive Evaluation of Multicategory Classification Methods for Microbiomic Data." *Microbiome* 1 (1): 11. https://doi.org/10.1186/2049-2618-1-11.

Swenson, Nathan G. 2011. "Phylogenetic Beta Diversity Metrics, Trait Evolution and Inferring the Functional Beta Diversity of Communities." Andrew Hector (ed.). *PLoS ONE* 6 (6): e21264. https://doi.org/10.1371/journal.pone.0021264.

Tsesmetzis, Nicolas, Michael J Maguire, Ian M Head, and Bart P Lomans. 2017. "Protocols for Investigating the Microbial Communities of Oil and Gas Reservoirs." In *Hydrocarbon and Lipid Microbiology Protocols: Field Studies*, Terry J McGenity, Kenneth N Timmis, and Nogales Balbina (eds.), 65–109. Berlin, Germany: Springer Berlin Heidelberg. https://doi.org/10.1007/8623_2016_212.

Weiss, Sophie, Zhenjiang Zech Xu, Shyamal Peddada, Amnon Amir, Kyle Bittinger, Antonio Gonzalez, Catherine Lozupone, et al. 2017. "Normalization and Microbial Differential Abundance Strategies Depend upon Data Characteristics," *Microbiome* 5: 27. https://doi.org/10.1186/s40168-017-0237-y.

Whittaker, R.H. 1960. "Vegetation of the Siskiyou Mountains, Oregon and California." *Ecological Monographs* 30 (3): 279–338. https://doi.org/10.2307/1943563.

Winkel, Matthias, Julia Mitzscherling, Pier P Overduin, Fabian Horn, Maria Winterfeld, Ruud Rijkers, Mikhail N Grigoriev, et al. 2018. "Anaerobic Methanotrophic Communities Thrive in Deep Submarine Permafrost." *Scientific Reports* 8: 1291.

Yazdani, Mehrdad, Bryn C Taylor, Justine W Debelius, Weizhong Li, Rob Knight, and Larry Smarr. 2016. "Using Machine Learning to Identify Major Shifts in Human Gut Microbiome Protein Family Abundance in Disease." *Proceedings - 2016 IEEE International Conference on Big Data, Big Data 2016*, 1272–80.

Index

C

N

O

Q

R

S